Non-covalent Interactions
Theory and Experiment

RSC Theoretical and Computational Chemistry Series

Editor in Chief:
Jonathan Hirst, *University of Nottingham, Nottingham, UK*

Series Editors:
Kenneth Jordan, *University of Pittsburgh, Pittsburgh, USA*
Carmay Lim, *Academia Sinica, Taipei, Taiwan*
Walter Theil, *Max Planck Institute for Coal Research, Mülheim an der Ruhr, Germany*

Titles in the Series:

How to obtain future titles on publication:
A standing order plan is available for this series. A standing order will bring delivery of each new volume immediately on publication.

For further information please contact:
Book Sales Department, Royal Society of Chemistry,
Thomas Graham House, Science Park, Milton Road, Cambridge,
CB4 0WF, UK
Telephone: +44 (0)1223 420066, Fax: +44 (0)1223 420247, Email: books@rsc.org
Visit our website at http://www.rsc.org/Shop/Books/

Non-covalent Interactions
Theory and Experiment

Pavel Hobza
Institute of Organic Chemistry and Biochemistry, Academy of Sciences of the Czech Republic, and Center for Biomolecules and Complex Molecular Systems, Prague, Czech Republic

Klaus Müller-Dethlefs
School of Chemistry and School of Physics and Astronomy, The University of Manchester, Manchester, UK

RSC Theoretical and Computational Chemistry Series No. 2

ISBN: 978-1-84755-853-4
ISSN: 2041-3181

A catalogue record for this book is available from the British Library

Published by The Royal Society of Chemistry,
Thomas Graham House, Science Park, Milton Road,
Cambridge CB4 0WF, UK

Registered Charity Number 207890

For further information see our web site at www.rsc.org

Foreword

Non-covalent (sometimes called van der Waals) interactions are involved in a vast number of phenomena related to the whole realm of molecular and macro-molecular science. Physical characteristics and reaction rates in solutions and at surfaces of solids are significantly influenced by them. The same is true about solid phase structures. On a long and tedious way aiming at understanding basic processes associated with life, these interactions do play a crucial role.

It is true that researchers in the area of molecular sciences have realized a key position of van der Waals interactions during the last 15 to 20 years. However, the knowledge of the most appropriate experimental and theoretical tools is still rather limited. Therefore, I welcome very much the text which follows, which assumes several pronounced positive features.

Specifically, I do appreciate the fact that one of the authors (P.H.) has been involved in theoretical studies of non-covalent interactions for about 40 years. The other author (K.M.-D.) has been very active in the area of experimental studies of essential features of those complexes with sophisticated, advanced methods for several dozens of years. Not only that: the first author has always been interested in relationships between calculated and experimental characteristics and the other one has been, already in the early stage of his investigations, aware of the fact that experimentalists can easily run into difficulties without assistance from the theoretical side. Simply speaking, the authors have always believed in the plausibility of the phrase "facts without theory are chaos, theory without facts is fantasy". The mentioned features of the authors manifest themselves throughout the reviews. I believe that this work will help to diminish the number of papers in which non-covalent interactions are interpreted not transparently enough, sometimes in terms of an unfortunate mixture of variational and perturbational energy differences superimposed with purely empirical terms.

In future, I mean in the future which is still beyond the horizon, more efficient and universal experimental techniques as well as theoretical (computational) tools will be available, suitable for a majority of tasks and problems in the realm of molecular and macromolecular science. At present, however, a broad spectrum of experimental procedures has to be considered together with an appropriate combination of computational methods in order to successfully tackle the problem under study. In this respect this work represents a very topical and useful guide-book to the whole area of non-covalent interactions.

Rudolf Zahradník
Prague

Preface

The book presented here is the result of a long-standing collaboration between an experimental (KMD) and a theoretical (PH) researcher. We met at the Technische Universität München, Institut für Physikalische Chemie, Garching, just after the opening of the Berlin Wall, enjoying the resulting breakdown of the established Cold War international political system. At that time, one of us (KMD) had just developed a new spectroscopic method, ZEKE spectroscopy, which is a high-resolution method of photoelectron spectroscopy, and now ventured into the investigation of non-covalent interactions in molecular complexes. The first study of hydrogen-bonded complexes, phenol...water, yielded quite spectacular results, crying out for a theoretical explanation by *ab initio* calculations. In Prague during the Cold War period, the second of us (PH) had, together with the future president of the Czechoslovak Academy of Sciences, Rudolf Zahradník, been able to expand his expertise of using the new *ab initio* computational methods and one important application became the study of non-covalent interactions, summarised in the book, *Intermolecular Complexes*, Elsevier, Amsterdam 1988 by PH and R. Zahradník.

In Garching, at the *Institut für Physikalische Chemie der Technischen Universität München* (Director: Professor E. W. Schlag), PH and KMD entered a very fruitful and enthusiastic, but also sometimes explosive discussion period, which laid the groundwork for the present book. We were looking for a more general understanding of the details of non-covalent interactions in molecular clusters beyond the limits of the electrostatic model and empirical potentials. These models were not really accurate enough to satisfactorily interpret highly resolved microwave, vibration-rotation-tunneling, REMPI, LIF, ZEKE, hole-burning and other spectra observed in the application of the, by then, widespread laser-based methods for the study of molecular clusters in supersonic jet expansions. The lesson we learned very early was essentially that the development of theory and experiment had to go hand in hand; the theory was not just the slave of the experiment but, on the contrary, very often theory could provide reasonable guidance provided the pitfalls of choosing an unsuitable

RSC Theoretical and Computational Chemistry Series No. 2
Non-covalent Interactions: Theory and Experiment
By Pavel Hobza and Klaus Müller-Dethlefs
© Pavel Hobza and Klaus Müller-Dethlefs 2010
Published by the Royal Society of Chemistry, www.rsc.org

level of theory were avoided, as discussed in the context of many of the methods described in this book.

After the move of KMD to the University of York in 1995 and the expansion of PH's group then at the Heyrovský Institute of Physical Chemistry in Prague, we saw a very fruitful summary of our research in addition to contributions from other leaders in the field in the *Chemical Reviews* centenary anniversary issue, *Chemical Reviews* 2000. Our review in this volume and a more recent one in *Collection of Czechoslovak Chemical Communications* in 2006 encouraged us to write this book. In the meantime, many new developments have taken place. PH became the distinguished Chair at The Institute of Organic Chemistry and Biochemistry in Prague and KMD moved to the University of Manchester as Director of The Photon Science Institute. The opportunities associated with these moves enabled us to expand on some of the problems that we both think need to be solved to understand some of the most vital problems of nature, of which a selected sample are illustrated in the book.

In the preparation of this book, we have to thank a variety of colleagues for providing material and these colleagues are mentioned and referred to in the corresponding sections of the book. In particular, we would like to thank the editor, Dr Merlin Fox of The Royal Society of Chemistry, Professor Masaaki Fujii (Yokohama) and Dr Mikko Riese (Manchester) for their comments and for their most valuable suggestions. Last but not least, we would like to thank all our coworkers over more than two decades of our scientific research for their most important and valued contributions, and we would like to thank our funding agencies for their support: the Institute of Organic Chemistry and Biochemistry, Prague, the Academy of Sciences of the Czech Republic, the Ministry of Education, Youth and Sports of the Czech Republic, the Korea Science and Engineering Foundation (World Class University Program), the UK Engineering and Physical Sciences Research Council and the Miller Foundation, Berkeley, California. Without the support of our wives the writing of this book would have been impossible: thank you Helen and thank you Pavla.

We do hope that the reader finds this book useful. We have tried to refer to the relevant literature in the most comprehensive manner. However, this book cannot cover this huge field completely and hence it presents a cross section of what we, as individual researchers, think is relevant and what may help others in stimulating and enhancing their enthusiasm about research in the field of non-covalent interactions. We think that this book is particularly suited to doctoral and postdoctoral researchers and also to undergraduate students on advanced courses. We do hope that our colleagues will also find this book useful and they will forgive us any omissions and inaccuracies they may find.

Klaus Müller-Dethlefs, Manchester, UK
Pavel Hobza, Prague, Czech Republic

Contents

RSC Theoretical and Computational Chemistry Series No. 2
Non-covalent Interactions: Theory and Experiment
By Pavel Hobza and Klaus Müller-Dethlefs
© Pavel Hobza and Klaus Müller-Dethlefs 2010
Published by the Royal Society of Chemistry, www.rsc.org

CHAPTER 1
Introduction

1.1 An Historical Remark

It was none other than van der Waals,[1] in the 1870s, who realised that the
discrepancies observed between the state function of a real gas and the ideal gas
law could be accounted for by the attracting forces between molecules or rare
gas atoms. Van der Waals introduced an equation of state suitable for
describing the behaviour of real (in contrast to ideal) gases. Although this law
does not provide the most accurate functional description for a real gas, it
nevertheless constituted a major breakthrough. Van der Waals made it explicitly obvious, for instance with respect to condensation of all real gases, that
significant attracting forces exist between gas molecules (or atoms, in the case
of monoatomic gases), which exhibit a tendency to *form a new type of bond.* An
important landmark in the history of understanding these attracting forces is
represented by the liquefaction of helium in experiments by Kamerlingh-
Onnes.[2] The very existence of liquid helium provides a most decisive argument
about the existence of attractive intermolecular forces acting even between
small spherical rare gas atoms such as helium, not bearing any charge or permanent electric multipole moment.

The formation of these special van der Waals bonds, compared to chemical
bonds, is not energetically demanding at all; these bonds are, under general
laboratory conditions, easily formed and just as easily split. What happens to
appear as a weakness represents, surprisingly, a strength of such bonds. In the
context of a scenario for the evolution of life on Earth it was necessary to find,
besides strong covalent bonding, another type of much weaker bonding
allowing easy reversibility of the formation process. The supermolecule formed
should allow for repeating opening and closing without changing any important structural feature.

Many years later, in 1930, Fritz London[3] (and soon afterwards Hans
Hellmann[4]) made a fundamental step in describing and interpreting these bonds.
This was only possible using the recently born quantum mechanics. Though
several contributions can be interpreted by classical physics, the most important

RSC Theoretical and Computational Chemistry Series No. 2
Non-covalent Interactions: Theory and Experiment
By Pavel Hobza and Klaus Müller-Dethlefs
© Pavel Hobza and Klaus Müller-Dethlefs 2010
Published by the Royal Society of Chemistry, www.rsc.org

ones giving rise to the repulsion and attraction between systems (exchange-repulsion and dispersion contributions, see later) that are of quantum origin and could be interpreted only by using the theoretical apparatus of quantum mechanics. Works of these and other pioneers are mentioned or outlined in the classic book on intermolecular interactions by J. O. Hirschfelder, C. F. Curtiss, and R. B. Bird,[5] and a survey of monographs and reviews up to the mid-1980s is presented in a book on intermolecular complexes.[6] Selected summarising works since about 1985 are presented in ref. 7. Specifically to be mentioned are three thematic issues of *Chemical Reviews* devoted to non-covalent interactions, which appeared in 1988, 1994 and 2000, and one thematic issue of *Phys. Chem. Chem. Phys.* devoted to the same subject in 2008; all these thematic issues were edited by one of us (PH). Besides these works three books need to be mentioned that supplement the present book. The first one by A. J. Stone[8] focuses on the theory of non-covalent interactions and perturbation calculations of the cluster inter-action energy. The second one by A. Karshikoff[9a] describes non-covalent interactions in proteins. The third one by I. G. Kaplan[9b] deals with the theory and computation of intermolecular interactions. The book presented here is largely based on our theoretical and experimental papers published in the last two decades, which are cited at the end of each chapter. A special place is held by our recent review[10] entitled, "The World of Non-covalent Interactions: 2006" by both present authors and Rudolf Zahradník.

1.2   A Remark on Nomenclature of Molecular Complexes

Why are molecular complexes, or molecular clusters, as they are most often called, of such interest? The main feature of molecular clusters is that they can be prepared experimentally in supersonic jet expansions and molecular beams as isolated systems exhibiting intermolecular bonds that originate from non-covalent interactions. From the theoretical point of view molecular clusters can also be studied using standard *ab initio* quantum-chemical methods, treating the cluster as a "supermolecule" composed of several moieties held together by non-covalent bonds.

An issue in the literature that sometimes is unclear relates to the definition of non-covalently bound complexes. A significant feature of such complexes is that the subsystems, of which they are constituted, are not bound by covalent interactions but solely by electric multipole–electric multipole interactions. We consider, however, not only permanent, but also inductive, and time-dependent multipoles. While it is possible to ascribe the stability of a complex to a bond, non-covalent in nature, it is not always easy to localise such a bond in space. When possible it is highly desirable to use another symbol for this bond than that which represents a covalent bond, *i.e.* a short full line – hence, three dots . . . may serve as a representation of a non-covalent bond. The hydrogen molecule and the helium (van der Waals) molecule are adequate representatives: H–H and He . . . He, or alternatively H_2 and $(He)_2$.

The second type of bond illustrated above still does not have a definite name. No doubt, it is possible to call it a non-covalent bond. Another label, which is sometimes used, is derived from the term weak interactions and therefore the name "weak bonds" is used. This is an unfortunate name, because it is derived from a designation that has been used for a long time in physics in a completely different context. We have favoured for years the designation "van der Waals" (vdW in abbreviated form), *e.g.*, vdW interactions, forces, bonds. It is unfortunately true that this designation has been corrupted – sometimes by poorly defined use – for a component of the empirical force field. In the case of empirical potentials the vdW term means a sum of London dispersion and exchange-repulsion terms. In our previous review[7a] we decided to use the term "*non-covalent*" to classify interactions that are *not covalent*. We are aware that this definition is again not straightforward and unambiguous since, for example, metallic interactions are also covered but we believe that the term non-covalent properly describes the origin and nature of these interactions. In the very broadest sense non-covalent interactions include *electrostatic interactions* between permanent multipoles (charge–charge, charge–dipole, charge–multipole, multipole–multipole. . .), *induction and/or polarisation interactions* between permanent and induced multipoles, *dispersion interactions* between instantaneous and induced multipoles and also charge-transfer, ionic and metallic interactions, and interactions leading to formation of H-bonding, halogen bonding and lithium bonding.

1.3   Purpose and Scope: Theory and Experiment

A purpose of this textbook is to illustrate why non-covalent interactions are of fundamental importance for chemistry and why their understanding is a *conditio sine qua non* for the molecular biodisciplines. Moreover, an attempt will be made to describe correct procedures for treating these interactions theoretically. Those who deal with this subject daily do not need such a recommendation. However, nowadays it is increasingly common for chemists to do the necessary calculations themselves and to develop experience in this field. In contrast to the realm where only chemical (*i.e.* covalent) bonds play a role, in the area of non-covalently bound complexes it is, in general, not trivial to assert how to proceed and which method and what level of theory guarantees obtaining reliable results. The situation is even more involved because problems of practical value in chemistry and still more in biology are – with respect to computer size – rather extensive. The choice of an appropriate method is especially challenging in these instances. Here, we will describe the main computational procedures to obtain static characteristics of non-covalent species; those characteristics are essential for the understanding of their dynamics.

1.4   Covalent *Versus* Non-covalent Bonds

The concept of covalent bonding belongs to the most successful concepts in modern science and is, at a certain level, a more or less closed chapter. After about eighty years of intense study, the processes of formation and breaking of covalent bonds are well understood and reliable descriptions of these processes can be performed at various theoretical levels. Calculated characteristic molecular properties agree well with the relevant experiments and there are no fundamental disagreements between state-of-the art theory and experiment. In contrast, the understanding of the nature of non-covalent interactions is far less clear and the respective calculations yield results that are frequently in conflict with experimental data. The basic principles of non-covalent interactions, for instance the hydrogen-bond (H-bond), by Linus Pauling,[11] were formulated in the 1930s. However, despite enormous progress made in theory as well as in experimental techniques in the last decades, we are still far from obtaining unambiguous and quantitatively satisfactory information about non-covalent complexes. Experiments do not yet yield full information on a complex being studied and combining various techniques introduces some ambiguity. Theory, on the other hand, is principally capable of providing full information about a non-covalent complex. For example, we can generate basic information such as structure and stabilisation energy and from the knowledge of the wavefunction we can obtain any other desired property.

A covalent bond is formed when two subsystems with unfilled electronic shells start to overlap. At that point, the electron density between them increases and a bond is created. (More specifically, electron density increases in the bonding region and this increase leads to strengthening of the bond. On the other hand, the increase of electron density in the antibonding region results in weakening of the bond.) The most efficient overlap arises at interatomic distances below $2\,\text{Å}$ and at distances larger than about $4\,\text{Å}$ overlap is negligible. Non-covalent interactions are, however, known to exist at much greater distances, sometimes at more than $10\,\text{Å}$ and in the case of biomacromolecules even at more than $100\,\text{Å}$, which points to the existence of some other source of attraction. The only possibilities are the electric, and to a lesser extent magnetic, properties of the systems. In particular, permanent, induced and instantaneous electric multipoles are sources of attraction (or repulsion) between systems of different types. *Electrostatic interaction*, using the usual terminology, applies to subsystems with permanent multipole moments each. For two subsystems this interaction energy between charges (= monopoles), dipoles, quadrupoles and higher multipoles is proportional to the product of the multipoles and the first or a higher power of reciprocal distance. In most cases electrostatic interaction dominates compared to other energy terms. *Induction (polarisation) interaction* represents the interaction between permanent and induced multipoles. For instance, if one subsystem has no permanent multipole moments, *i.e.* it is neutral and spherically symmetric (*i.e.* an atom), the electrostatic interaction term is zero and induction is responsible for attraction. If one system is of near central symmetry, such as SF_6 (whose lowest permanent moment is the hexadecapole) or CH_4 (whose lowest permanent moment is the octopole) electrostatic interaction and induction may be of comparable magnitude. Systems with vanishing permanent moments, for instance, two nonpolar and spherically symmetric systems, exhibit attraction as well. This experimental finding manifested by the liquefaction of noble gases in the early 20th century was at that time very surprising. Only on the basis of quantum mechanics was it possible to theoretically derive the attraction between noble-gas atoms in terms of the London *dispersion energy*. At that time this explanation represented an important success of the recently born quantum mechanics. London's theory took into account the oscillations of electron clouds and nuclei leading to the generation of *instantaneous multipoles*, which vanish when integrated over time. For this case one talks about *instantaneous multipole–induced multipole moment interactions*. The dispersion energy, which is of quantum origin, is proportional to the product of the polarisabilities of the subsystems and a sixth (or higher) power of reciprocal distance. It was believed for a long time that dispersion energy was always smaller than the two previously mentioned energy contributions and that it stabilised mainly noble gas atoms. It has now been shown that dispersion energy between aromatic systems with delocalised electrons is substantial, and stabilisation of these structures is comparable to that of other, for example, hydrogen-bonded structures. This finding has cast new light on the nature of the stabilisation of such biomacromolecules as DNA and proteins. All three energy contributions (electrostatic, induction, dispersion) can be of attractive nature and as such should be balanced by some repulsive force. The so-called exchange-repulsion energy is of quantum-chemical nature too and becomes important when two subsystems overlap. Unlike for the covalent bond, where the electron density between subsystems having unpaired electrons increases in the bonding region, here (subsystems with occupied electron shells) the electron density increases in an antibonding region, which results in mutual repulsion. Most non-covalent complexes exhibit nonzero overlap of their electron clouds. Consequently, some charge transfer between both subsystems exists, giving rise to a small covalent contribution. This is covered by the charge-transfer energy contribution realised not only between the electron donor and the electron acceptor (*e.g.* benzene...tetracyanoethylene), but also, more generally, between the proton donor and the proton acceptor (*e.g.* hydrogen bonding). Only a very small portion of the electron (≈ 0.01) is transferred between the subsystems. It must be kept in mind that even chemical processes are connected with only slight changes of the total density and that in the case of non-covalent interactions these changes should be even smaller.

Stabilisation of all the mentioned types of non-covalent complexes is due to favorable energy. This means that the energy of a complex is lower than the sum of the energies of its separated subsystems, which occurs systematically if a complex is formed *in vacuo*. The situation in an environment and mainly in the water phase is different. This is due to the fact that equilibrium of any non-covalent (as well as covalent) process is not determined by the change of energy (enthalpy) but by the change of the Gibbs energy ($\Delta G = \Delta H - T\Delta S$). The binding can be now realised not only because of favorable energy (enthalpy) but also because of favorable entropy (ΔS is positive, *i.e.* entropy at the right side of a reaction is higher than that at the left side); in this case the enthalpy (energy) change can be even unfavorable ($\Delta H > 0$). This is the so-called hydrophobic interaction, for which the term "hydrophobic bond" is sometimes (albeit incorrectly) used. Evidently, the origin of the stabilisation is completely different. Let us again mention that hydrophobic interactions never occurred for systems *in vacuo* and is mostly connected with the water environment.

Comparison of theoretical and experimental results is of vital importance for theory as well as for experiments because it allows for the testing of the ability and accuracy of newly developed procedures and techniques. The combination of experiment and theory also gives a deeper insight into the problem studied and so leads to a deeper understanding.

1.5   Experimental Observables

The first question is which properties of non-covalent complexes are observable explicitly? The surprising answer is not so many of them! The structure is not directly observable and can only be determined by measuring the rotational constants thus providing the three principal moments of inertia. Rotational constants, however, do not provide an unambiguous answer concerning structure and geometry (see Section 2.1). A similar situation exists for the determination of stabilisation energies and of the various experimental techniques available only zero-electron kinetic energy (ZEKE) spectroscopy[12–18] provides directly measurable high-accuracy information on stabilisation energies[19] (see Section 2.3). In addition, directly observable characteristics of a non-covalent complex are vibrational frequencies, not all of which may be seen due to Franck–Condon factors or symmetry selection rules.

In the following we venture into a comparison of theory and experiment. Before going into any further detail of the theoretical and computational procedures, we would like to briefly summarise what can be learned from experiments with regard to understanding the intermolecular interactions that lead to non-covalent binding.

The first information that can be obtained comes from vibrational frequencies. There are essentially two methods for vibrational spectroscopy, based on infrared (IR) absorption or the Raman effect. Though some efforts have been made towards the development of stimulated Raman population transfer as a method of vibrationally resolved spectroscopy[20–23] this method has not been utilised by many other groups. Most studies of vibrational spectroscopy for the determination of vibrational energy levels rely on ionisation detection.[24–28] For molecular clusters mass selection is essential and the detection methods are often based on the most sensitive ionisation techniques. The determination of vibrational frequencies is often straightforward and provides information on the structure and strength of a non-covalent complex.[29–33] The structural information, however, is only indirect and is mostly based on the fact that the stretching vibration frequency of an X–H bond is changed when the X–H . . . Y hydrogen bond is formed. In general, direct absorption methods are unsuitable for molecular beam studies (except for recently developed IR cavity ring-down spectroscopy in slit nozzles[34–38]).

Second, very important information comes from the experimental determination of rotational constants. These are obtained from high-resolution spectroscopy with rotational level resolution: microwave spectroscopy (MW) as the method of highest resolving power,[39–47] vibration-rotation-tunnelling spectroscopy(VRT),[48–50] rotational electronic spectroscopy[51–62] and deconvolution of only partially resolved rotational structure.[63–67]

Most suitable are methods based on mass spectrometry, particularly time-of-flight methods, which allow a mass signature to be obtained and thus allow for identification of certain molecular clusters and, in addition, provide single-molecule detection efficiency. For vibrational spectroscopy, a most successful method, the detection of infrared absorption by population depletion, is infrared-UV hole burning with ionisation detection, first utilised by Brutschy and coworkers.[68]

Contemporary research into molecular interactions aims at obtaining accurate intermolecular potential energy surfaces through a combination of experiment and theory.[7a] Within this arena, molecular clusters are useful model systems for investigating intermolecular forces that also operate in bulk systems, and have been actively studied over the last 25 years since they can yield high-resolution spectroscopic data and be treated with state-of-the-art theory. For closed-shell systems in their electronic ground states, the goal of obtaining spectroscopically accurate potential surfaces has been achieved for several systems including $Ar_2 \ldots DCl$[69] and $(H_2O)_2$,[70] but the situation is quite different for ionic clusters. Progress in this area has been hampered by the experimental challenge of producing sufficient quantities of ionic complexes under the molecular beam conditions that are crucial for performing high-resolution spectroscopy. The detailed characterisation of ion–solvent interactions is clearly an important goal, since they control fundamental chemical processes such as the solution and solvation of salts.

Zero electron kinetic energy (ZEKE) photoelectron spectroscopy is one of the spectroscopic techniques that has greatly contributed to improving our knowledge of the (ro)vibronic structure of cationic molecular complexes. Benzene $\ldots Ar$[71] and phenol $\ldots H_2O$[72] were the first van der Waals (vdW) and hydrogen-bonding complexes, respectively, to be studied using ZEKE spectroscopy.[12–14,] Unlike traditional photoelectron spectroscopy, the technique offers sufficient spectral resolution to allow the identification of the low-frequency intermolecular modes that characterise the vdW potential energy surface. Since its introduction in 1984, the ZEKE technique has been reviewed in a number of articles,[73–81] and here we will focus on its application to non-covalent bonding.

Resonance-enhanced multiphoton ionisation (REMPI) spectroscopy provides spectroscopic information on the excited states of neutral systems.[82–84] The technique is considered here along with ZEKE spectroscopy for two reasons. Firstly, the two spectroscopies have frequently *both* applied to the chemical systems we consider and provide complementary chemical information on the structure (and occasionally, the dynamics) of the molecular cluster. Secondly, ZEKE spectra are often recorded using resonant two-photon ionisation (R2PI), where the first photon accesses an intermediate electronic state of the neutral system and a second photon effects ionisation. This approach has many benefits including (i) facile generation of the laser photons if ionisation is achieved in this two-step approach, (ii) state selection of a specific molecular cluster so that the resulting ZEKE spectrum is not contaminated by a coincidentally overlapping signal that results from an entirely different cluster, (iii) improved spectral resolution and (iv) the opportunity to gain a fuller picture of the ionic potential energy surface by effecting ionisation *via* different intermediate states. Furthermore, the R2PI scheme is central to IR-UV double-resonance spectroscopies which have recently provided a wealth of information on vdW and hydrogen-bonded complexes.[85–87]

Extensive IR-UV hole-burning studies were carried out by the groups of Zwier,[88,89] Mikami[25,90–93] and Kleinermanns,[94–96] which have even been applied to molecules of biological interest in the gas phase. The principle of IR-UV hole burning is shown in Figure 1.1. An infrared laser is used to shift the population of the molecule of interest from its vibrational ground state into a vibrationally excited state. The molecule is also ionised from its vibrational

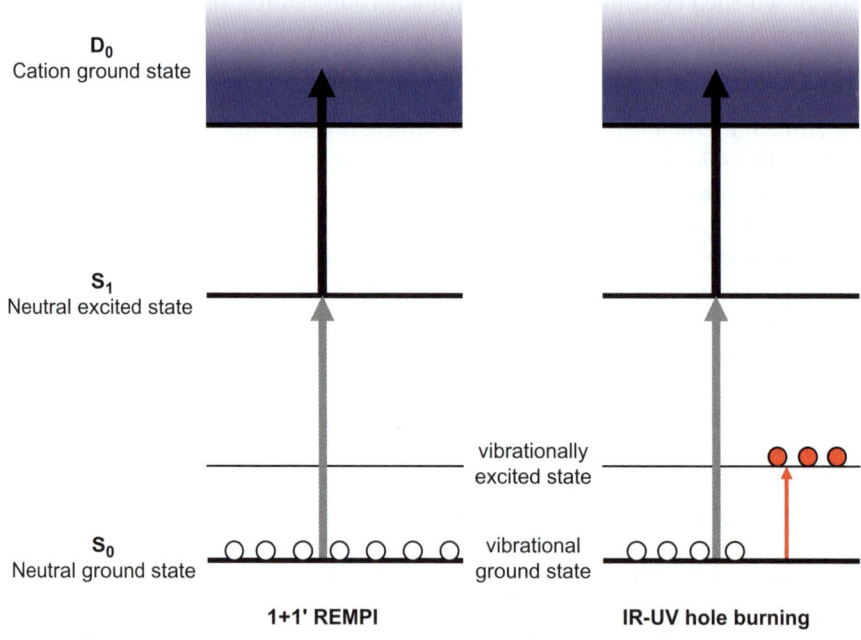

Figure 1.1   Principle of IR-UV hole burning. An infrared laser is used to shift the population of the molecule of interest from its vibrational ground state into a vibrationally excited state. The molecule is also ionised from its vibrational ground state by a UV laser pulse in a two-photon ionisation process. Due to the depletion of the ground state by the infrared absorption a depletion in the ion signal is observed in the experiment, *i.e.* an infrared absorption is characterised by a dip in the ion signal.

ground state by the UV laser in a two-photon ionisation process. Due to the depletion of the vibrational ground state by the infrared absorption, a depletion in the ion signal is observed in the experiment, *i.e.* an infrared absorption is characterised by an ion dip. This type of experiment has been very successful, and with the advent of new optical parametric IR laser sources the OH and NH[97] vibration region of various species has been extensively probed.

Most recently, the availability of infrared lasers in the region above 3.5 μm, based on difference frequency mixing and parametric amplification in nonlinear silver selenide crystals has extended the IR tuning range substantially towards probing lower vibrational frequencies, particularly in the CO region.[98–100] For the infrared region above 6 μm the big challenge comes from producing infrared laser sources that are suitable for laboratory use and that emit pulses of high enough energies (a few mJ) to be capable of inducing a population depletion that can easily be detected by photoionisation. This challenge has been resolved by Gerhards who has been able to extend the IR laser tuning range from 2 to 16 μm, using three different crystals, while producing pulse energies around the mJ range.[98,101] The IR free electron laser,[102] in particular FELIX[103] at the FOM Institute Rujnhuisen provides a very high-powered (albeit not of very high resolution) laser source with IR frequencies down to about 100 cm^{-1}. Several groups have now investigated molecules of biological interest in the gas phase using the infrared depletion technique;[63,104,105] for example, de Vries, Kleinermanns *et al.* have investigated the guanine . . . cytosine cluster using IR depletion, generating a wealth of new vibrational information compared with previously recorded electronic excitation spectra.[106–111]

It should be noted that the "fingerprint" vibrational frequency region below 1000 cm^{-1} is of particular interest. It is obvious from observed spectra that an assignment can only be made with the help of quantum-chemical *ab initio* calculations. In this spectral region, where the vibrations are much softer than in the high-frequency region, it is imperative to perform anharmonic frequency calculations in order to interpret spectra. Let us add here that anharmonic effects are also of prime importance, for hydrogen-bonded clusters, for a rather low barrier for proton transfer between proton donor and proton acceptor. In both cases the perturbative treatment of anharmonic motion is not adequate and the procedure fails to provide reasonable frequencies. The other newly developed tools, for instance, Gerber's anharmonic self-consistent field method,[112] or recently introduced adiabatic separation[113] allow full assignment of the currently available spectra of complex systems such as guanine . . . cytosine[114] or the stretch vibration modes of various tautomers of guanine not far from spectroscopic accuracy (a few cm^{-1}).[113]

Time-resolved picosecond pump-probe,[115–125] and hole-burning experiments are also a very useful tool for the study of the dynamics in molecular clusters. This spectroscopic method has been carried out for investigating the intracluster vibrational energy redistribution (IVR) and subsequent dissociation of molecular clusters. It provides information about the dynamics of energy transfer from one moiety to a non-covalent bond, often in a two-stage process, and reveals that the intramolecular vibrational energy redistribution takes place within the chromophore site, creating a hot moiety (or compound). Then the energy flows from the hot moiety to the intermolecular vibrational modes of the cluster and finally, the molecular cluster dissociates. A picosecond pump-probe method was employed to investigate the dynamics of competition between H-bonding and π-stacking[126,127] in the phenol . . . Ar_2 cation radical (see Figure 1.2).[128] This can also be viewed as competition between H-bonding (hydrophilic) and π-stacking (hydrophobic) binding sites, which is an example of a chemical recognition process at the molecular level. Reference 128 describes the first direct observation of a hydrophobic $\rightarrow$ hydrophilic site switching induced by resonant ionisation in the phenol . . . Ar_2 trimer cation radical. When the cation radical is prepared by photoionisation, it is produced in the π-bound geometry of the neutral precursor, with the Ar atom binding to the hydrophobic ring site. On the time scale of a few picoseconds, one of the Ar atoms switches from the hydrophobic ring site to the hydrophilic OH site, thus creating a hydrogen bond. The dynamics of this isomerisation process is monitored in real time by a change in the OH-stretching vibrational wave

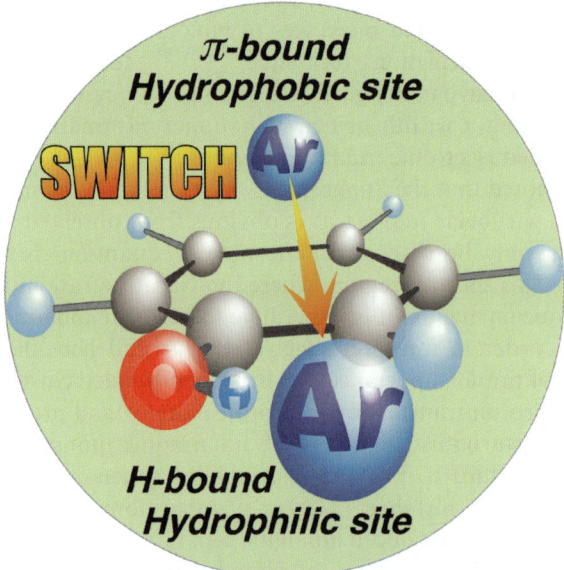

Figure 1.2   The first direct observation of a hydrophobic $\rightarrow$ hydrophilic site switching induced by resonant ionisation in the phenol . . . Ar_2 trimer cation radical. By photoionisation the cation radical is produced in the π-bound geometry of the neutral precursor with Ar binding to the hydrophobic ring site. On the time scale of a few picoseconds, one of the Ar atoms switches from the hydrophobic ring site to the hydrophilic OH site thus creating a hydrogen bond. The dynamics of this isomerisation process are monitored in real time by a change in the OH-stretching vibrational wave number using time-resolved picosecond UV-IR pump-probe ionisation depletion spectroscopy (reproduced with permission from ref. 128).

number using time-resolved picosecond UV-IR pump-probe ionisation depletion spectroscopy.[128]

1.6 Covalent and Non-covalent Interactions in Nature

Broadly, chemistry means covalent bonding. The covalent description is fully adequate when a molecule is considered in free space, *i.e.* isolated from any surroundings. Experimental conditions in molecular beams made from a supersonic jet expansion are close to these conditions. However, once the molecule is surrounded by other molecules, such as in solution or in the bulk, these surroundings affect the covalent bonding and the electronic system of the molecule is perturbed. This perturbation depends on the strength and extent of non-covalent interactions with the most pronounced changes occurring in ionic and H-bonded systems. For certain cases the H-bonding interactions, for instance between an anion and a polar neutral system, and covalent interactions can be of comparable strength. Mostly, however, non-covalent interactions are considerably weaker (by one to three orders of magnitude) than covalent bonds. Despite, or probably because of this, non-covalent bonds play a subtle but decisive key role in nature.

First, the very existence of a liquid phase, and also all related effects like solvation phenomena, can be attributed to non-covalent interactions. The existence of a condensed phase probably represents the most important example of non-covalent interactions and it must be stated that theory still has serious problems in describing adequately the role of solvents in general and, more specifically, the role of liquid water. Whereas covalent systems can now be investigated with chemical accuracy ($\sim$ 1 kcal/mol), the error in evaluation of hydration energies is much larger and, even worse, results from various theoretical methods differ significantly. Development of new procedures allowing determination of hydration energy is an important task for today's theoretical chemists.

Second, non-covalent interactions are responsible for the structure of biomacromolecules such as DNA, RNA, peptides and proteins. It must be recalled that the double-helical structure of DNA, which is of key importance for the transfer of genetic information, is mainly due to non-covalent interactions in which the interactions of nucleic acid bases play a dominant role. These bases are polar, aromatic heterocycles that interact either *via* planar H-bonds or vertical π–π interactions, resulting in two structural motifs, planar H-bonding and π-stacking; both are important not only in determining the architecture of nucleic acids but in a much more general sense. The basic question is: what is the relative strength of these interactions? It was long believed that specific H-bonding originating from electrostatic effects was the dominant term, while the nonspecific stacking originating due to London dispersion effects was considered to be energetically much less significant. Only the most accurate calculations (see Section 2.3) have revealed that stacking can be associated with surprisingly large stabilisation energies comparable with those of strong H-bonding.[129–131]

Third, non-covalent interactions play a key role in molecular recognition processes. This molecular recognition is most important in life processes, where it ensures an extremely high fidelity in the formation of biomacromolecules.

The key role of non-covalent interactions in biodisciplines is also conditioned by easy formation and also easy decomposition of non-covalent complexes. For example, DNA should be stiff enough to be able to store genetic information, yet simultaneously soft enough to allow, after receiving enzymatic information, for unwinding and thus reproduction of genetic information. Opening and closing of the molecule should be perfectly reproducible. Nature has evidently selected non-covalent interactions over the too strong (approximately by one to two orders of magnitude) covalent bonds.

Finally, non-covalent interactions play a role not only in microscopic but also in the macroscopic world. The effect of purely dispersion-driven non-covalent interactions can be seen in the fantastic ability of geckos to rapidly climb up smooth vertical surfaces, even flat glass. It was found that even a macroscopic, quite large animal such as a gecko can fully support its substantial body weight using the non-covalent interactions between the few hundred thousand keratinous hairs, or setae, on their feet and the surface. The adhesive force values observed support the hypothesis that individual setae operate *via* non-covalent interactions, or as the authors put it, van der Waals forces.[132]

1.6.1   Quantum-Chemical Methods for Non-covalent Complexes

Non-covalent interactions can be studied in a manner similar to covalent interactions, that is, by standard methods of quantum chemistry based either on perturbation or variation theory. While the former approach separates the overall stabilisation energy into various physically well-defined contributions such as electrostatic, induction, dispersion and exchange-repulsion energies, the latter separates the stabilisation energy only into Hartree–Fock (HF) and post-HF (correlation) contributions. Furthermore, the Hartree–Fock or correlation energy is not observable, nor are the separate energy contributions (however physically meaningful they may be) defined in the perturbation theory expansion. An important advantage of this approach (over variational) remains, namely the perturbation treatment is free of the most serious problem of the variation calculation – the basis set inconsistency. This inconsistency is associated with the fact that in a variational calculation the supersystem is described in the sum over the basis sets of all subsystems, leading to a more complete description compared to the isolated subsystems (being described only by their own individual basis sets) and thus to an artificially lowered energy. The effect was first independently recognised already at the end of the 1970s by Kestner[133a], and Jansen and Ros[133b] and the latter authors even suggested (correctly) how to eliminate it. One year later Boys and Bernardi, probably completely independently, came to the same solution of the problem and called the respective procedure the function counterpoise (CP) method and the respective error the basis set superposition error, BSSE.[134] This effect, which is purely mathematical in origin, has no physical meaning and can be eliminated either by describing the subsystems in the basis set of the whole supersystem, or by using infinite AO basis sets. Using the first approach one can obtain BSSE-corrected energy as well as any complex property that can be derived as a first- or second-order derivative of the energy; this means optimised geometries or vibrational frequencies. The counterpoise correction is realised by introducing so-called ghost orbitals, *i.e.* the ghost subsystem has basis set functions placed at the position of its atoms but no electrons and protons. This means the BSSE is geometry dependent, different orientations of the ghost system (toward the real one) yield different values of the BSSE. The BSSE-corrected characteristics converge much faster to the limit values than the BSSE-uncorrected ones. The second approach (the use of an infinitive basis set), which seemed quite unrealistic until recently, is now available through an extrapolation procedure to the complete basis set (CBS) limit (see Section 2.3.2), for which, by definition, the BSSE disappears. The BSSE-corrected and BSSE-uncorrected energy curves cross at one point and this fact is used in one extrapolation technique (see later) thus allowing the CBS limit to be obtained. Any variational treatment, which supposedly eliminates this basis set inconsistency, can be easily tested on the basis of obtaining the stabilisation energy by perturbation theory, which is, by definition, free of this problem.

The BSSE originates from a non-adequate description of a subsystem that then "tries" to improve it by "borrowing" functions from the other subsystem(s). Is this effect limited only to molecular clusters? The answer is no and the same effect should take place also within an isolated system where one part is improving its description by "borrowing" orbitals from the other one. The existence of the "intramolecular" BSSE was for a long time not accepted and it was "discovered" only recently. The role of intramolecular BSSE was first shown for aromatic–backbone interactions occurring in peptides.[135–137] It was shown that various peptide structures and their relative energies are dramatically affected by this intramolecular BSSE. Elimination of this BSSE is (contrary to the intermolecular case) more involved and, as the easiest way, the use of as large a basis set as possible is recommended. An elegant solution to the problem was suggested by Asturiol *et al.*[138] who removed the problem of nonplanarity of benzene and other arenes simply by correcting the total energy (of benzene) for the intramolecular BSSE using twelve atomic fragments, or six C–H fragments, or three H–C–C–H fragments. In all cases, correcting for the intramolecular BSSE fixes the erroneous behaviour of correlated methods such as MP2 or CCSD.

Splitting the stabilisation energy into components is important mainly because it allows for the nature of stabilisation to be understood. In the majority of cases we are, however, interested only in the total stabilisation energy, which is without doubt more accurately and easily provided by the variational calculation. Further, the variational calculations are easily combined with gradient optimisation of the whole complex, thus allowing us to determine not only the intermolecular coordinates but also the intramolecular ones. Problems associated with basis set inconsistency (which is the main obstacle of the variational calculations) have now become less serious, which is due to: (i) hardware and software development allowing the use of much larger AO basis sets that, by definition, are consistent with lower BSSE; (ii) development of procedures that allow gradient optimisation that takes the basis set inconsistency effect into account *a priori*, in each gradient cycle; and (iii) development of extrapolation techniques allowing for the estimation of the complete basis set limit that should, by definition, be free of BSSE.

1.7   Aims of this Book

This book aims to understand the main aspects of non-covalent chemistry (mainly in the gas phase) and specifically compares experimental and theoretical data available for non-covalent complexes and subsequent problems associated with this comparison. We stress that the book reflects the personal views of both authors: we are considering both experimental (KMD) and theoretical (PH) aspects of non-covalent chemistry and we believe our views might be of help to researchers engaged in both theoretical and experimental areas of this subject area. We are certainly aware that two authors working in quite different areas will bring two different languages and despite our effort for their convergence some differences remain. We feel, however, that experiment and theory is now so far away that their consistent description by one author is practically impossible. Finally, and this is the main philosophy of our approach, we are both strongly convinced that any textbook on non-covalent interactions cannot be limited either to theory or experiment. Both approaches are nowadays so closely connected that one cannot exist without the other and *vice versa* and their mutual connection provides the consistent description of non-covalent processes in our world.

References

1. J. D. van der Waals, Over de continuiteit van den gas- en vloeistof toestand. Ph.D. Thesis, Leiden 1873, quoted according to L. M. Brown, A. Pais, B. Pippard (ed.) Twentieth Century Physics, Vol. I. American Institute of Physics Press, New York, 1995.
2. H. Kamerlingh-Onnes, *Proc. Sec. Sci.* 1908, KNAW, **XI**, 68.
3. F. London, *Z. Phys. Chem.*, 1930, **B11**, 222.
4. H. Hellmann, *Acta Physicochim.*, 1935, **2**, 273.
5. J. O. Hirschfelder, C. F. Curtiss and R. B. Bird, *Molecular Theory of Gases and Liquids*, Wiley, New York, 1954.
6. P. Hobza and R. Zahradník, *Intermolecular Complexes*, Elsevier, Amsterdam, 1988.
7. (a) K. Müller-Dethlefs and P. Hobza, *Chem. Rev.*, 2000, **100**, 143; (b) Van der Waals Complexes I, II, III: *Chem. Rev.* 1988, **88**; 1994, **94**; 2000, **100**; (c) Stacking Interactions: *Phys. Chem. Chem. Phys.* 2008, **10**.

8. A. J. Stone, *The Theory of Intermolecular Forces*, Oxford University Press, Oxford, 2002.

9. (a) A. Karshikoff, *Non-covalent Interactions in Proteins*, Imperial College Press, Singapore, 2006; (b) I. G. Kaplan, *Intermolecular Interactions; Physical Picture, Computational Methods and Model Potentials*, Wiley, Chichester, 2006.

10. P. Hobza, R. Zahradník and K. Müller-Dethlefs, *Collect. Czech. Chem. Commun.*, 2006, **71**, 443.

11. L. Pauling, *J. Am. Chem. Soc.*, 1931, **53**, 1367; L. Pauling, The Nature of the Chemical Bond, Cornell University Press, Ithaca, 1939.

12. K. Müller-Dethlefs, M. Sander and E. W. Schlag, *Z. Naturforsch.*, 1984, **39a**, 1089.

13. K. Müller-Dethlefs, M. Sander and E. W. Schlag, *Chem. Phys. Lett.*, 1984, **112**, 291.

14. K. Müller-Dethlefs, M. Sander and L. A. Chewter, in: *Laser Spectroscopy VII*, ed. T.W. Hänsch and Y.R. Shen, Springer Verlag, 1985, p. 118.

15. G. Reiser, W. Habenicht, K. Müller-Dethlefs and E. W. Schlag, *Chem. Phys. Lett.*, 1988, **152**, 119.

16. W. Habenicht, G. Reiser and K. Müller-Dethlefs, *J. Chem. Phys.*, 1991, **95**, 4809.

17. K. Müller-Dethlefs, *J. Chem. Phys.*, 1991, **95**, 4822.

18. C. E. H. Dessent, S. R. Haines and K. Müller-Dethlefs, *Chem. Phys. Lett.*, 1999, **315**, 103.

19. C. E. H. Dessent and K. Müller-Dethlefs, *Chem. Rev.*, 2000, **100**, 3999.

20. W. Kim, M. W. Schaeffer, S. Lee, J. S. Chung and P. M. Felker, *J. Chem. Phys.*, 1999, **110**, 11264.

21. S. Lee, J. S. Chung, P. M. Felker, J. L. Cacheiro, B. Fernandez, T. B. Pedersen and H. Koch, *J. Chem. Phys.*, 2003, **119**, 12956.

22. S. Lee, J. Romascan, P. M. Felker, T. B. Pedersen, B. Fernandez and H. Koch, *J. Chem. Phys.*, 2003, **118**, 1230.

23. C. Tanner, D. Henseler, S. Leutwyler, L. L. Connell and P. M. Felker, *J. Chem. Phys.*, 2003, **118**, 9157.

24. B. C. Garrett, D. A. Dixon, D. M. Camaioni, D. M. Chipman, M. A. Johnson, C. D. Jonah, G. A. Kimmel, J. H. Miller, T. N. Rescigno, P. J. Rossky, S. S. Xantheas, S. D. Colson, A. H. Laufer, D. Ray, P. F. Barbara, D. M. Bartels, K. H. Becker, H. Bowen, S. E. Bradforth, I. Carmichael, J. V. Coe, L. R. Corrales, J. P. Cowin, M. Dupuis, K. B. Eisenthal, J. A. Franz, M. S. Gutowski, K. D. Jordan, B. D. Kay, J. A. LaVerne, S. V. Lymar, T. E. Madey, C. W. McCurdy, D. Meisel, S. Mukamel, A. R. Nilsson, T. M. Orlando, N. G. Petrik, S. M. Pimblott, J. R. Rustad, G. K. Schenter, S. J. Singer, A. Tokmakoff, L. S. Wang, C. Wittig and T. S. Zwier, *Chem. Rev.*, 2005, **105**, 355.

25. A. Abou El-Nasr, A. Fujii, T. Ebata and N. Mikami, *Mol. Phys.*, 2005, **103**, 1561.

26. E. A. El-Hakam, A. Ei-Nasr, A. Fujii, T. Yahagi, T. Ebata and N. Mikami, *J. Phys. Chem. A*, 2005, **109**, 2498.

27. P. Imhof, D. Krugler, R. Brause and K. Kleinermanns, *J. Chem. Phys.*, 2004, **121**, 2598.
28. R. H. Wu and B. Brutschy, *J. Phys. Chem. A*, 2004, **108**, 9715.
29. C. Jacoby, W. Roth, M. Schmitt, C. Janzen, D. Spangenberg and K. Kleinermanns, *J. Phys. Chem. A*, 1998, **102**, 4471.
30. W. Roth, M. Schmitt, C. Jacoby, D. Spangenberg, C. Janzen and K. Kleinermanns, *Chem. Phys.*, 1998, **239**, 1.
31. M. Schmitt, C. Jacoby, M. Gerhards, C. Unterberg, W. Roth and K. Kleinermanns, *J. Chem. Phys.*, 2000, **113**, 2995.
32. M. Schmitt, C. Ratzer and W. L. Meerts, *J. Chem. Phys.*, 2004, **120**, 2752.
33. A. Westphal, C. Jacoby, C. Ratzer, A. Reichelt and M. Schmitt, *Phys. Chem. Chem. Phys.*, 2003, **5**, 4114.
34. R. N. Casaes, J. B. Paul, R. P. McLaughlin, R. Saykally and T. van Mourik, *J. Phys. Chem. A*, 2004, **108**, 10989.
35. A. J. Huneycutt, R. N. Casaes, B. J. McCall, C. Y. Chung, Y. P. Lee and R. J. Saykally, *ChemPhysChem*, 2004, **5**, 321.
36. L. Biennier, F. Salama, L. J. Allamandola and J. J. Scherer, *J. Chem. Phys.*, 2003, **118**, 7863.
37. X. F. Tan and F. Salama, *J. Chem. Phys.*, 2005, **122**, 084318.
38. E. Witkowicz, H. Linnartz, C. A. de Lange, W. Ubachs, A. Sfounis, M. Massaouti and M. Velegrakis, *Int. J. Mass Spectrom.*, 2004, **232**, 25.
39. G. C. Cole and A. C. Legon, *J. Chem. Phys.*, 2004, **121**, 10467.
40. P. W. Fowler, A. C. Legon, J. M. A. Thumwood and E. R. Waclawik, *Coord. Chem. Rev.*, 2000, **197**, 231.
41. B. M. Giuliano and W. Caminati, *Angew. Chem. Int. Ed.*, 2005, **44**, 603.
42. B. M. Giuliano, P. Ottaviani, W. Caminati, M. Schnell, D. Banser and J. U. Grabow, *Chem. Phys.*, 2005, **312**, 111.
43. J. L. Alonso, S. Antolinez, S. Blanco, A. Lesarri, J. C. Lopez and W. Caminati, *J. Am. Chem. Soc.*, 2004, **126**, 3244.
44. W. Caminati, J. C. Lopez, J. L. Alonso and J. U. Grabow, *Angew. Chem. Int. Ed.*, 2005, **44**, 3840.
45. R. Sanchez, S. Blanco, A. Lesarri, J. C. Lopez and J. L. Alonso, *Chem. Phys. Lett.*, 2005, **401**, 259.
46. A. Lesarri, E. J. Cocinero, J. C. Lopez and J. L. Alonso, *Angew. Chem. Int. Ed.*, 2004, **43**, 605.
47. S. Blanco, J. C. Lopez, A. Lesarri, W. Caminati and J. L. Alonso, *Mol. Phys.*, 2005, **103**, 1473.
48. N. Goldman, C. LeForestier and R. J. Saykally, *Philos. Trans. Royal Soc. London Series A-Math. Phys. Eng. Sci.*, 2005, **363**, 493.
49. F. N. Keutsch, J. D. Cruzan and R. J. Saykally, *Chem. Rev.*, 2003, **103**, 2533.
50. J. D. Cruzan, L. B. Braly, K. Liu, M. G. Brown, J. G. Loeser and R. J. Saykally, *Science*, 1996, **271**, 59.
51. S. Chervenkov, P. Q. Wang, J. E. Braun, S. Georgiev, H. J. Neusser, C. K. Nandi and T. Chakraborty, *J. Chem. Phys.*, 2005, **122**, 244312.

52. S. Georgiev and H. J. Neusser, *J. Electron Spectrosc. Relat. Phenom.*, 2005, **142**, 207.
53. K. Siglow and H. J. Neusser, *Chem. Phys. Lett.*, 2001, **343**, 475.
54. F. Aguirre and S. T. Pratt, *J. Chem. Phys.*, 2004, **121**, 9855.
55. M. R. Hockridge, E. G. Robertson, J. P. Simons, D. R. Borst, T. M. Korter and D. W. Pratt, *Chem. Phys. Lett.*, 2001, **334**, 31.
56. G. Berden, W. L. Meerts, M. Schmitt and K. Kleinermanns, *J. Chem. Phys.*, 1996, **104**, 972.
57. M. Gerhards, M. Schmitt, K. Kleinermanns and W. Stahl, *J. Chem. Phys.*, 1996, **104**, 967.
58. C. Kang and D. W. Pratt, *Int. Rev. Phys. Chem.*, 2005, **24**, 1.
59. C. W. Kang, J. T. Yi and D. W. Pratt, *J. Chem. Phys.*, 2005, **123**, 094306.
60. T. V. Nguyen, T. M. Korter and D. W. Pratt, *Mol. Phys.*, 2005, **103**, 2453.
61. D. W. Pratt, *Science*, 2002, **296**, 2347.
62. J. A. Reese, T. V. Nguyen, T. M. Korter and D. W. Pratt, *J. Am. Chem. Soc.*, 2004, **126**, 11387.
63. Y. H. Lee, J. W. Jung, B. Kim, P. Butz, L. C. Snoek, R. T. Kroemer and J. P. Simons, *J. Phys. Chem. A*, 2004, **108**, 69.
64. D. R. Borst, P. W. Joireman, D. W. Pratt, E. G. Robertson and J. P. Simons, *J. Chem. Phys.*, 2002, **116**, 7057.
65. M. S. Ford, X. Tong, C. E. H. Dessent and K. Müller-Dethlefs, *J. Chem. Phys.*, 2003, **119**, 12914.
66. X. Tong, M. S. Ford, C. E. H. Dessent and K. Müller-Dethlefs, *J. Chem. Phys.*, 2003, **119**, 12908.
67. M. S. Ford and K. Müller-Dethlefs, *Phys. Chem. Chem. Phys.*, 2004, **6**, 23.
68. C. Riehn, C. Lahmann, B. Wassermann and B. Brutschy, *Chem. Phys. Lett.*, 1992, **197**, 443.
69. M. J. Elrod, R. J. Saykally, A. R. Cooper and J. M. Hutson, *Mol. Phys.*, 1994, **81**, 579.
70. R. S. Fellers, C. Leforestier, L. B. Braly, M. G. Brown and R. J. Saykally, *Science*, 1999, **284**, 945.
71. L. A. Chewter, K. Müller-Dethlefs and E. W. Schlag, *Chem. Phys. Lett.*, 1987, **135**, 219.
72. G. Reiser, O. Dopfer, R. Lindner, G. Henri, K. Müller-Dethlefs, E. W. Schlag and S. D. Colson, *Chem. Phys. Lett.*, 1991, **181**, 1.
73. K. Müller-Dethlefs, *J. Electr. Spectrosc. Rel. Phenom.*, 1995, **75**, 35.
74. K. Müller-Dethlefs and E. W. Schlag, *Angew. Chem. Int. Ed.*, 1998, **37**, 1346.
75. K. Müller-Dethlefs, O. Dopfer and T. G. Wright, *Chem. Rev.*, 1994, **94**, 1845.
76. K. Müller-Dethlefs, *High Resolution Photoionization and Photoelectron Studies*, ed. I. Powis, T. Baer, C.Y. Ng, J. Wiley & Sons Ltd, Chichester, 1995, p.21.
77. M. C. R. Cockett, K. Müller-Dethlefs and T. G. Wright, *Ann. Rep. Royal Soc. Chem. Sect. C*, 1998, **94**, 327.

78. K. Müller-Dethlefs and M. C. R. Cockett, Chapter 7 in *Nonlinear Spectroscopy for Molecular Structure Determination*, Blackwell Science, Oxford, England, 1998.

79. K. Müller-Dethlefs, E. W. Schlag, E. R. Grant, K. Wang and B. V. McKoy, *Advances in Chemical Physics XC*, Wiley, Chichester, 1995.

80. F. Merkt, *Ann. Rev. Phys. Chem.*, 1997, **48**, 675.

81. F. Merkt and T. P. Softley, *Int. Rev. Phys. Chem.*, 1993, **12**, 205.

82. D. A. Beattie and R. J. Donovan, *Prog. React. Kin.*, 1998, **23**, 281.

83. R. B. Bernstein, *J. Phys. Chem.*, 1982, **86**, 1178.

84. S. Leutwyler and J. Bosinger, *Chem. Rev.*, 1990, **90**, 489.

85. C. J. Gruenloh, J. R. Carney, C. A. Arrington, T. S. Zwier, S. Y. Fredericks and K. D. Jordan, *Science*, 1997, **276**, 1678.

86. H.-D. Barth, K. Buchhold, S. Djafari, B. Reimann, U. Lommatzsch and B. Brutchy, *Chem. Phys.*, 1998, **239**, 49.

87. Ch. Janzen, D. Spangenberg, W. Roth and K. Kleinermanns, *J. Chem. Phys.*, 1999, **110**, 9898.

88. B. C. Dian, J. R. Clarkson and T. S. Zwier, *Science*, 2004, **303**, 1169.

89. J. R. Clarkson, B. C. Dian, L. Moriggi, A. DeFusco, V. McCarthy, K. D. Jordan and T. S. Zwier, *J. Chem. Phys.*, 2005, **122**, 214311.

90. S. Tanabe, T. Ebata, M. Fujii and N. Mikami, *Chem. Phys. Lett.*, 1993, **215**, 347.

91. M. Miyazaki, A. Fujii, T. Ebata and N. Mikami, *Science*, 2004, **304**, 1134.

92. M. Miyazaki, A. Fujii and N. Mikami, *J. Phys. Chem. A*, 2004, **108**, 8269.

93. V. Venkatesan, A. Fujii and N. Mikami, *Chem. Phys. Lett.*, 2005, **409**, 57.

94. M. Gerhards, A. Jansen, C. Unterberg and K. Kleinermanns, *Chem. Phys. Lett.*, 2001, **344**, 113.

95. I. Hunig, K. A. Seefeld and K. Kleinermanns, *Chem. Phys. Lett.*, 2003, **369**, 173.

96. R. Linder, M. Nispel, T. Haber and K. Kleinermanns, *Chem. Phys. Lett.*, 2005, **409**, 260.

97. M. Honda, A. Fujii, E. Fujimaki, T. Ebata and N. Mikami, *J. Phys. Chem. A*, 2003, **107**, 3678.

98. A. Gerlach, C. Unterberg, H. Fricke and M. Gerhards, *Mol. Phys.*, 2005, **103**, 1521.

99. M. Gerhards, C. Unterberg, A. Gerlach and A. Jansen, *Phys. Chem. Chem. Phys.*, 2004, **6**, 2682.

100. M. Gerhards, *Opt. Commun.*, 2004, **241**, 493.

101. M. Gerhards, private communication.

102. A. Doria, R. Bartolini, J. Feinstein, G. P. Gallerano and R. H. Pantell, *IEEE J. Quantum Electron.*, 1993, **29**, 1428.

103. M. S. Ding, H. H. Weits and D. Oepts, *Nucl. Instrum. Methods Phys. Res. Sect. A*, 1997, **393**, 504.

104. P. Carcabal, R. A. Jockusch, I. Hunig, L. C. Snoek, R. T. Kroemer, B. G. Davis, D. P. Gamblin, I. Compagnon, J. Oomens and J. P. Simons, *J. Am. Chem. Soc.*, 2005, **127**, 11414.

105. R. A. Jockusch, R. T. Kroemer, F. O. Talbot, L. C. Snoek, P. Carcabal, J. P. Simons, M. Havenith, J. M. Bakker, I. Compagnon, G. Meijer and G. von Helden, *J. Am. Chem. Soc.*, 2004, **126**, 5709.

106. E. Nir, C. Plutzer, K. Kleinermanns and M. de Vries, *Eur. Phys. J. D*, 2002, **20**, 317.

107. K. A. Seefeld, C. Plutzer, D. Lowenich, T. Haber, R. Linder, K. Kleinermanns, J. Tatchen and C. M. Marian, *Phys. Chem. Chem. Phys.*, 2005, **7**, 3021.

108. M. Abd El Rahim, R. Antoine, L. Arnaud, M. Barbaire, M. Broyer, C. Clavier, I. Compagnon, P. Dugourd, J. Maurelli and D. Rayane, *Rev. Sci. Instrum.*, 2004, **75**, 5221.

109. R. Brause, M. Schmitt, D. Krügler and K. Kleinermanns, *Mol. Phys.*, 2004, **102**, 1615.

110. W. Chin, I. Compagnon, J. P. Dognon, C. Canuel, F. Piuzzi, I. Dimicoli, G. von Helden, G. Meijer and M. Mons, *J. Am. Chem. Soc.*, 2005, **127**, 1388.

111. N. C. Polfer, B. Paizs, L. C. Snoek, I. Compagnon, S. Suhai, G. Meijer, G. von Helden and J. Oomens, *J. Am. Chem. Soc.*, 2005, **127**, 8571.

112. S. K. Gregurick, G. M. Chaban and B. Gerber, *Biophys. J.*, 2001, **80**, 1256.

113. D. Nachtigallová, P. Hobza and V. Špirko, *J. Phys. Chem. A*, 2008, **112**, 1854.

114. G. M. Chaban, J. O. Jung and R. B. Gerber, *J. Chem. Phys.*, 1999, **111**, 1823.

115. Y. Yamada, T. Ebata, M. Kayano and N. Mikami, *J. Chem. Phys.*, 2004, **120**, 7400.

116. M. Miyazaki, A. Fujii, T. Ebata and N. Mikami, *Chem. Phys. Lett.*, 2004, **399**, 412.

117. Y. Yamada, N. Mikami and T. Ebata, *J. Chem. Phys.*, 2004, **121**, 11530.

118. T. Ebata, M. Kayano, S. Sato and N. Mikami, *J. Phys. Chem. A*, 2001, **105**, 8623.

119. M. Kayano, T. Ebata, Y. Yamada and N. Mikami, *J. Chem. Phys.*, 2004, **120**, 7410.

120. J. D. Pitts and J. L. Knee, *J. Chem. Phys.*, 1998, **109**, 7113.

121. J. D. Pitts, J. L. Knee and S. Wategaonkar, *J. Chem. Phys.*, 1999, **110**, 3378.

122. X. Zhang and J. L. Knee, *Discuss. Faraday Soc.*, 1994, **97**, 299.

123. S. M. Bellm, P. T. Whiteside and K. L. Reid, *J. Phys. Chem. A*, 2003, **107**, 7373.

124. J. A. Davies, K. L. Reid, M. Towrie and P. Matousek, *J. Chem. Phys.*, 2002, **117**, 9099.

125. A. K. King, S. M. Bellm, C. J. Hammond, K. L. Reid, M. Towrie and P. Matousek, *Mol. Phys.*, 2005, **103**, 1821.

126. N. Solcá and O. Dopfer, *J. Phys. Chem. A*, 2001, **105**, 5637.

127. N. Solcá and O. Dopfer, *J. Mol. Struct.*, 2001, **241**, 563.

128. S. Ishiuchi, M. Sakai, Y. Tsuchida, A. Takeda, Y. Kawashima, M. Fujii, O. Dopfer and K. Müller-Dethlefs, *Angew. Chem. Int. Ed.*, 2005, **44**, 6149.

129. P. Jurečka and P. Hobza, *Chem. Phys. Lett.*, 2002, **365**, 89.
130. P. Hobza and J. Šponer, *J. Am. Chem. Soc.*, 2002, **124**, 11802.
131. P. Jurečka and P. Hobza, *J. Am. Chem. Soc.*, 2003, **125**, 15608.
132. K. Autumn, Y. A. Liang, S. T. Hsieh, W. Zesch, W. P. Chan, T. W. Kenny, R. Fearing and R. R. Full, *Nature*, 2000, **405**, 681.
133. (a) N. R. Kestner, *J. Chem. Phys.*, 1968, **48**, 252; (b) H. B. Jansen and P. Ros, *Chem. Phys. Lett.*, 1969, **3**, 140.
134. S. F. Boys and F. Bernardi, *Mol. Phys.*, 1970, **19**, 553.
135. N. Y. Palermo, J. Csontos, M. C. Owen, R. F. Murphy and S. Lovas, *J. Comput. Chem.*, 2007, **28**, 1208.
136. L. F. Holroyd and T. van Mourik, *Chem. Phys. Lett.*, 2007, **42**, 442.
137. H. Valdés, V. Klusák, M. Pitoňák, O. Exner, I. Starý, P. Hobza and L. Rulíšek, *J. Comput. Chem.*, 2008, **29**, 861.
138. D. Asturiol, M. Duran and P. Salvador, *J. Chem. Phys.*, 2008, **128**, 144108.

CHAPTER 2

Characteristics of Non-covalent Complexes and Their Determination by Experimental and Theoretical Techniques

2.1 Structure and Geometry

The primary property of any non-covalent complex is not just its "equilibrium structure" but its potential-energy surface (PES) with its stationary points. Determination of the structure and geometry of a complex cannot be separated from discussing its PES. To illustrate this point consider the intramolecular hydrogen atom dynamics of acetylacetone, for which a hydrogen-shifting keto–enol tautomerisation and an interconversion of the enol in two indistinguishable enolone structures was proposed.[1,2] Ultrafast electron-diffraction experiments on acetylacetone by Zewail and coworkers[3,4] have shown that of the two tautomeric forms in dynamic equilibrium the C_s-enolic structure is the dominant one. This was attributed to some π-delocalisation, leading to a type of intramolecular interaction called "resonance-assisted hydrogen bond". The net result is that the hydrogen prefers to be close to one of the oxygens (the enolone C_s structure) and not in between the two oxygens (C_{2v}). They also observed the interconversion of the enol between the two equivalent enolone minima on the PES. The observation of a C_s structure seems to contradict results from microwave spectra of acetylacetone that are compatible with a C_{2v} structure.[5] However, this "contradiction" is only a rhetorical one; it can be easily understood that the microwave experiment will show an averaged C_{2v} structure provided the interconversion barrier between the two enolones is sufficiently low. The microwave experiment averages the tunnelling motion of the hydrogen atom between the two enolones in a similar way to donor–acceptor tunnelling in the ammonia dimer (see Section 2.2.1). Similar results

RSC Theoretical and Computational Chemistry Series No. 2
Non-covalent Interactions: Theory and Experiment
By Pavel Hobza and Klaus Müller-Dethlefs
© Pavel Hobza and Klaus Müller-Dethlefs 2010
Published by the Royal Society of Chemistry, www.rsc.org

and conclusions were obtained for comparable systems exhibiting intramolecular hydrogen bonds such as malondialdehyde.[6,7]

For the nonspecialist reader who might think experimental spectroscopic data can always be interpreted explicitly it must be stated that structural information resulting from rotational constants is not unambiguous since one set of rotational constants can be assigned to various structures.[8,9] Hence, some additional information, sometimes referred to as chemical intuition, or another technique (mostly theoretical calculations) should be combined with the evidence from rotational constants to generate reliable structural information. Rotational constants for larger complexes can be obtained by using various experimental techniques: microwave (MW), terahertz, (THZ), infrared (IR) spectroscopies. It is of very high interest to extend this to the determination of rotational constants in electronic excitation spectra, for instance in REMPI spectra[10–24] and also, for the cation complex, from ZEKE spectra,[25] or from IR predissociation direct absorption in the cation monitored by predissociation.[26–29]

In our book we will focus mainly on larger non-covalent complexes having more than 24 atoms (the benzene dimer, with 24 atoms, is considered here as an arbitrary boundary between small, medium and extended complexes). Direct experimental determination of structure and geometry of extended non-covalent complexes is impractical: they can presently be obtained only indirectly from rotational constants. It must, however, be mentioned that there exist only a very limited set of non-covalent complexes for which reliable rotational constant exist.

2.2 Microwave and Terahertz Spectroscopy

2.2.1 Ultrasoft Potentials: The Riddle of the Ammonia Dimer

One of the most powerful methods in molecular sciences, one could even call it the classical method of molecular spectroscopy, is microwave spectroscopy. Microwave spectroscopy features superior, very high spectral resolution and allows the obtaining of – besides rotational constants – hyperfine splittings and other interactions in molecules and molecular clusters with unprecedented precision. This is a mature technique that is applied in a routine way in many laboratories to study a large variety of molecular properties, including non-covalent interactions in molecular clusters and biomolecules in the gas phase.[30–38] What has to be said, however, is that structural determination is not as straightforward as one might believe, particularly for molecular clusters, in spite of the precision and accuracy of the spectral lines and the corresponding rotational constants and moments of inertia obtained. For rigid molecular systems there is a very clear recipe for molecular spectroscopy to reduce the rotational constants to moments of inertia, thus constructing a molecular structure. However, for nonrigid systems that exhibit large-amplitude motion – this is a particular feature of many molecular clusters held together by non-covalent interactions – and that are, furthermore, particularly susceptible to centrifugal distortion effects, the connection between spectroscopic constants determined from experiment and structural determination is not so straight-forward.

A prime example of this comes from the problem of the structure of ammonia dimer, $(NH_3)_2$, initiated by the seminal microwave spectroscopic experiments conducted by Klemperer and coworkers in the early 1980s.[39] Though this work has already been reviewed by us before[40] we are mentioning it here once more because this molecular cluster shows two important features: large-amplitude vibration and tunnelling motion. The original spectra of $(NH_3)_2$ and $(ND_3)_2$ were interpreted in terms of donor–acceptor tunnelling and rotational tunnelling of one NH_3 moiety (but not inversion tunnelling, which was assumed to be quenched by the dimer formation), according to the molecular symmetry group G_{16}.[41,42] Much to their surprise – and the surprise of the scientific community – their result implied a dimer structure that seemed not to be hydrogen-bonded at all. Their structure, shown in Figure 2.1, seemed to resemble a cyclic structure with the nitrogen atoms more or less facing each other in a distorted structure that did not at all seem to be hydrogen-bonded. However, it was quite apparent that the result from the microwave experiment would only be fully conclusive if a full interpretation of the rotational tun-nelling structure could be determined from the starting point of a high-level potential-energy surface calculated with a reliable *ab initio* method. This pro-cedure took more than a decade and resulted in a most decisive contribution from van der Avoird and coworkers, who finally computed the potential-energy surface to a sufficient precision to allow numerical determination of the full rotational tunnelling spectrum of the NH_3 dimer, and also the perdeuterated species, within the molecular symmetry group G_{144}.[43] The result shows con-vincingly that, for an extremely flat potential with respect to the dimer-bending motion, a normal microwave spectroscopy experiment would result in the determination of moments of inertia that represent an averaged vibrational-rotational structure. This shows clearly that the interpretation drawn from the first microwave spectra of the Klemperer group had, of course, been both correct but not fully conclusive at the same time. Correct in so far that – of

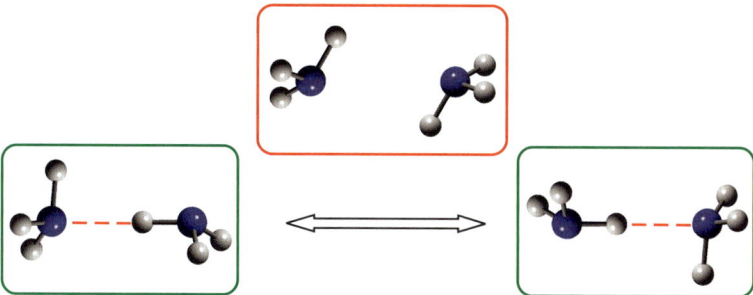

Figure 2.1 Ammonia dimer structure as it appears after vibrational averaging (centre) and two hydrogen-bonded structures (left and right) undergoing large-amplitude motion.

course – the moments of inertia determined were correctly deduced from the experimental data, but incomplete in the context that only with a high-quality potential-energy surface and a complete quantum calculation of energy states on that surface can the spectra be fully interpreted. These, naturally, were not available at the time when the first microwave measurements were carried out.

2.2.2   From Water Clusters to a Potential for Liquid Water

The condensed phase of water is interesting for different reasons, the most important being the fact that life probably originated in an aqueous phase. In chemistry, water plays also a key role as solvent. The unique position of water, *i.e.* what makes it different from isoelectronic or similar systems like HF, NH_3, CH_4, H_2S, is due to its various anomalous physical properties such as:[44]

- High boiling point
- Large heat of vaporisation
- Solid (ice) floats on liquid (water)
- Ice has many different phases
- Liquid density maximum at $4\,°C$
- Compressibility decreases with increasing T up to $46.5\,°C$
- Large specific heat C_p with minimum at $36\,°C$
- Large dielectric constant
- High surface tension (highest except metals)

Despite enormous progress in experiment and theory a united description of bulk water starting from the basic molecular interactions is still missing. Before starting the description of water clusters let us stress that vibrational spectroscopy of the OH-stretch mode is a sensitive indicator of the strength and coordination of the hydrogen bond.[45] This mode is shifted in water clusters up to $800\,cm^{-1}$ to the red, compared to the vibrational stretch modes of the isolated water molecule: the symmetric one at $3655\,cm^{-1}$, and the asymmetric one at $3756\,cm^{-1}$, respectively. (For a definition of H-bonding see Section 4.1.)

A story similar to that of the ammonia dimer, albeit of higher complexity, can be developed for water clusters. Water clusters are of substantial interest to understand not only the pairwise additive interactions, such as in the water dimer, but also the three-body (and many-body) interactions, such as those present in larger clusters like the trimer, tetramer, *etc.* The pioneering work of Saykally and coworker,[46] inspired by the first spectrum of the water dimer by Miller,[47] has led to a very significant improvement in the understanding of these systems.[48–50]

The method developed by Saykally is vibration–rotation–tunnelling (VRT) spectroscopy carried out in the terahertz region of the electromagnetic spectrum, *i.e.* around $100\,cm^{-1}$. In this spectral region, transitions are seen that involve not only rotations but also low-frequency vibrational and librational motions and, in particular, tunnelling motions. The water clusters all exhibit very significant tunnelling motions and the corresponding large-amplitude motions and passages through the associated transition states resemble complicated energy landscapes.[51] A concerted effort by several theoretical groups, from early *ab initio* calculations by Kim,[52,53] Kim and Jordan,[54] to extensive high-level *ab initio* explorations by Xantheas,[55] to full six-dimensional calculations by Leforestier,[56] and the use of the diffusion Monte Carlo method by Clary[57,58] has now left us with a clearer picture of what is happening in water clusters and how their dynamics relate to the structure and dynamics of liquid water. The first conclusion that may be drawn is that when you build up the clusters to a certain size, the first three-dimensional structure is the hexamer, all other clusters are planar (or, more precisely, pseudoplanar because some hydrogens point out of plane) as can be seen in Figure 2.2. Taking into account the zero-point energy, ZPE, it is now believed that the cage structure for the water hexamer is the lowest in energy,[59] in contrast to earlier work, which proposed the prism structure. However, this might not yet be the final definitive answer for the hexamer structure since the full consideration of the zero-point energy level on such potentials of such high dimensionality is extremely demanding. In other words, for such systems entropy plays a most important role and it is necessary to consider not only the PES, but rather the free (Gibbs) energy surface (FES).

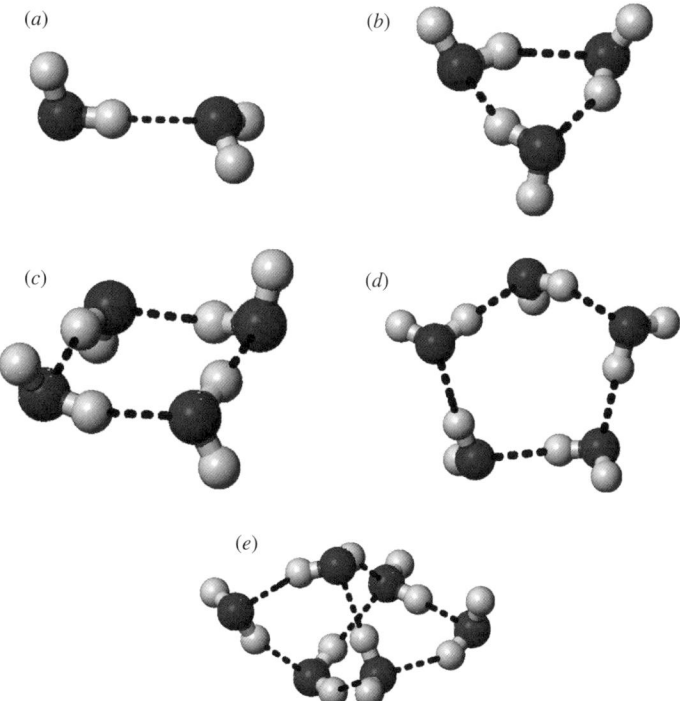

Figure 2.2 Water-cluster geometries from $n = 2$ (linear), $n = 3$ to $n = 5$ (planar), to $n = 6$ for the most probable three-dimensional water hexamer structure. Redrawn from Ref. 62.

With the present state-of-the-art it is now clear that the spectroscopy of water clusters and the corresponding theoretical efforts have led to very substantially improved potentials that will be useful for the simulation of bulk liquid water. These new potentials developed by different groups[60,61] are based on the knowledge gained from the cluster experiments: the pair potential must be known with extreme precision and the three-body interactions contribute about 20% to the potential energy; higher-body interactions give a rather negligible contribution.

A detailed description of small (2–10) water clusters based on terahertz VRT spectroscopy and *ab initio* quantum-mechanical calculations can be found in Ref. 62. These gargantuan efforts have led the way to stimulate the development of highly accurate water–water potentials based solely on *ab initio* calculations. Van der Avoird *et al.*[63] developed a force field for water entirely from first principles that contains both pairwise and many-body interactions. This force field predicts the properties of the water dimer and of liquid water in excellent agreement with experiments.

The pair polarisable analytical potential (CC-pol) was developed on the basis of the CCSD(T)/CBS (CBS = complete basis set limit) calculations and the same set of 2510 carefully selected grid points as in the previously used SAPT-5s potential[64] was used. Similarly as in this work the monomers were kept rigid. The MP2/CBS was constructed from aug-cc-pVTZ and aug-cc-pVQZ basis sets supplemented by midbond functions. The CCSD(T) correction term was determined with the cc-pVTZ basis set. The uncertainty of the computed interaction energies is about 0.07 kcal/mol, while that of the mentioned SAPT-5s potential was much higher (0.3 kcal/mol). The root-mean-square deviation of the fit for the points with negative energies was 0.09 kcal/mol. The functional form of CC-pol represents a compromise between the accuracy of reproducing the computed CCSD(T)/CBS energies and the simplicity needed for molecular simulations.

The VRT dimer levels for states with overall rotation quantum numbers J and K equal to 0, 1, and 2 computed with the CC-pol potential agree nicely with the levels deduced from the measured spectra, both for the ground state and for the excited intermolecular vibrations. The only substantial error occurs for the interchange splitting $a(0) + a(1)$, which is 19% too large. This observable is particularly sensitive to monomer flexibility effects that decrease its value.[65] Thus, rigid-monomer potentials such as CC-pol should overestimate this splitting. The present CC-pol potential performs considerably better than earlier *ab initio* potentials (ASP-W, ASP-S, SAPT-5s, TTM2.1, and SDFT-5s). The empirical TIP4P potential, which is widely used in simulations of biomacromolecules, fails completely.

Recently, Cencek *et al.*[66] refitted the *ab initio* water dimer interaction energies obtained from coupled cluster calculations and used in the CC-pol water pair potential to a site–site form containing eight symmetry-independent sites in each monomer, denoted as CC-pol-8s.

The water dimer vibration–rotation–tunneling spectrum predicted by the CC-pol-8s potential[67,68] agrees substantially and systematically better with experiment than the already very accurate spectrum predicted by CC-pol, while specific features that could not be accurately predicted previously now agree very well with experiment. It can be anticipated that the time is now ripe to see a very significant improvement in the modelling of the properties of liquid water from these results.

In this context it should also be mentioned that a completely different method for resolving the issue of modelling a complex hydrogen-bonded network such as water could come from the Carr–Parrinello method.[69] This method has been extensively tested on large complex systems and though it is based on density-functional theory it is extremely powerful since the potential is calculated on-the-fly during the propagation of the dynamics, rather than using a static approximation.

On lowering the temperature to $0\,^{\circ}C$ water freezes to ice, which is a well-known process. However, the modelling of the water freezing process with atomistic resolution is surprisingly difficult. The main problem is not the quality of the potential describing the water–water interactions but the extremely complicated potential-energy landscape possessing a large number of possible configurations. Consequently, very long simulation time is required. To the best of our knowledge there still exists only one study[70] where freezing of the bulk water from scratch was successfully modelled. In this extremely long simulation (several months on a supercomputer) an ice nucleus was finally formed spontaneously and, consequently, the whole system froze. This ice nucleation occurred once a sufficient number of relatively long-lived H-bonds (longer than 2 ns) developed spontaneously at the same location resulting in the formation of a compact initial nucleus. The authors performed 6 trajectory calculations (512 water molecules in a cubic box with a periodic condition) of the order of microseconds but only one successfully crystallised. During the study the standard TIP4P empirical potential was utilised. In order to facilitate the simulations of the freezing process a box of atomistic water was placed in contact with a prebuilt ice patch.[71] This procedure avoids simulating the rather rare event of the spontaneous formation of an ice nucleus, which is extremely demanding on simulation time.

2.2.3 Rotational Coherence Spectroscopy

Another, still quite new technique for the determination of rotational constants, using ultrafast laser pulses, is based on rotational coherence and the measurement of its recurrence.[72,73] Most recently, this method has been substantially refined by Brutschy, Riehn and coworkers,[74,75] using a two-colour, transform-limited picosecond laser system and ionisation detection. Due to the very high quality of their spectra the analysis and the determination of rotational constants (which is not unambiguous) have been substantially improved and it is fair to state that rotational coherence spectroscopy is now a complementary method compared with energy-level-resolved spectroscopy.

2.2.4 Quantum-Chemical *ab initio* Methods

The "inversion" of experimental data to obtain the structure and geometry of a molecular cluster is normally not possible and fitting procedures using a suitable model are generally employed. In contrast, theoretical quantum-chemical methods, using gradient-optimisation techniques, yield reliable structural and geometrical information even for extended complexes. Over the last half century, *ab initio* computational techniques have been developed and extensively used for covalent systems. The recipe for *ab initio* computations of non-covalent complexes is essentially the same: use basis sets as large as possible and take into account as much of the correlation energy as you can afford. The use of the Hartree–Fock (HF) method is not recommended since the method does not cover the correlation energy. The situation with DFT methods is more involved since these methods cover some part of correlation energy (see later). In the past geometry and structure determination was considered as less demanding than energy determination and the use of lower-level theoretical procedures was recommended. It seems now that for obtaining highly accurate structural data the state-of-the art theoretical procedures should be applied. For small complexes it is possible to optimise the geometry by using the most reliable (and most expensive) CCSD and CCSD(T) methods and for complexes with no more than a few atoms even the CCSDT methods.[76] For extended complexes the use of these higher-level coupled cluster methods for geometry optimisation is difficult and too expensive and mostly only a limited number of internal coordinates is optimised. For a long time it was believed that coupled-cluster and MP2 methods provide similar characteristics. This view should be, however, modified in the light of the fact that MP2 for some structures strongly overestimates (in comparison with CCSD(T)) the binding energy. Consequently, the MP2 and CCSD(T) optimised geometries can differ substantially.

The choice of basis set is critical for a reasonably reliable description of any structural type of non-covalent complex: the basis sets required should have at least two sets of first-polarisation functions and one set of second-polarisation functions, conditions satisfied by, *e.g.* the cc-pVTZ basis set (for atoms of the second period and hydrogen: [4s3p2d1f/3s2p1d]). A plethora of work provides evidence that the MP2/cc-pVTZ gradient optimisation or counterpoise-corrected (cc) gradient optimisation at the same level yield correct geometries for the neutral ground state of molecular complexes and that MP2/cc-pVTZ is the level of theory required to distinguish between, for instance, stacking and hydrogen bonding in nuclear base pairs (see later). The cc-gradient optimisation yields generally longer intersystem separations and in the case of flat PES the difference (with respect to standard gradient optimisation) can be rather large (0.3–0.6 Å). Specifically, geometries of our benchmark database of accurate interaction energies (set S22 in Ref. 77; for a detailed description of the S22 set see Section 3.6) were mostly determined at the latter level. It must, however, be mentioned that this is only because of compensation of errors (extension of the basis set in the MP2 and consideration of the CCSD(T) procedure).

The problem mentioned can be best demonstrated with the case of phenol dimer. The system is interesting from several viewpoints. First, the complex belongs to very few (extended) complexes for which the rotational constants were determined by time-resolved rotational[78,79] coherence spectroscopy.

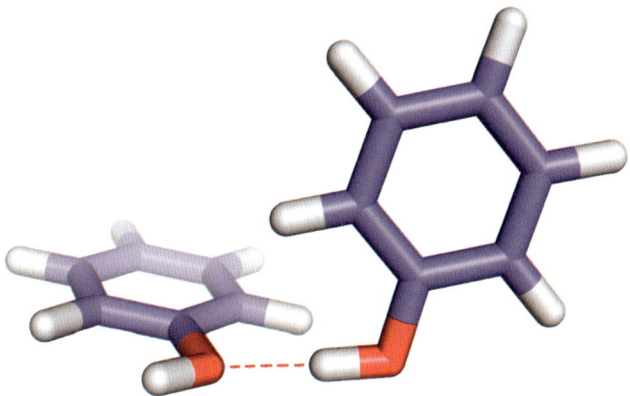

Figure 2.3 Optimised structure of the phenol dimer.

Further, its structure (see Figure 2.3) is not only stabilised by OH...O hydrogen bonding but also by substantial (stacking-like) interaction of the benzene rings. The correct description of such a structure thus requires not only the description of the H-bond, which is not so particularly demanding, but also the description of the π–π interaction, which is much more involved. The reliability of the MP2/cc-pVDZ method was tested[80,81] and a reasonably good agreement between experimental and theoretical rotational constants (with an average relative deviation from experimental value smaller than 7%) was obtained. We expected the error to be reduced significantly when passing to considerably larger cc-pVTZ basis set. To our surprise, the opposite was true, and the error increased dramatically (relative deviation from experimental values was larger than 60%). The problem is connected with the above-mentioned overestimation of the MP2 stabilisation energy (when the extended basis set is used) between phenyl rings and only when passing to the CCSD(T) level can more reliable geometry be obtained. Since the CCSD(T) gradient optimisation of such a complex is and will be in the near future impractical, other methods should be used. In the part devoted to the calculation of interaction energies we will see that density-functional theory (DFT) methods cannot be generally used for geometry determination since these methods do not cover the London dispersion energy. When using the modified DFT method properly covering the dispersion energy (the method provides very similar results to the much more expensive CCSD(T) procedure, see later) an excellent agreement between experimental and theoretical rotation constants (relative deviation from experimental values was smaller than 1.5%) was found.

For extended complexes even the MP2/cc-pVTZ level of theory is computationally challenging and becomes increasingly impractical with increasing size of the complex. We have explored[82] the applicability of the resolution of the identity MP2 (RI-MP2) approximation method[83–85] and shown that when combined with extended basis sets containing f-functions, RI-MP2 is capable of accurate descriptions of H-bonded and stacked DNA-base interactions. The

RI-MP2 method implemented in the TURBOMOLE code[86] yields almost identical absolute as well as relative (interaction) energies as the exact MP2 method for the nucleic acid bases and base pairs studied,[82] whilst the computational time saving can be as large as one order of magnitude.

2.2.5   Gradient Optimisation and Basis Set Superposition Error

The use of the gradient-optimisation techniques for non-covalent complexes is associated with a serious problem: the basis set superposition error[87] is normally not *a priori* included in the geometry optimisation cycles. In contrast, the BSSE is mostly included *a posteriori* in order to improve the determination of stabilisation energy. So, normally, this important correction is only taken into account when determining the stabilisation energy but it is ignored in the geometry optimisation of the complex. Only several years ago, was a counterpoise-corrected gradient-optimisation procedure introduced, which covers the basis set extension effect in each optimisation cycle.[88,89] Until very recently, numerical applications were limited to small and medium basis sets as well as to structural motifs where the BSSE is rather small. Recently, we have used the counterpoise-corrected gradient-optimisation procedure for the phenol dimer:[81] it was found that the relative deviation of rotational constants from experimental values was the same as in the case of standard MP2 gradient optimisation, which we believe was due to the use of a small basis set. To test the procedure further we optimised the structure of stacked and planar H-bonded uracil dimers and the stacked adenine...thymine;[77] besides the counterpoise-corrected gradient optimisation, we also optimised the geometry of these clusters with a step-by-step procedure based on the CCSD(T) method, including the counterpoise correction in every step. The results can be summarised as follows: (i) standard geometry optimisation at the MP2 level with small basis sets (*e.g.* 6-31G**) provides fairly reasonable intermolecular separation (this is due to compensation of errors); (ii) geometry optimisation with extended basis sets at the MP2 level underestimates the intermolecular distances compared with the reference CCSD(T) results, whereas the MP2/cc-pVTZ counterpoise-corrected optimisation agrees well with the reference geometries.

It is possible to conclude that the MP2 method when combined with basis set of TZ quality provides reasonable geometries for various structural types of a molecular cluster. Care should be taken with the use of DFT methods that are considerably less demanding on the CPU time. These methods yield good geometries for one structural type (*e.g.* H-bonding) but might fail for the other one (*e.g.* stacking). Considerably better results are obtained with DFT-D methods covering the dispersion energy and these methods can be recommended for extended clusters possessing various structural motifs. A very strong point of the latter method is the fact that BSSE is less important than in the case of wavefunction theory (WFT) methods.

Geometrical data obtained from quantum-chemical calculations correspond normally to optimised geometries at stationary points on the PES. These data cannot be directly compared with experiment since the experimental characteristics involve the vibrational energy. Even at 0 K, the effect of the zero-point energy, to which all vibrations contribute, has to be considered. In order to compare to experiment, vibrationally averaged geometries have to be obtained around the stationary points on the PES. The spectacular effect of vibrational averaging was already demonstrated for the ammonia dimer in Section 2.2.1. In general, the less harmonic the PES is around a stationary point, the more pronounced is the effect of this vibrational averaging. When considering the zero-point vibrational energy level for very flat potentials, for which the harmonic approximation becomes meaningless, the position of the zero-point energy level can be rather difficult to estimate. For instance, for very flat potentials with two or more (adjacent) minima it is quite possible for the ZPE level to lie above a barrier.

The optimisation procedure described above concerns rigid systems where the concept of structure is meaningful. In the case of floppy complexes, large-amplitude motions play a significant role such that the concept of a definite structure must be replaced by the more generalised concept of the potential-energy surface (see discussion of the ammonia dimer in Section 2.2.1 and benzene dimer in Section 3.1).

2.3 Stabilisation Energy

We define stabilisation energy as the negative of the dissociation energy measured from the dissociation asymptote to the minimum of the PES. Stabilisation energy thus has a positive value, while the other frequently used term, interaction energy, has a negative value. The determination of stabilisation energies is of key importance since it relates to fundamental thermodynamics. Stabilisation energy can be obtained only from theoretical calculations, while experiments yield stabilisation enthalpy (this corresponds to the inclusion of the zero-point energy). For experiments at very low temperatures (*e.g.* in a supersonic expansion molecular beam), the data refer to the stabilisation enthalpy at 0 K, which is obtained theoretically by adding the zero-point energy to the stabilisation energy. However, for higher temperatures, the inclusion of temperature-dependent enthalpy terms, derived from the partition function, is required.

Experimental determination of the stabilisation enthalpy of a cluster is difficult and relevant data exist only for a few complexes. The classical method involves measuring the temperature dependence of the equilibrium constant, which yields the stabilisation enthalpy of complex formation. Using the field-ionisation mass spectrometry method, Sukhodub *et al.*[90] measured stabilisation enthalpies of methylcytosine...methylcytosine, methylguanine...methylcytosine, methyl-adenine...methylthymine and methylthymine...methylthymine complexes. It must be stressed that these, almost 25-year-old data are still the only data on stabilisation energies (enthalpies) of DNA base pairs *in vacuo*.

2.3.1 Experimental Methods for the Determination of the Binding Enthalpy

The energetics of a dissociation reaction without reverse activation energy of a heterodimer consisting of the two molecules A and B (for illustration see Figure 2.4) can be described, by invoking a Haber–Bosch cycle, as simple relations:

$$D_0 = AE - IE(A) \tag{2.1}$$

$$D_0^+ = AE - IE(A \ldots B) \tag{2.2}$$

where D_0 represents the binding enthalpy of the neutral dimer $A \ldots B$ and D_0^+ the binding enthalpy of the charged cluster $A^+ \ldots B$. $IE(A)$ and $IE(A \ldots B)$ are the (adiabatic) ionisation energies of the monomer A and the dimer $A \ldots B$, respectively. AE is the so-called appearance energy, *i.e.* the energy at which an ionic fragment appears for the first time by fragmentation of the dimer cation. So the binding enthalpies D_0 and D_0^+ can be obtained from the measured values of AE, $IE(A)$ and $IE(A \ldots B)$.

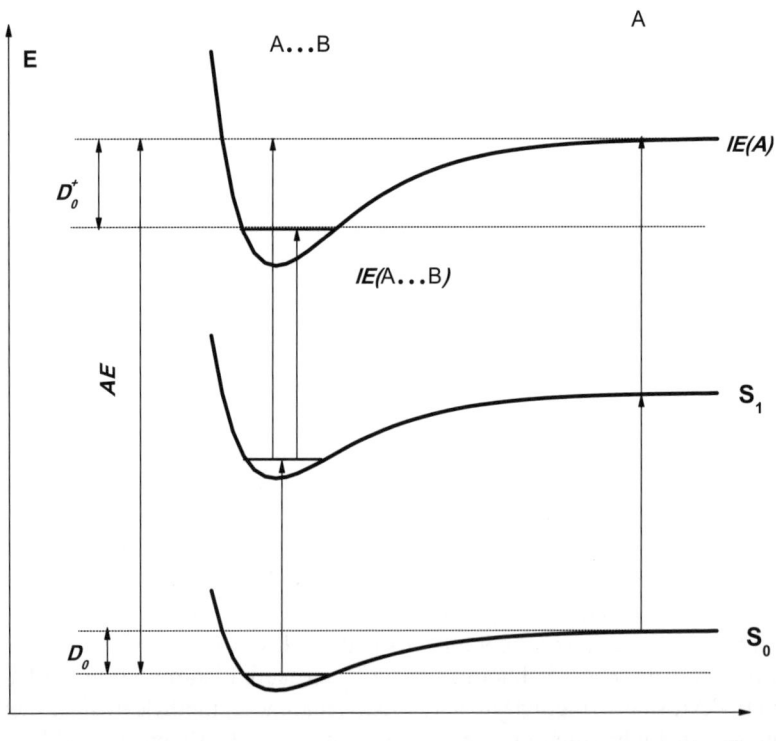

Figure 2.4 Haber–Bosch cycle for a dimer dissociation

The determination of binding enthalpies (BE) such as D_0 and D_0^+ has been of continuous interest since BE values are important microscopic parameters for the description of the properties of condensed matter. High-pressure mass spectrometry,[91] IR absorption spectroscopy in gas mixtures,[92] or bolometric methods[93] are successful techniques but have not yet been applied to isolated, weakly bound clusters containing larger polyatomic molecules.

2.3.1.1 Breakdown Method

The breakdown technique is based on photo-ion-efficiency (PIE) spectro-scopy[94–100] invented in 1954 by Watanabe,[101] which can be used to determine the ionisation energy as well as vibrational levels in ionic molecules and clusters. Using the R2PI method already described, the molecule is excited by a first photon into an intermediate (resonant) state and it is ionised by the absorption of a second photon. Scanning the energy of the ionisation laser and recording the resulting ion current yields the PIE spectrum, from which the ionisation energy and some vibrational levels can be determined as steps in the PIE spectrum.

For the breakdown technique the energy of the first photon is tuned to a transition into an intermediate state of the cluster, normally the vibrationless ground state of the first electronically excited state. If the total photon energy is below the appearance energy only the complex ion $(A \ldots B)^+$ is observed. Once the energy exceeds the appearance energy, fragmentation of the complex cation occurs and a fragment ion A^+ appears in the mass spectrum. Recording the ion signal of the A^+-fragment, while scanning the wavelength of the ionisation laser yields the breakdown spectrum from which the appearance energy of the fragment can be determined.

As shown by Grover *et al.*[97] the appearance energy AE can be determined with the breakdown technique. A least-squares fit of the curves is applicable to determine AE as the intersection of the two straight lines. Figure 2.5 shows a

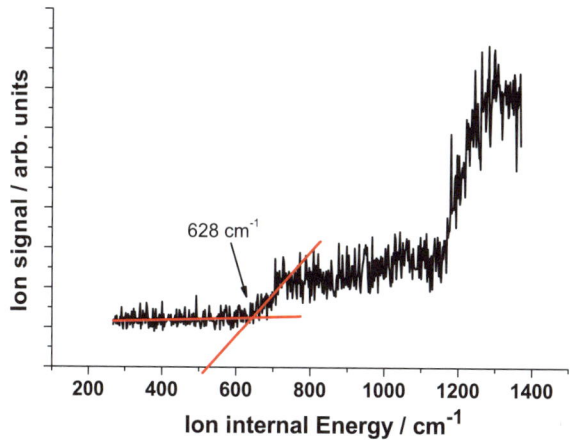

Figure 2.5 Breakdown spectrum for pyrazine ... argon. Redrawn from Ref. 102.

breakdown spectrum for the pyrazine...Ar complex[102] recorded by Riese *et al.* In the first part of the spectrum the energy of the second laser is high enough to ionise the cluster but not high enough to fragment it – no fragment signal is observed. At $628 \pm 20 \, \text{cm}^{-1}$ above the ionisation energy of the complex IE(A...B) the fragment signal (pyrazine$^+$) starts to appear as a step in the PIE spectrum. This fragment PIE curve is normally called a *partial PIE* spectrum. According to eqn (2.2) this appearance energy is assigned to the binding enthalpy in the ionic ground state.

A second step in the partial PIE curve is observed at $1130 \pm 20 \, \text{cm}^{-1}$, corresponding to a vibrational state of the cluster assigned to the $5^1 6a^1$ or the $10a^2$ vibration. Generally, a step in the PIE curve arises when a new ionisation channel, for instance, a new vibrational state is opened. This occurs when the photon energy of the ionisation laser exceeds the energy of a vibrational state of the cluster ion resulting in an increase in the ion current of the cluster and the daughter ion.

Though the breakdown method is broadly applicable and a lot of dissociation energies of neutral and ionic dimers and larger clusters have been determined[103–106] it suffers from several disadvantages. First, the ionisation energy often cannot be determined unambiguously since the Franck–Condon factors for transitions to the vibrationless cationic ground state are often very weak. This is especially true for clusters. Second, the dissociation threshold is not well determined since a defined photon energy does not lead to a defined internal energy and no sharp dissociation threshold can be found. Another problem arises from the use of a one-colour excitation scheme in contrast to a two-colour one. In such one-colour experiments it can be quite difficult to distinguish between monomer ions resulting from the dissociation of the cluster and direct ionisation of the monomer.

2.3.1.2 *Mass-Resolved ZEKE Spectroscopy*

The best method so far to determine stabilisation energies with spectroscopic precision comes from ZEKE spectroscopy.[107–114] The ZEKE spectroscopy method has been extensively reviewed,[115–125] also in the context of Rydberg state[126] and photoionisation dynamics[127] and applied to molecular clusters.[128–133] Let us recall that ZEKE spectroscopy is based on the pulsed-field ionisation of very long-lived Rydberg states[134,135] of very high principle quantum numbers ($n > 200$). The pulsed-field ionisation of a ZEKE Rydberg state produces both an electron and an ion. When this electron is detected we call the method zero electron kinetic energy (ZEKE) photoelectron spectroscopy; when the corresponding ion is selectively measured, we call this mass-selected ZEKE, or mass-analysed threshold ionisation (MATI).[136–138] Since the motion of the Rydberg electron is completely decoupled from the internal motion of the ion core, a ZEKE spectrum results in the observation of vibrational and rotational structure of the corresponding molecular cation. This is fully equivalent to the spectrum one would obtain from photoelectron detection if one could improve the resolution to the required level. Mass-selected ZEKE provides the additional advantage of having a mass signature, which is particularly useful for molecular clusters and the observation of fragmentation processes. For molecular clusters, the most useful application of MATI comes from observation of the dissociation of the Rydberg state ion core that allows the determination of dissociation and

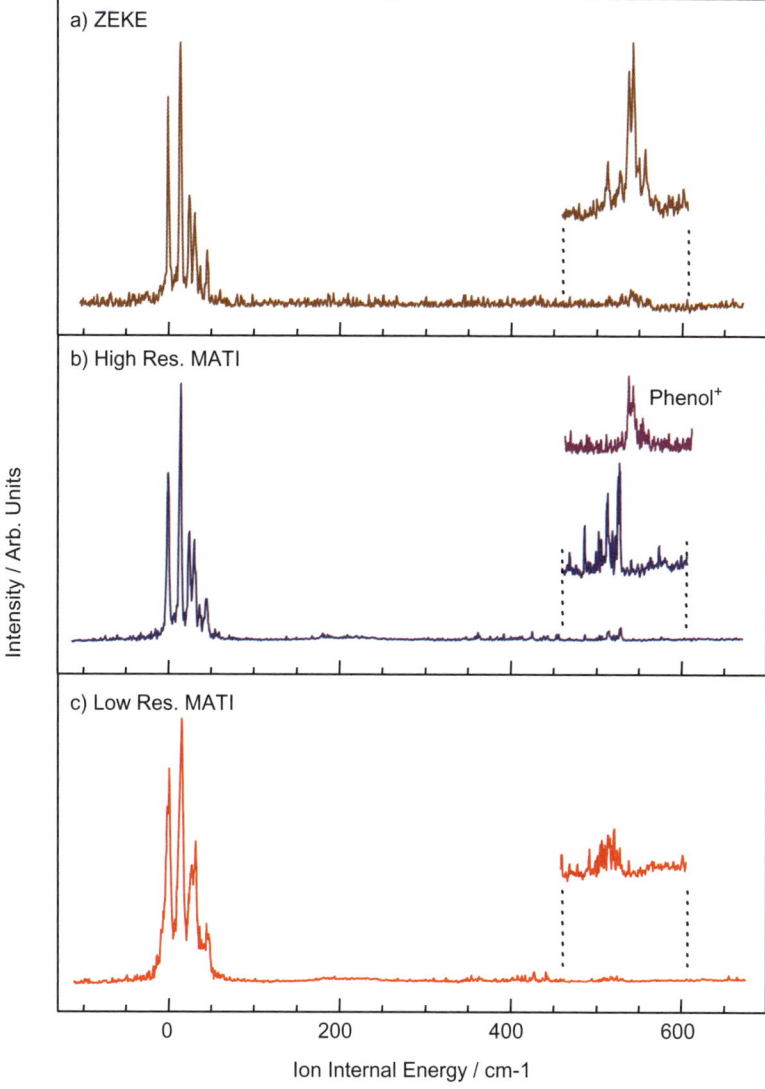

Figure 2.6 ZEKE spectroscopy of phenol . . . argon: (a) high-resolution ZEKE spectrum (electron detection); (b) high-resolution and (c) low-resolution MATI spectra measured for the phenol$^+$ . . . argon (parent) and phenol$^+$ (daughter) dissociation product. The dissociation threshold is enlarged in the inserts in (b) and (c). The dissociation energy can be very precisely determined in (b) at the point of disappearance of the parent ion and the appearance of the daughter fragment ion. Redrawn from ref. 10.

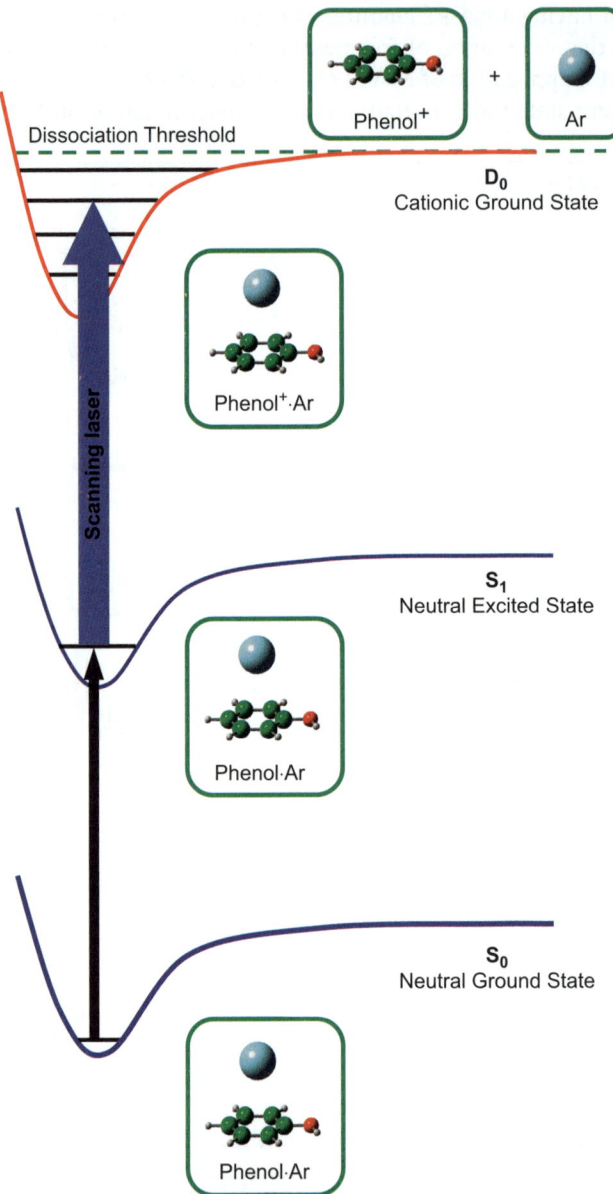

Figure 2.7   Energy diagram for the MATI dissociation process of phenol . . . argon.

binding energies of the molecular cluster cation. Once the cation dissociation energy is known, the dissociation energy of the neutral cluster is also known, as we have shown for the phenol . . . N_2 and phenol . . . CO complexes.[139–147] In terms of spectral resolution, the MATI resolution can be better than $0.1\,\text{cm}^{-1}$ and ZEKE and MATI are fully equivalent.[148] For the phenol . . . argon complex,

Figure 2.6 shows a comparison between ZEKE electron (a) and MATI detection with high (b) and lower (c) spectral resolution. The insert in Figure 2.6(b) shows an enlarged section of both the phenol$^+$ ... Ar (parent) and the phenol$^+$ (daughter) ion around the dissociation threshold of the phenol$^+$ ... Ar cluster. It can be seen that the dissociation energy can be very precisely determined as the difference between the disappearance of the parent ion and the appearance of the daughter fragment ion. The corresponding energetics for the cation ground state, the neutral excited state and the neutral ground state are illustrated in Figure 2.7 along the intermolecular dissociation reaction coordinate.

The way that this high-precision dissociation spectroscopy works is depicted in Figure 2.8. The ZEKE Rydberg electron circles around the ion core at a very large distance. Any dynamic process, such as vibrational predissociation, even electronic excitation in the cation in the core, will leave the Rydberg electron untouched. So, if the ion core dissociates, the ionic fragment will continue to support the Rydberg electron, whereas the neutral fragment will escape virtually unnoticed. Upon pulsed-field ionisation the ion generated will then be the daughter fragment ion and no longer the parent ion. In the time-of-flight spectrum, the daughter ion will be measured and if this is done simultaneously with the measurement of the parent ion then a spectrum such as those shown in Figures 2.6 and 2.9 will be obtained. It cannot be stressed enough that presently the only viable method of high spectroscopic precision for the determination of dissociation and stabilisation energies of molecular complexes comes from this mass-resolved variant of ZEKE spectroscopy.

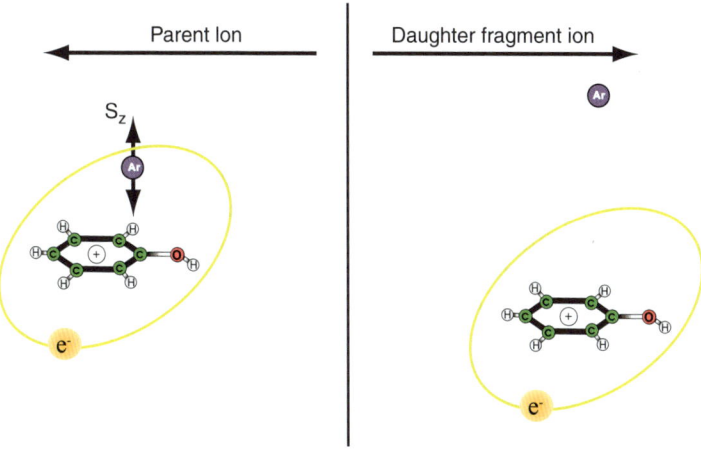

Figure 2.8 Dynamics of the MATI dissociation process. The ZEKE Rydberg electron is not perturbed by any ion-core dynamics, including vibrational predissociation. When the ion core dissociates, the ionic fragment will continue to support the Rydberg electron due to the dominant Coulomb interaction. Upon pulsed-field ionisation, the ion produced is the (daughter) fragment ion. The neutral fragment is not detected.

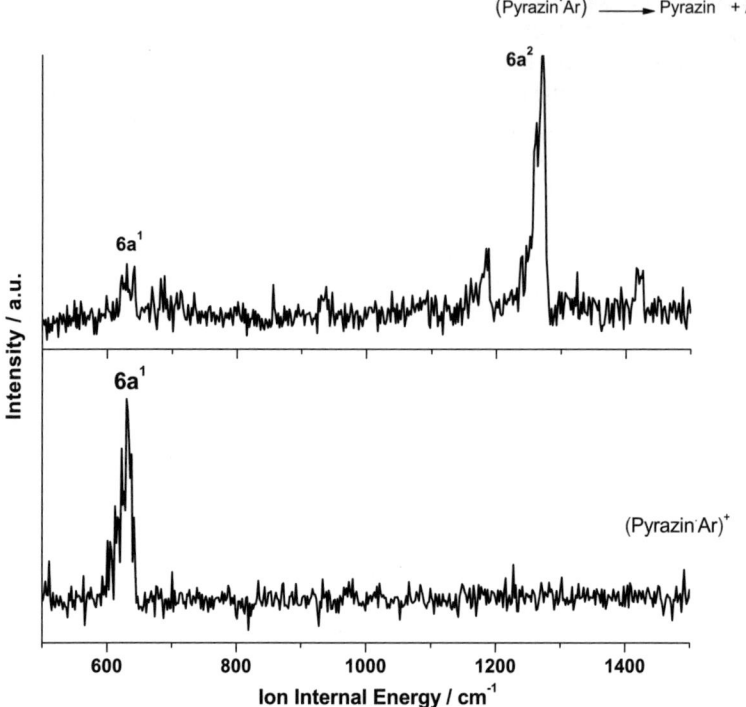

Figure 2.9 MATI spectrum of pyrazine . . . argon. Redrawn from Ref. 102.

If the two-photon energy exceeds the appearance energy the threshold signal changes from the cluster mass to the fragment mass, so normally a signal should only be observed on one mass channel at the same time. Figure 2.9 shows the MATI spectra of the pyrazine . . . Ar cluster recorded on the cluster and the monomer mass, by exciting via the $6a^1$ intermediate state of the cluster. The $6a^1$ vibration can be clearly observed in both spectra. The reason for the simultaneous appearance of bands in the cluster as well as the fragment mass channel will be discussed in the following.

Grebner *et al.* demonstrated for the case of fluorobenzene . . . Ar clusters that the measured value of the fragment-ion appearance energy depends on the strength of the applied ionisation field.[149] For low electric fields (125 V/cm) the appearance of peaks in the fragment channel was shifted towards higher internal energy (some few 10ths of a cm^{-1}) compared to high electric fields (975 V/cm). This observation was explained by a coupling of the originally excited high Rydberg levels of series converging to a vibrational state below the field-free dissociation threshold to lower Rydberg states converging to a vibrational state above the dissociation threshold of the ionic core. This coupling is induced by the pulsed electrical ionisation field. The process is illustrated in Figure 2.10. The coupling results in a transfer of electronic energy from the highly excited high Rydberg electron to the vibrational degrees of

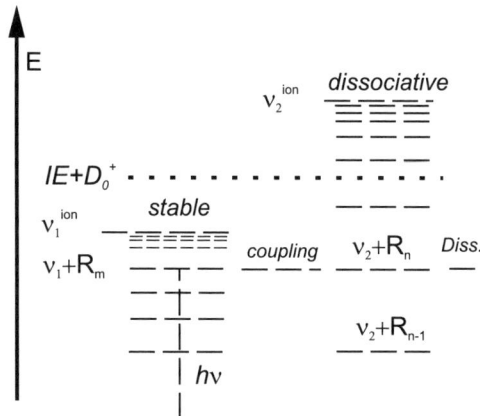

Figure 2.10 Coupling introduced by pulsed electric ionisation field.

freedom of the cluster ion core, causing a dissociation of the cluster, even when the total energy is less than IE $(A \ldots B) + D_0^+$. Here, it is assumed that the Rydberg states survive the dissociation, as shown in ref. 150. The dissociation can be detected if the ionising field strength is sufficient to ionise the low Rydberg states of the fragment Rydberg molecule. MATI spectra that are simultaneously recorded at the cluster and the fragment mass channel show an overlap region of approximately $50 \, \mathrm{cm}^{-1}$ for the fields used in these experiments. In this internal energy range some clusters dissociate, whereas others remain stable, since not all clusters gain vibrational energy sufficient for dissociation through the coupling. This leads to a broadening of the appearance energy range for the daughter ions as described in detail in ref. 149.

2.3.2 Computation of Stabilisation Energy

Perturbation and variation supermolecular methods. The perturbation method is a natural solution to the evaluation of the interaction energy since it is determined directly as the sum of the various perturbation contributions (not as a difference between very large quantities as in the variation method, see below). Using symmetry adapted perturbation treatment (SAPT),[151] the interaction energy can be expressed as a sum of first-, second- and higher-order perturbation contributions. The first-order contribution contains electrostatic and exchange energies, while the second-order term includes induction and dispersion energies. The charge-transfer energy is included in the second-order induction energy and higher-order contributions. Provided an extended atomic orbital (AO) basis set is used, the SAPT treatment yields accurate values of single energy terms (each term corresponding to a well-defined and physically meaningful phenomenon) as well as of the total interaction energy. The main advantage of the perturbation SAPT treatment is the fact that the interaction energy can be determined in a straightforward manner and without any theoretical problems (see, *e.g.* the basis set superposition problem in the case of the variation method). A broader application of the method is, however, complicated by the fact that it was based on the assumption that the Schrödinger equation for monomers is solved exactly or nearly exactly. The original SAPT was also highly demanding on CPU time, making its use for larger complexes prohibitive. This is especially true when utilising correlated monomer wavefunctions. An alternative that is less demanding is the variant of the method where the monomer wavefunctions are described at the HF level. A significant improvement has been reached by utilising a combination of SAPT and DFT theories,[152–154] and having been combined with extended basis sets, the method has been used for extended complexes such as the benzene dimer and DNA base pairs.[155,156] The (DFT) SAPT (or SAPT-DFT) total interaction energy E^{int} is calculated as the sum of the polarisation (electrostatic), exchange, induction and dispersion components, the induction and dispersion components also having their exchange counterparts.

$$E^{int} = E^1_{Pol} + E^1_{Ex} + E^2_{Ind} + E^2{}_{Ex\text{-}Ind} + E^2_{Disp} + E^2_{Ex\text{-}Disp} + \delta(HF) \qquad (2.3)$$

The $\delta(HF)$ term denotes the estimated higher-order Hartree–Fock contributions (induction, exchange-induction and charge-transfer); the higher-order dispersion energy components are not included. This term is determined as a difference of HF (variation) interaction energy and sum of electrostatic, exchange-repulsion, induction and exchange-induction energies. Evidently, it includes one serious problem – the HF interaction energy is namely determined with elimination of basis set superposition error (BSSE, see later). Hence, one of the most important advantages of the perturbation treatment – its independence of BSSE, no longer applies. The (DFT) SAPT method is substantially faster than the standard SAPT (where monomer correlated wavefunctions were adopted), because intramolecular correlation is treated fully by the DFT, whereas the intermolecular interaction is left to the intermolecular perturbation theory. Significant speed-up was obtained by using the density-fitting procedure. The (DFT) SAPT method scales worse with the complex size than supermolecular DFT (as N^5 *vs.* N^3) but it scales much better than other highly accurate wavefunction theories. Since perturbation theory exploits orbital energies, the inherently incorrect DFT orbitals (both occupied and virtual) need to be corrected in some way. Hesselmann and Jansen used a gradient-controlled shift procedure,[157] which needs a difference (shift) between the vertical ionisation potential (IP) and the HOMO energy of the DFT method used as an input. To obtain reliable energy components, an extended AO basis set should be used. The "smallest" basis set yielding correct energies is the aug-cc-pVDZ one (or a similarly constructed basis set), and when this basis set is used, all energy components with the exception of dispersion energy reach correct values (within less than 5% of the accurate values obtained with the aug-cc-pVQZ basis set); the dispersion term is underestimated by approximately 20%.[158] The (DFT) SAPT procedure depends much less on the DFT functional than supermolecular DFT and all recently developed functionals provide comparable results. This is especially true for hybrid potentials and among them the PBE0 yields particularly accurate results. Another important advantage of the (DFT) SAPT is the fact that it converges faster with the basis set than the regular SAPT.[159]

The variation (also called supermolecular) method determines the interaction energy indirectly as a difference between the energy of a complex (supersystem) and the energies of all subsystems

$$\Delta E = E^{R...T} - (E^R + E^T) \tag{2.4}$$

The BSSE is eliminated using the function counterpoise procedure, *i.e.* the subsystem energies are determined not in their own basis sets (E^R, E^T) but in the basis set of the whole complex ($E^{R(T)}$, $E^{T(R)}$)

$$\Delta E = E^{R...T} - (E^{R(T)} + E^{T(R)}) \tag{2.5}$$

As mentioned above, the gradient optimisation is to be performed using the counterpoise-corrected procedure and in this case the resulting interaction energy is determined on the basis of eqn (2.5). However, in most cases the optimisation is performed with a standard procedure and only the final interaction energy (determined for the energy minimum) is corrected for the BSSE. The full gradient optimisation certainly also modifies the subsystem coordinates and the final interaction energy thus does not refer to isolated subsystems but to subsystems modified during the optimisation procedure. This inconsistency is removed by correcting the final interaction energy by a so-called deformation term that is defined as a difference between the energies of the subsystems in geometries resulting from the optimisation procedure and the energies of the isolated subsystems. Since the latter energies are always lower than the former ones the deformation energy is always positive (*i.e.* repulsive). Let us again recall that when performing the counterpoise-corrected gradient optimisation this problem does not occur.

In wavefunction theories (WFT), the energy of a system is expressed as a sum of Hartree–Fock (HF) and correlation (COR) energies. Consequently, the interaction energy is also expressed as a sum of HF interaction energy and correlation interaction energy

$$\Delta E = \Delta E^{HF} + \Delta E^{COR} \tag{2.6}$$

In principle, any WFT can be applied, but due to low values of stabilisation energies of non-covalent complexes, only the most accurate methods should be utilised. First, the HF level is not acceptable, because the correlation interaction energy is always important. Second, the use of the size-consistent, variational full configuration interaction (FCI) method is impractical. The other (cheaper) variants of the CI method such as CI-SD (CI covering single and double electron excitations) have serious problems (*e.g.* size inconsistency), which prevents them from being very useful. Thirdly, the most promising methods are based on the couple-cluster approach, and the CCSDT method, covering iteratively the single, double and triple electron excitations, yields very accurate energies. In 2004 the first study appeared[160] where the stabilisation energies of model H-bonded and stacked systems were determined at the CCSDT level, and it has been shown that the method (which is highly demanding in terms of CPU time) yields energies practically identical to those of the much cheaper CCSD(T) method. When triples play such an important role for the interaction energy a natural question is raised about the role of quadruple electron excitations. In the case of benzene dimer and a moderately large basis set it was shown that their role is negligible.[161] As a compromise between economy and accuracy, the CCSD(T) method, where the triple excitations are covered in a noniterative way, is the gold standard of single-reference calculations. It covers a significant portion of correlation energy and can be applied even for extended complexes with more than 24 atoms. For larger complexes, however, the CCSD(T) method cannot be used, and here the MP2 method should be employed covering double electron excitations at second order of perturbation theory. If combined with an extended basis set, the method overestimates the correlation interaction energy. Here the critical role is played by diffuse polarisation functions and this is why the first reliable basis set is represented by Dunning's aug-cc-pVDZ (or similar) basis set. Due to some error compensation (see below), the MP2 method in combination with smaller basis sets provides surprisingly reliable values of stabilisation energies. A very special position is held here by the small 6-31G* (0.25) basis set that was used in the early days of correlated quantum-chemical calculations on DNA base pairs. This basis set contains one set of d-polarisation functions but instead of a standard exponent of about 0.8 it contains more diffuse ones with an exponent of 0.25. The MP2 stabilisation energies of extended clusters determined with this basis set agree surprisingly well (see later) with much more expensive CCSD(T)/CBS values. Nevertheless, it should be mentioned that this is due to error compensation and care should be taken with relying on this compensation.

Grimme[162] introduced a promising modification to the standard MP2 method. This newly developed method, called spin-component-scaled MP2 (SCS-MP2), in most cases improves the description of molecular ground-state energies when compared to the unmodified MP2. The method is based on a separate scaling of the correlation-energy contributions from antiparallel- and parallel-spin pairs of electrons by two new scaling factors. Because of the appearance of these parameters, SCS-MP2 is no longer a "pure" *ab initio* method. The modification when introduced effectively corrects the overestimation of the dispersion contribution to the interaction energy, thus offering higher-accuracy interaction energies for non-covalent complexes. As, however, mentioned above, standard MP2 overestimates the dispersion energy and that is a most important topic for stacked structures. Consequently, the original SCS-MP2 yields reliable stabilisation energies for stacked structures while these energies for H-bonded structures are underestimated (MP2 and CCSD(T) interaction energies for H-bonded structures evaluated with the same basis set are very similar). Improvement for stacking is, however, significant and the method can be safely used even for geometry optimisation.[163–165]

The original SCS-MP2 procedure was modified by several authors and SCSN-MP2 and SCS(MI)-MP2 procedures were suggested.[166–169] The SCSN-MP2 method relies on the scaling of only the antiparallel spin component of the MP2 energy, where the scaling parameter was fitted against the benchmark interaction energies of base pairs from work of Jurečka and coworkers.[77] SCS(MI)-MP2 is based on scaling of both spin components, which were optimised against the S22 test set.[77] Both of these methods perform surprisingly well compared to the SCS-MP2 method, and are capable of delivering interaction energies with deviations of only a few tenths of a kcal/mol, both for H-bonded and dispersion-dominated complexes. However, the quantitative agreement for the H-bonded complexes compared with CCSD(T)/CBS values is still not quite satisfactory.

The separate scaling technique was very recently applied also at the CCSD level and the proposed SCS-CCSD method was shown to provide very good stabilisation energies for methane and benzene dimers.[170] Although this nicely supports the general SCS idea, the wider use of the method is, however, hampered by the fact that CCSD calculations are computationally much more demanding than MP2 ones (for simplified alternatives to CCSD based on the CEPA approximation, see Ref. 171).

As mentioned above, the theoretical determination of stabilisation energies requires using high-level techniques since the stabilisation energy is very sensitive to the theoretical level applied. A compromise between economy and accuracy is the CCSD(T) method where the triple excitations are covered in a noniterative way. Because of the strong dependence of stabilisation energy on the AO basis-set size, it is important to perform the calculation with a basis set as large as possible or, preferably, at the complete basis set (CBS) limit. The direct determination of the CBS limit for CCSD(T) calculations is difficult due to the need for two energy points generated with systematically improved AO basis sets. Let us repeat that the first reliable basis set (in the cc-pVXZ series) is the Dunning aug-cc-pVDZ basis, followed by aug-cc-pVTZ. The CCSD(T) calculations for the complexes mentioned are extremely expensive for basis sets larger than DZ + P (*e.g.* 6-31G** or cc-pVDZ). There exists, however, a single method by which one may overcome this problem. This is based on the fact that CCSD(T) and MP2 energies have a similar dependence on the size of the basis set used (see Figure 2.11). Assuming the difference between CCSD(T) and MP2 interaction energies ($\Delta E^{\text{CCSD(T)}} - \Delta E^{\text{MP2}}$) exhibits a relatively small basis set dependence compared with MP2 CBS determination, the CBS CCSD(T) interaction energy can be approximated as:[172]

$$\Delta E_{\text{CBS}}^{\text{CCSD(T)}} = \Delta E_{\text{CBS}}^{\text{MP2}} + (\Delta E^{\text{CCSD(T)}} - \Delta E^{\text{MP2}})|\text{medium basis set} \qquad (2.7)$$

The CCSD(T)-MP2 difference has been investigated for H-bonded as well as stacked model complexes.[77] It was found that even rather small basis sets like

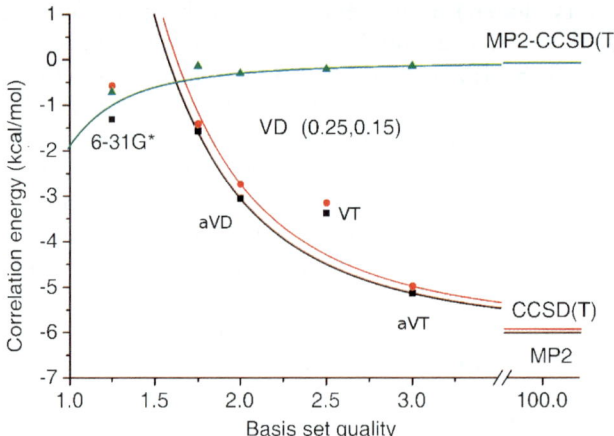

Figure 2.11   The dependence of MP2, CCSD(T) and (MP2 – CCSD(T)) interaction
energies on the basis-set size.

cc-pVDZ, or even the 6-31G**, yield satisfactory values for this difference.
Among various small and medium basis sets the best performance was exhibited by the 6-31 + G** basis set and if there is no chance to use larger basis sets
(*e.g.* aug-cc-pVDZ) this basis set can be recommended for evaluation of the
CCSD(T) correction term. The need to work with a small basis set is under-
standable, because CCSD(T) scales as N^7 with the complex size N (MP2 and
CCSD scale much better as N^5 and N^6). Let us recall here that (DFT) SAPT,
which provides highly accurate interaction energies, well comparable to
CCSD(T) ones, scales only as N^5 (see above).

As mentioned, the MP2 procedure strongly overestimates the stabilisation
energy. Higher-order MP contributions, however, behave differently. Typically
MP3 (and also MP4(D) and MP4(SDQ)) underestimate the stabilisation
energy and, what is important, the underestimation is roughly the same as the
overestimation of MP2. If the behavior of MP2 and MP3 is systematic for
non-covalent interactions, these considerations lead us (see Ref. 173) to a linearly interpolated MP method that damps the third-order correction by 50%,
eqn (2.8),

$$E(\text{MP2.5}) = E(\text{MP2}) + 1/2\,E^{(3)} \tag{2.8}$$

where $E(\text{MP2})$ is the total MP2 energy and $E^{(3)}$ is the third-order MP correlation energy correction. For obvious reasons this method is termed MP2.5 and
represents a simple average of MP2 and MP3. A more empirical generalisation
is to introduce a scaling factor "c", which defines the SMP3 (scaled MP3)
method in eqn (2.9),

$$E(\text{SMP3}) = E(\text{MP2}) + c\,E^{(3)} \tag{2.9}$$

The empirical scaling factor "*c*" was determined by a fit to CCSD(T) reference data. From the discussion given above it is already clear that when "*c*" is about a half, the problem of MP2 with stacked aromatic complexes will be more or less solved, while for the systems for which MP2 is already good, the relative magnitude of the relative third-order energy will be small and MP2 quality is recovered. Note, that these SMP3 and MP2.5 really include new correlation effects since coupling of the first-order pair excitations is introduced at third order. Furthermore, the nonadditivity of the dispersion energy is also recovered at third order (see Ref. 174). It is obvious that by setting $c=0$, we recover MP2 and by $c=1$ we arrive at MP3 so that our proposal can be regarded as a linear interpolation scheme.

The total SMP3 (or MP2.5 for $c=0.5$) energy in a large basis (or at the CBS limit) can similarly be defined as in eqn (2.9), giving eqn (2.10):

$$E(\text{SMP3/Large basis}) \approx E(\text{MP2/Large basis}) + c\,E^{(3)}/\text{Small basis} \quad (2.10)$$

Scaled MP3 interaction energies calculated as a sum of MP2/CBS interaction energies and scaled third-order energy contributions obtained in small or medium size basis sets agree very closely with the estimated CCSD(T)/CBS interaction energies for the 22 H-bonded, dispersion-controlled and mixed non-covalent complexes from the S22 data set. Performance of this so-called MP2.5 (third-order scaling factor of 0.5) method has also been tested for 33 nucleic acid base pairs and two stacked conformers of porphine dimer. In all the test cases, performance of the MP2.5 method was shown to be superior to the scaled spin-component MP2 based methods, *e.g.* SCS-MP2, SCSN-MP2 and SCS(MI)-MP2. In particular, a very balanced treatment of hydrogen-bonded compared to stacked complexes is achieved with MP2.5. The main advantage of the approach is that it employs only a single empirical parameter and is thus biased by two rigorously defined, asymptotically correct *ab initio* methods, MP2 and MP3. The method is proposed as an accurate but computationally feasible alternative to CCSD(T) for the computation of the properties of various kinds of non-covalently bound systems.

The MP2 part of the stabilisation energy is extrapolated to the CBS limit as follows. Whereas the HF interaction energy can be considered to converge with respect to the one-electron basis already for relatively small basis sets, the MP2 part of the interaction energy converges unsatisfactorily slowly to its complete basis-set limit. In order to correct the computed results for the basis-set incompleteness error, several extrapolation schemes have been successfully employed in literature. These are the schemes of Helgaker and coworkers:[175]

$$E_X^{\text{HF}} = E_{\text{CBS}}^{\text{HF}} + Ae^{-\alpha X} \quad \text{and} \quad E_X^{\text{corr}} = E_{\text{CBS}}^{\text{corr}} + BX^{-3} \quad (2.11)$$

and of Truhlar:[176]

$$E_X^{\text{HF}} = E_{\text{CBS}}^{\text{HF}} + AX^{-\alpha} \quad \text{and} \quad E_X^{\text{HF}} = E_{\text{CBS}}^{\text{HF}} + BX^{-\beta} \quad (2.12)$$

where E_X and E_{CBS} are energies for the basis set with the largest angular momentum X and for the complete basis set, respectively; α and β are parameters fitted by the authors. These schemes were chosen because (i) both approaches extrapolate HF and correlation energy separately and (ii) both use the two-point form (they extrapolate two successive basis sets results). The two-point extrapolation form is preferable as it was shown[177] that inclusion of additional (lower-quality basis set) results in the extrapolation often spoils the quality of the fit, especially when the smallest basis set (cc-pVDZ) is used. For non-covalent complexes it is recommended to use augmented Dunning's basis sets rather than nonaugmented ones to reduce the extrapolation error (note that the aug-cc-pVDZ basis set gives absolute energies as well as interaction energies comparable with those calculated with the TZVPP basis).

In addition to the extrapolation techniques mentioned above that require a knowledge of few parameters (determined for smaller systems where higher-level calculations are feasible) an alternative method suggested in our previous paper should be mentioned.[178] The previously mentioned extrapolation techniques ignore the BSSE-uncorrected energies and use for extrapolation only the BSSE-corrected energies. The present procedure is based on knowledge of both BSSE-corrected and BSSE-uncorrected interaction energies and the "interpolation-like" procedure relies on the fact that BSSE-corrected and -uncorrected quantities converge to the same asymptotic value in the CBS limit. An evident advantage of the procedure is the fact that knowledge of any parameter or scaling factor is not required.

Extrapolation to the complete basis-set limit is also important considering the basis-set inconsistency problem. The BSSE gives rise to a better description of the supersystem compared with the subsystems; the supersystem uses basis sets of both subsystems (and all subsystems in more extended species), forming the larger dimer-centred basis set (DCBS). The BSSE is largest for small basis sets and its value reduces when the basis-set size increases. When working with infinite basis sets, the BSSE should converge, by definition, to zero. Figure 2.12 shows the dependence of stabilisation energies of the adenine...thymine H-bonded pair on the AO basis-set size. It is evident that corrected as well as uncorrected stabilisation energies converge to the same complete basis-set limit, the first from above, the latter from below (though, in the second instance, not necessarily in a monotonic fashion). From Figure 2.12 it is further evident that when using small and medium AO basis sets, the BSSE-corrected stabilisation energy is closer to the CBS limit and as such justifies the use of the BSSE correction when working with these basis sets.

2.4 Is Density-Functional Theory Capable of Describing Non-covalent Interactions?

In the DFT methods, the energy is determined by using exchange and correlation functionals, and an important advantage of the method (which is size consistent) is its computational economy. Contrary to WFT methods,

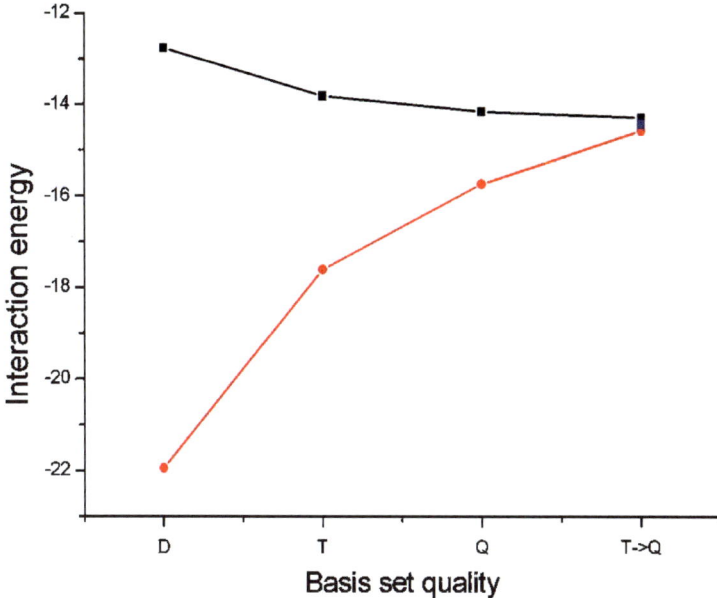

Figure 2.12 Dependence of uncorrected (red line) and BSSE-corrected (black line) stabilisation energies (in kcal/mol) of hydrogen-bonded structure of the adenine . . . thymine pair on the AO basis-set size (D, T, Q denote double-, triple- and quadruple-zeta basis set; T → Q means CBS extrapolation based on triple-zeta and quadruple-zeta single-point energies).

which strongly depend on the quality of the basis set, the DFT methods are generally much less dependent on the size of the basis set and reliable results are mostly already obtained with medium basis sets. The DFT method provides very good results on properties of isolated systems like geometry, vibration frequencies, *etc.* The same is true for selected non-covalent complexes. This concerns geometry and energetics of H-bonded complexes, geometry and spectral features of classic charge-transfer complexes,[179] the magnetic exchange coupling between transition-metal ions in binuclear complexes[180] and metal–ligand aromatic cation–π interactions (topical, for example, in metalloproteins).[181] The application of DFT methods with current density functionals (based on local density, its gradient, and the local kinetic-energy density) to non-covalent complexes is, however, limited as the method fails to describe the (nonlocal) dispersion energy. The fact that dispersion energy is not included in the local DFT energy has not been accepted by the DFT community readily, although the first papers showing this drawback were published in the mid-1990s.[182,183] Unfortunately, this problem is shared by all LDA and GGA functionals, including even the most advanced meta-GGA functionals. For hydrogen-bonded complexes, DFT is known to offer results that are often in fair agreement with the reference data, both experimental and WFT-based.[184,185] For further discussion it is necessary to mention that these good results are obtained as a consequence of error cancellation. The missing dispersion attraction (which actually plays an important role even in hydrogen-bonded complexes) is partly simulated in some density functionals by the (erroneous) attraction of the exchange functional.[186] Further, this missing dispersion contribution is effectively compensated in many DFT calculations by the basis set superposition error, which amounts to several kcal/mol for the medium-sized dimers and double-zeta quality basis sets. (The BSSE in DFT methods is in general smaller than that in WFT methods.) The same problems apply even for single molecules, whose dimensions are comparable to or larger than the sums of the vdW radii of their constituent atoms (and especially if highly polarisable groups like phenyl rings or peptide bonds are present). In such cases, we obviously cannot rely on the BSSE compensation. This renders DFT-based methods highly unreliable as they often fail to reproduce accurate geometries of molecules with the mentioned moieties, such as peptides.[187,188] As previously stated, error compensation (which is principally not reliable and hence not acceptable) is able to produce sufficient additional non-physical attraction for the H-bonded complexes but not for the case of the dispersion-bonded complexes, which results in unacceptable errors. Again, the use of the present DFT methods for non-covalent complexes (*i.e.* for most biological applications) is not justified, because dispersion energy plays a crucial role, and thus the results obtained with DFT methods must be taken with caution. Numerous attempts were made to include dispersion interaction into DFT, and three main directions can be followed when solving this problem: (i) theoretically based approaches with as little empiricism as possible (nonempirical approaches), (ii) attempts to reparametrise the existing density functionals so that they describe dispersion properly or cover the dispersion energy at least qualitatively, (iii) empirical approaches, based on the force-field-like terms, where the empirical expression for dispersion energy is simply applied and the total energy is constructed as a sum of DFT and dispersion energies.[189–191] Owing to the low number of dispersion-bonded complexes included in the widely used training sets, good performance of the developed functionals can simply not be expected. In an attempt to improve this weakness, a training set of 22 smaller molecular complexes (S22) and a rather extended set of 143 real-size complexes (JSCH2005) containing the geometries and accurate intermolecular interaction energies (for details see later) adopted from our previous papers[192–195] were introduced.[77] The nonempirical ways of including dispersion into DFT will be mentioned only in brief. A molecular complex is usually subdivided into two subsystems, and the dispersion interaction between them is calculated either based on intermolecular perturbation theory,[196–198] on the dynamic polarisabilities,[199,200] on the ground-state densities only,[201–206] or in another way.[207–209] Note that the division of the complex into subsystems is in fact one of the major drawbacks of these methods, because intramolecular dispersion in larger molecules (peptide chains, proteins, nucleic acids) cannot be accounted for. However, methods relying on molecular fragmentation as a principle are an exception, and a promising solution has recently been presented.[197] Density functionals not requiring the splitting of the system (seamless methods) have been suggested by Kohn *et al.*[210] and Dion *et al.*[211] While the density functional of Kohn *et al.* is prohibitively demanding in terms of computation, the method of Dion *et al.* is not unreasonably expensive and may appear useful after the necessary testing has been done. However, the evaluation of dispersion energy using these methods is either very tedious or is based on simplifications that lead to some inaccuracies, typically over 10% and sometimes even over 20%. Moreover, in most cases relatively large basis sets are required to describe correctly the dynamic polarisabilities of the molecules. Therefore, the errors in the computed dispersion energies are usually not much smaller than the errors that appear in the empirical dispersion calculations due to approximate dispersion coefficients and damping. The construction of a computationally inexpensive density functional accounting for dispersion in a general and seamless way thus remains an immense challenge. Regarding the attempts to modify the existing density functionals, the mainstream is represented by modifying the exchange functional.[212–215] The exchange functionals certainly need to be improved, because the long-range exchange that codetermines the vdW bonding is often only poorly predicted.[213] However, the dispersion interaction would be better described by the correlation functional as it is wholly a correlation effect. Moreover, whereas dispersion corresponds to $1/r^6$, $1/r^8$, … values, its inclusion in the exchange functional results in a different, incorrect dependence, namely $1/r^3$ values. Nevertheless, these contradictions do not preclude that some partial success may be achieved also using the exchange functional modification. In the work of Xu and Goddard,[214] the exchange functional was optimised considering among others the rare-gas dimers; the results on the real systems, however, turned out to be poor.[216] Grimme[217–219] introduced and tested a modified version of the B97 functional, which performs well for the complexes contained in our set of structures. Truhlar *et al.*[220–227] have tested their kinetics-optimised functionals on several larger dispersion-bonded as well as H-bonded complexes (from the S22 set) with reasonable success. The M06-2X that contains HF exchange provides very good estimates for various non-covalent complexes and can be even used (without any further parametrisation) for isolated systems like peptides. The local version of the potential, M06-L which is computationally much less demanding, also provides reasonable characteristics for various non-covalent complexes. Both functionals represent a new generation of DFT functionals that successfully describe all possible energy components including the London dispersion. These authors have also quite recently introduced the very tempting possibility of performing excited-state calculations employing their functionals, using the time-dependent DFT machinery. It seems possible to use one of these functionals (M06-HF) even to describe properly the excited states of a charge transfer character. Promising results for π-stacking interactions have been obtained[228] with the BH&H functional, which contains an equal mixture of the exact Hartree–Fock and local-density approximation for the description of exchange energy, coupled with Lee, Yang and Parr's expression for the correlation energy. Unfortunately, the performance of the functional for H-bonded systems is considerably worse. Kurita *et al.*'s modified PW91 functional performs relatively well for the stacked cytosine dimers. Unfortunately, while providing reasonable data for π-stacking, the latter two procedures[215,228] fail for H-bonding. This feature, namely that certain functionals work either for the hydrogen-bonded or for the stacked complexes but not for both, seems to be shared by all the approaches based on modifying the exchange functional. It must be clearly stated that this is not acceptable as only rarely can we distinguish between H-bonding and stacking. Interactions of nucleic acid bases in DNA represent such a case, but, *e.g.* interaction of amino acids in proteins is different and here one cannot unambiguously assign a structural motif either to H-bonding or stacking.

In the empirical dispersion approach, the dispersion energy is represented by the well-known C_6/r^6 formula. Dispersion energy is calculated separately from the DFT calculation, damped by an appropriate distance-dependent damping function to compensate for the overlap effects, and simply added to the DFT energy. Inclusion of the damping function is inevitable, because the C_6/r^6 dispersion energy reaches unrealistically high values at small interatomic distances. It is assumed that since dispersion acts mostly at long range, it has only a small effect on the electron densities, and its separate (uncoupled) calculation should not be problematic. Therefore, it forces a reliance on dispersion influencing the chemistry of the system mainly through the geometries, which is a great simplification. This may not be true in some specific situations, such as dipole bound states[229,230] or excited states, which are, however, beyond the scope of this study.

Among the first, already in the 1970s, Scoles *et al.*[231,232] combined empirical dispersion, damped by a sophisticated multiparameter damping function, with Hartree–Fock calculations. This method was fairly successful as it dealt with particularly weak rare-gas atom complexes, where the intermolecular repulsion is fairly well described by the HF non-local treatment. A similar approach was successfully used later by Hobza *et al.* for the description of larger complexes.[233,234] Since then, this initial success has stimulated several attempts to apply this to DFT. The works of Meijer and Sprik,[235] Mooij *et al.*,[236,237] and Wu *et al.*[238] showed significant improvements to the pure DFT, but also pointed to several questions regarding the reliability of DFT for the long-range exchange repulsion. Surprisingly accurate stabilisation energies of various types of non-covalent complexes were also obtained when the semiempirical tight-binding SCC-DFTB-D method by Elstner *et al.*[189] and Zhechkov *et al.*[239] was employed. A unique property of this method is that it allows us to study extended clusters having several thousands of atoms and the procedure can even be used for *ab initio* MD simulations. A successful application of the method for biomolecular systems (nucleic acid base pairs, amino-acid pairs, oligopeptides, and hydrophobic core of proteins) has been reviewed recently.[240] Wu and Yang[241] and Zimmerli *et al.*[242] touched on the question of the damping function. Unfortunately, all these studies share the drawback that only one or very few molecules were tested, which casts some doubt on the transferability of the results. Significant progress in this direction was made in a study by Grimme[190] in which various density functionals were tested on a set of molecules. The basis-set dependence was examined, and basis sets of at least triple-zeta quality were recommended. Notably, Grimme recognised the need for the dispersion to be adjusted for a given functional form and introduced a simple scaling factor optimised for each particular density functional. Hillier *et al.*[243] tested this method on our JSCH2005 set of complexes with reasonable success.

The basic idea of the above-mentioned approach is that DFT energy does not contain the dispersion term. The missing energy part (dispersion energy) has some functional form and parameters of this form are fitted to accurate CCSD(T) calculations. A step further was made by Bludský *et al.*[244] who proposed a novel DFT/CCSD(T) correction scheme for the precise calculations of non-covalent interactions. The difference between DFT/aug-cc-pVQZ and CCSD(T)/CBS energies was expressed as a sum of atom–atom corrections that are functions of interatomic distance. The correction functions were obtained from calculations on model systems and their transferability was proven. The method was recently successfully used for the study of benzene ... naphthalene and naphthalene ... naphthalene complexes.[245] The global energy minimum for both complexes corresponds to parallel-displaced structures and T-shaped structures were less stable by 0.2 and 1.6 kcal/mol.

We have seen that addition of dispersion energy to HF or DFT procedures highly improves their performance in general and in the case of stacked structures this improvement is of key importance. There is a good reason for this – in both methods the dispersion energy is completely missing. In the case of the HF method the addition of dispersion energy is fully justified. The situation with DFT is, however, different since here some part of the correlation energy is covered. There exist opinions that DFT + D overcorrects the dispersion energy. We believe that the results presented in this book contradict these opinions.

In the case of semiempirical QM methods the situation is even more difficult since here some portion of the correlation energy (*via* empirical parameters) is covered. The first attempt to include the dispersion energy into semiempirical QM methods was made by Martin and Clark.[246] The authors introduced an additional term to treat dispersion in NDDO-based semiempirical molecular orbital techniques. The dispersion energy was calculated using additive "atomic orbital" polarisability tensors. Dispersion energies are best calculated using the Slater–Kirkwood modification of the London formula, although they can also be obtained using the Casimir–Polder approach with frequency-dependent polarisabities. Recently, McNamara and Hillier combined semiempirical AM1 and PM3 methods with an empirical dispersion term.[247] After reparametrisation of both the semiempirical method and dumping function, considerably improved characteristics of non-covalent complexes resulted. The semiempirical methods of the OMx family[248] are known to describe complex systems better than AM1 or PM3 that is mainly due to the use of orthogonalisation corrections. These corrections, which are missing in other semiempirical methods, improve the description of Pauli repulsion. Tuttle and Thiel[249] augmented the OM2 and OM3 methods by empirical dispersion energy but contrary to the above-mentioned case they use the semiempirical method in the original version (*i.e.* no reparametrisation of the semiempirical method was required). Similarly as in the case of AM1 and PM3, significant improvements of the original methods for non-covalent interactions are achieved upon inclusion of the dispersion term. The use of semiempirical QM methods covering the dispersion energy is without doubt very attractive since they are genuine QM methods with all the advantages and similarly with still reasonable CPU demands. Evidently, careful investigations of their performance before these methods will be widely applied, is required.

We have recently developed an empirical correction for DFT[191] and since the procedure was quite successful in application to biomacromolecules and also nanostructures the method will be described in detail. Let us also mention that very similar conclusions can be drawn about the DFT + D procedure suggested by Grimme. The critical point represents the use of an improved scheme for the damping function. In this (as well as similar procedures) only the damping function is parametrised, while the dispersion term is untouched. The damping function is optimised on a small, well-balanced set of 22 vdW complexes (S22 dataset) and verified on a validation set of 58 vdW complexes. The results can be summarised as follows: (1) a simple empirical dispersion correction can significantly improve the description of the intermolecular interactions. (2) Damping of the empirical dispersion term is apparently the most critical point in DFT-D, since it concerns questions of double counting of correlation energy (see above). (3) Extending the basis set, the importance of the empirical correction becomes more and more obvious. It is recommended to add the empirical dispersion correction only to sufficiently large basis sets (of at least a TZVP quality). (4) In the case of hydrogen-bonded systems the BSSE for the basis sets of DZ quality is of the same direction and of almost the same size as the dispersion interaction. Therefore, the BSSE uncorrected DFT calculations provide reasonable agreement with experiment for these complexes. (5) The quality of the results is considerably affected by the quality of functional used. In general, the more advanced functionals (hybrid functionals, meta-GGA functionals) give better results than LDA or simple GGA approximations when combined with empirical dispersion. (6) The best results were obtained with the TPSS functional combined with an extended basis set containing f- and d-functions on heavy atoms and hydrogens, respectively.

As mentioned above, the procedure is similar to that used by Grimme[190] and the major change is that we discarded the coefficient scaling the dispersion strength to various density functionals. The main reason is that it scales the dispersion strength also far from the overlap region where the dispersion interaction is not affected by the choice of a particular density functional. This behaviour is unphysical and it introduces some error that will be important for extended clusters. This was nicely demonstrated in Ref. 250, where the cohesive energy of a set of 14 molecular crystals ranging from hydrogen-bonded to dispersion-bonded ones was determined using the DFT + D procedure. The authors have shown that the mean absolute deviation of the theoretical values (from the experimental data) was considerably larger for the Grimme and coworkers approach than for the procedure of Jurečka and coworkers (9.5 *vs.* 5.5 kJ/mol). The authors concluded that "the overall results are very promising and the application of this empirical dispersion term deserves to be further investigated".

This DFT-D method (where the damping function was parametrised to molecular clusters) works very well not only for molecular clusters but also for extended single molecules such as oligopeptides[188,251,252] or even simple alkanes,[253] for which the damping function was not even parametrised. Our intention is to use the DFT-D method mainly for calculations and simulations of peptides and nucleic acids. For such large-scale calculations, we prefer to perform the calculation using a nonhybrid XC functional, because the density-fitting approximation can be exploited, bringing significant speed-ups (by about one order of magnitude). Unfortunately, density fitting cannot be applied to the hybrid functionals (containing the explicit HF exchange). Among other works, von Lilienfeld *et al.*[254] simulated dispersion by reparametrising the nonlocal part of the pseudopotential in the plane-wave DFT.

Owing to the problems with dispersion energy, the DFT geometries of non-covalent complexes are not accurate. In the case of stacking, the standard DFT methods fail completely and usually no minimum is detected for stacking motifs. But also in the case of H-bonding, where the dispersion energy is definitely not negligible, the DFT geometries are not accurate. Due to the compensation of errors, however, the geometries of H-bonded clusters are often acceptable. As shown above, the easiest way to remove the problems with dispersion energy is by adding it to the DFT energy (the gradients for the DFT-D method can easily be determined). The DFT-D geometries are surprisingly accurate. In the case of the phenol dimer,[81] the DFT-D geometries agree very well with experimental data. We have reason to believe that the DFT-D procedure for now represents the most accurate method for determining the structure and geometry of extended molecular clusters.

We conclude this part devoted to DFT by stating that when the method properly includes the dispersion energy its applicability increases considerably and the method has a chance to become a "universal" technique suitable for the treatment of complex molecular systems (*i.e.* extended molecular complexes as well as extended isolated systems like peptides or proteins).

2.5 Quantum Monte Carlo

All previously mentioned methods were based on the application of different WFT or DFT schemes. In the case of WFT, both the quality of the AO basis set and the amount of the correlation energy covered play a key role. The electronic structure of molecular complexes can also be determined using a completely different procedure, namely the quantum Monte Carlo (QMC) method.[255] This method is attractive in that it includes the correlation energy explicitly in the multielectron wavefunction (partitioning of the total energy is not necessary); the effect of the finite basis set is either none or very small (since it is used only for the construction of the guiding function); it is size consistent and variational; the current cubic scaling of the method is much more favourable than for WFT methods (and it is believed that this will further improve); and finally the QMC algorithm is intrinsically parallel, scaling linearly with the number of processors. QMC of course has some drawbacks too. First, one can in fact obtain only the energy: there are no easily applicable ways to calculate energy derivatives, which means that the chance to, for example, optimise the geometry of the studied systems using QMC is very small. The second problem is connected with the quality of the nodal system used, which affects the quality of the obtained results. Fortunately, in the case of non-covalent complexes, this error cancels out.

The diffusion quantum Monte Carlo method is based on the similarity of the Schrödinger equation to Fick's diffusion equation. The differential equation is simulated using a procedure in which many "random walkers" are allowed to diffuse and multiply in a series of finite time steps until the normalised distribution of the walkers approaches a "steady-state" distribution. This distribution fluctuates around an average steady-state distribution that corresponds to the lowest-energy wavefunction satisfying the Schrödinger equation. Because the distribution of walkers cannot recognise nodes in the wavefunction, it is necessary to somehow restrict this distribution in such a way that it does not diffuse to the nodeless ground-state wavefunction. The most common way of restricting the walker distribution is to enforce nodes contained within analytic variational wavefunctions (typically the Hartree–Fock wavefunction). When this technique is used along with the quantum Monte Carlo procedure, the resulting method is referred to as fixed-node diffusion Monte Carlo.[256] The QMC method has been successfuly applied to various small complexes.[257]

2.6 Vibrational Frequencies

A very important feature of vibrational frequencies is the fact that they are observable and the formation of a non-covalent complex can be easily detected by measuring its intermolecular vibrational frequencies. Upon formation of a non-covalent complex, new intermolecular frequencies arise, which are generally much smaller than intramolecular frequencies and are typically observed in the sub-$100\,cm^{-1}$ region of the vibrational spectrum; frequently, they are below $50\,cm^{-1}$. Generally, six intermolecular frequencies exist for a dimer, corresponding to the loss of three translational and three rotational degrees of freedom upon its formation. Intermolecular frequencies are mostly rather similar for various types of non-covalent complexes and thus cannot be used for identification of a specific complex. For the identification, one can use changes of intramolecular frequencies upon formation of a complex, the observed shift of which usually correlates with the strength of molecular interactions. Well known is the shift of X–H stretching frequencies to lower values upon formation of a hydrogen bond of the type X–H . . . Y. This redshift can be very large, even several hundreds of wave numbers, affording an easy and unambiguous way of proving the formation of an H-bonded complex. In recent years an opposite shift, *i.e.* a shift to higher frequencies (blueshift), was also observed for a similar class of systems and the so-called improper blueshifting H-bond was detected in many complexes in nature. More detailed information on both types of H-bonding will be provided later.

Harmonic frequencies are easily determined even for large non-covalent clusters using the Wilson FG analysis and the procedure is now routinely available in quantum-chemical codes. The frequencies calculated, however, are in reality nonharmonic, and this should be taken into consideration. Mostly, this is covered *via* scaling and there exist recommended values of scaling factors, which are different for Hartree–Fock and correlated calculations. The use of scaling, without doubt, results in better agreement between experimental and theoretical values but its use cannot be generally recommended since it sometimes leads to the assumption that a small correction for anharmonicity is acceptable for describing the system, whereas in reality it is nonharmonic, as for the large-amplitude motion in the ammonia dimer. The standard approach to the nonharmonic vibrational problem for large non-covalent clusters is represented by perturbation theory.[258,259] If the zero-order Hamiltonian (usually a harmonic oscillator) is a good approximation to the true vibrational Hamiltonian, perturbation theory is a very efficient and reliable tool for calculating vibrational frequencies. In the traditional approach, the matrix representation of the molecular Hamiltonian is diagonalised by successive contact transformations. This procedure, however, fails in the case of accidental resonances, which is often the case if we deal with large systems with many vibrational modes. In this case the terms connecting the resonant levels have to be treated variationally[260] and considerable progress has been achieved in developing new procedures based on perturbative treatment for the calculation of anharmonic frequencies.[261,262] The potential-energy function is constructed as a low-order polynomial (up to fourth order) expressed in normal coordinates. The force constants are obtained by least-squares fitting of energies, gradients, and Hessians calculated at geometries close to the global minimum on the PES. The main advantage of this approach stems from its computational efficiency as the number of the required Hessians scales linearly with the number of vibrational modes. Thus, the method can even be used for large systems, while respecting the full dimensionality of the problem. However, the applicability of the procedure is less straightforward for non-covalent clusters since they are nonrigid systems. The vibrational dynamics of floppy systems cannot be described in the framework of a single-reference Hamiltonian and, therefore, the perturbation series used are necessarily strongly divergent. In such a case the only alternative is a more exact treatment of the large-amplitude vibrational modes including all relevant parts of the coordinate space. This requires calculation of the global PES, which becomes computationally prohibitive even for systems with only a few degrees of freedom. However, the number of large-amplitude motions is usually a small fraction of the total number of vibrations. Consequently, the large-amplitude vibrations can be removed from the perturbative treatment and the Schrödinger equation for the effective large-amplitude Hamiltonian solved variationally. However, a coupling between a large-amplitude vibrational mode and other modes is not considered and can lead to rather large errors.

The fully variational method is free of any limitation but it is prohibitively expensive even for problems of low dimensionality. The literature on higher-dimensional anharmonic vibrational calculations of non-covalent clusters based on *ab initio* correlated calculations includes a variational six-dimensional intermolecular vibrational frequency calculation for the adenine...thymine Watson–Crick base pair[263] and a twelve-dimensional vibrational frequency calculation for the water dimer by perturbation theory.[264] Also, a six-dimensional frequency calculation for the water dimer based on various empirical potentials was reported by LeForestier *et al.*[265]

Other rigorous treatments of vibrations in clusters have also been reported.[266–269] Harmonic and anharmonic vibration frequencies were determined for the guanine...cytosine complex and compared with gas-phase IR-UV double resonance spectral data. Harmonic frequencies were obtained at RI-MP2/cc-pVDZ and RI-MP2/TZVPP levels and anharmonic frequencies were obtained by the CC-VSCF method based on improved semiempirical PM3 results.[270] Comparison of the data with experimental results indicates that the average absolute percentage deviation for the method is 2.6% for harmonic RI-MP2/cc/pVDZ, 2.5% for harmonic RI-MP2/TZVPP and 2.3% for the adopted PM3 CC-VSCF. The use of an empirical scaling factor for the *ab initio* harmonic calculations improves the stretching frequencies but decreases the accuracy of the other mode frequencies.

A serious problem with floppy non-covalent complexes is represented by the conformational instabilities of the studied systems. The presence of surpassable barriers on the molecular PES (this is a very typical feature of not only all molecular clusters but also of large isolated systems like peptides or proteins) prevents various theoretical treatments going beyond the harmonic approximation as discussed above and from being used safely, such as molecular dynamics, perturbation theory, and vibrational self-consistent techniques. A popular way of overcoming these limits is based on an empirical scaling of the calculated harmonic frequencies (see above and also Refs. 271 and 272). However, in the case of vibrational motions opposed by strongly anharmonic potentials (for instance, motions involved in hydrogen bondings, internal rotations, and ring deformations), the approach is no longer reliable[273] and, as these motions constitute very sensitive probes for structural assignments (see, *e.g.* Refs. 274 and 273 and references therein), a more realistic method is highly desirable. A possible way of meeting this requirement may lie in an adiabatic separation of the probing of molecular modes from the "bath" of the remaining molecular motions, while disregarding the non-adibatic couplings (*i.e.* a Born–Oppenheimer-type approximation). Being only few-dimensional, the resulting dynamical problems are tractable in a numerically exact way for practically any shape of the corresponding effective potentials (see, *e.g.* Ref. 275). This approach is also very economical in terms of *ab initio* calculations as it requires energy optimisations for only very few molecular geometry coordinates. Apparently, it is ideally suited for the case of single, highly characteristic molecular vibrations, which are only weakly coupled to the remaining molecular motions – typically the stretch vibrations. Let us recall that the majority of recent gas-phase experiments on molecular clusters and peptides detect just stretch vibrations in the region 3000–4000 cm^{-1}. Although formally one-dimensional, the approach allows for all the important interaction terms from the potential energy surface and even rotation–vibration interaction terms originating from kinetic energy by allowing for the molecular valence coordinates to vary with the reference coordinate.[276,277]

The procedure was tested[278] for the assignment of the NH stretches of the keto/enol and 7/9 NH guanine tautomers, which has been the subject of many discussions.[279–282] The main reason for the contradictory results of several groups is the close resemblance of the corresponding vibrational frequencies.[279–281] In addition, the tautomers are very close in their energy contents, four of them being within 1.2 kcal/mol. The assignment of the IR-UV double resonance spectra is further complicated by the very short lifetime of some of the tautomers because of ultrafast nonradiative decay.[283,284] Despite being relatively large, guanine is still tractable by means of highly accurate *ab initio* procedures and thus allows for benchmark calculations. Moreover, the NH frequencies of the four lowest-energy tautomers are conclusively assigned by comparing experimental vibration transition moment angles (obtained using the helium nanodroplet isolation technique) with their *ab initio* counterparts.[282] The main outcome of the study was the comparison of the calculated and experimental HN and NH$_2$ fundamental frequencies of guanine tautomers. All the calculated "adiabatic" frequencies are in very close agreement with their experimental counterparts and by applying only a very small linear shift this agreement becomes practically perfect. This is especially true when most accurate CCSD(T) calculations are utilised. The resulting vibrational frequencies for all tautomers have been determined with "spectroscopic" accuracy and the sum of squares of the deviations to experiment is unprecedently small, less than 4 cm^{-1}. It is thus evident that the method can be used for analytical purposes, allowing spectral separations of less than a few cm^{-1} to be distinguished. Despite deviating much more from the experiment than the "adiabatic" predictions, even the harmonic vibrational frequencies are still in a one-to-one correspondence with the experiment, thus evidencing the reliability of the standard harmonic approach in the probed case. On the contrary, the frequencies evaluated using the standard second-order perturbation theory exhibit strong disharmony with experiment. This approach is clearly not suitable for the purposes of assignment. Its failure is not surprising: the standard polynomial quartic force-field representation is inadequate for describing hydrogen-containing stretching motions,[285] and moreover, the perturbation series in such cases strongly diverge and their correct summation requires accounting for very high-order corrections.[286] Importantly, the frequency differences resulting from fairly "cheap" MP2 and DFT procedures exhibit only slightly larger (theory *vs.* experiment) dispersions than their "expensive" CCSD(T) counterparts. The former methods thus offer a promising potential for the adiabatic treatment of systems that are prohibitive even for the standard harmonic normal coordinate analysis.

The soft intermolecular vibrational modes may serve for structural elucidation. Their total number ranges from one (vdW system consisting of two atoms) up to six, which is available for any vdW system consisting of two polyatomic subsystems. The hydrogen fluoride dimer possesses four inter-molecular modes for obvious reasons. The vibrational predissociation[287] and the vibrational second overtones[288] of $(HF)_2$ were studied thoroughly. Counterpoise-corrected *ab initio* analytical potential-energy and dipole hypersurfaces were computed and the dimer dissociation energy, $D_e=19.1\,kJ/mol$ was obtained.[289] The energy barrier to the hydrogen-bond exchange amounts to $4.2\,kJ/mol$; the disrotatory in-plane bending vibration involved in this process was studied earlier.[290] A careful study of the near-infrared spectra of $(DF)_2$ permitted determination of all four intermolecular modes.[291] Extensive attention was paid to the role of the basis set superposition error (BSSE) in connection with oligomers $(HF)_n$ ($n = 3, 4$); the consequences are of greater importance.[292]

The high-resolution IR[293] and vibrational excitation[294] spectra were also recorded and analysed. Information on the coupling between the intra- and intermolecular modes can be obtained from the redshift of the HF stretching mode; an analysis of the rotational constants[287] can be used for the same purpose. The CCSD(T) calculations were carried out in a study of vibrational predissociation of the Ne...Br_2 system.[295] Determination of a high-quality PES for $(N_2)_2$ leads[296] to a T-shaped structure with a well depth of $107\,cm^{-1}$ and a distance of $4.03\,Å$. A rotationally resolved IR spectrum for C_2H_4...HCl (with ^{35}Cl and ^{37}Cl) was interpreted with the assistance of CCSD(T) calculations.[297]

A π-type complex between 2,3-dihydrobenzofuran (coumaran) and Ar was studied by combining resonance-enhanced multiphoton ionisation and zero electron kinetic energy spectroscopy.[298]

The OH...CO vdW radical represents an intermediate in the OH + CO → $H + CO_2$ process, which assumes an important role in combustion and atmospheric chemistry.[299] Intermolecular excitations in the region 50–250 cm^{-1} were interpreted with the assistance of the CCSD(T) method.

Multiphoton IR photodissociation spectroscopy was used for the investigation of the solvation effect with the $(HBr)_nBr^-$ ($n=1, 2, 3$) systems. The harmonic approximation is not sufficient for the interpretation of experimental IR spectra.[300] The intermolecular π-bond between protonised benzene $(C_6H_7^+)$ and Ar, N_2, CH_4, and H_2O was studied by means of IR photodissociation spectra of mass-selected clusters and MP2 calculations.[301]

References

1. G. Buemi and C. Gandolfo, *J. Chem. Soc. Faraday Trans. 2*, 1989, **85**, 215.
2. J. J. Dannenberg and R. Rios, *J. Phys. Chem.*, 1994, **98**, 6714.
3. R. Srinivasan, J. S. Feenstra, S. T. Park, S. J. Xu and A. H. Zewail, *J. Am. Chem. Soc.*, 2004, **126**, 2266.
4. S. J. Xu, S. T. Park, J. S. Feenstra, R. Srinivasan and A. H. Zewail, *J. Phys. Chem. A*, 2004, **108**, 6650.

5. W. Caminati, *private communication.*
6. W. F. Rowe, R. W. Duerst and E. B. Wilson, *J. Am. Chem. Soc.*, 1976, **98**, 4021.
7. G. Karlstrom, B. Jonsson, B. Roos and H. Wennerstrom, *J. Am. Chem. Soc.*, 1976, **98**, 6851.
8. D. Gerhard, A. Hellweg, I. Merke, W. Stahl, M. Baudelet, D. Petitprez and G. Wlodarczak, *J. Mol. Spectrosc.*, 2003, **220**, 234.
9. M. Schnell, J. U. Grabow, H. Hartwig, N. Heineking, M. Meyer, W. Stahl and W. Caminati, *J. Mol. Spectrosc.*, 2005, **229**, 1.
10. M. S. Ford, S. R. Haines, I. Pugliesi, C. E. H. Dessent and K. Müller-Dethlefs, *J. Electron Spectrosc. Relat. Phenom.*, 2000, **112**, 231.
11. S. Chervenkov, P. Q. Wang, J. E. Braun, S. Georgiev, H. J. Neusser, C. K. Nandi and T. Chakraborty, *J. Chem. Phys.*, 2005, **122**, 244312.
12. S. Georgiev and H. J. Neusser, *J. Electron Spectrosc. Relat. Phenom.*, 2005, **142**, 207.
13. K. Siglow and H. J. Neusser, *Chem. Phys. Lett.*, 2001, **343**, 475.
14. F. Aguirre and S. T. Pratt, *J. Chem. Phys.*, 2004, **121**, 9855(b).
15. M. R. Hockridge, E. G. Robertson, J. P. Simons, D. R. Borst, T. M. Korter and D. W. Pratt, *Chem. Phys. Lett.*, 2001, **334**, 31.
16. G. Berden, W. L. Meerts, M. Schmitt and K. Kleinermanns, *J. Chem. Phys.*, 1996, **104**, 972.
17. M. Gerhards, M. Schmitt, K. Kleinermanns and W. Stahl, *J. Chem. Phys.*, 1996, **104**, 967.
18. C. Kang and D. W. Pratt, *Int. Rev. Phys. Chem.*, 2005, **24**, 1.
19. C. W. Kang, J. T. Yi and D. W. Pratt, *J. Chem. Phys.*, 2005, **123**, 094306.
20. T. V. Nguyen, T. M. Korter and D. W. Pratt, *Mol. Phys.*, 2005, **103**, 2453.
21. D. W. Pratt, *Science*, 2002, **296**, 2347.
22. J. A. Reese, T. V. Nguyen, T. M. Korter and D. W. Pratt, *J. Am. Chem. Soc.*, 2004, **126**, 11387.
23. Y. H. Lee, J. W. Jung, B. Kim, P. Butz, L. C. Snoek, R. T. Kroemer and J. P. Simons, *J. Phys. Chem. A*, 2004, **108**, 69.
24. D. R. Borst, P. W. Joireman, D. W. Pratt, E. G. Robertson and J. P. Simons, *J. Chem. Phys.*, 2002, **116**, 7057.
25. M. Ford, R. Lindner and K. Müller-Dethlefs, *Mol. Phys.*, 2003, **101**, 705.
26. H. S. Andrei, N. Solca and O. Dopfer, *J. Phys. Chem. A*, 2005, **109**, 3598.
27. U. Lorenz, N. Solca and O. Dopfer, *Chem. Phys. Lett.*, 2005, **406**, 321.
28. N. Solca and O. Dopfer, *Chem. Phys. Lett.*, 2003, **369**, 68.
29. N. Solca and O. Dopfer, *J. Am. Chem. Soc.*, 2004, **126**, 9520.
30. G. C. Cole and A. C. Legon, *J. Chem. Phys.*, 2004, **121**, 10467.
31. P. W. Fowler, A. C. Legon, J. M. A. Thumwood and E. R. Waclawik, *Coord. Chem. Rev.*, 2000, **197**, 231.
32. B. M. Giuliano, P. Ottaviani, W. Caminati, M. Schnell, D. Banser and J. U. Grabow, *Chem. Phys.*, 2005, **312**, 111.

33. B. M. Giuliano and W. Caminati, *Angew. Chem. Int. Ed.*, 2005, **44**, 603.
34. J. L. Alonso, S. Antolinez, S. Blanco, A. Lesarri, J. C. Lopez and W. Caminati, *J. Am. Chem. Soc.*, 2004, **126**, 3244.
35. W. Caminati, J. C. Lopez, J. L. Alonso and J. U. Grabow, *Angew. Chem. Int. Ed.*, 2005, **44**, 3840.
36. R. Sanchez, S. Blanco, A. Lesarri, J. C. Lopez and J. L. Alonso, *Chem. Phys. Lett.*, 2005, **401**, 259.
37. A. Lesarri, E. J. Cocinero, J. C. Lopez and J. L. Alonso, *Angew. Chem. Int. Ed.*, 2004, **43**, 605.
38. S. Blanco, J. C. Lopez, A. Lesarri, W. Caminati and J. L. Alonso, *Mol. Phys.*, 2005, **103**, 1473.
39. D. D. Nelson, G. T. Fraser and W. Klemperer, *J. Chem. Phys.*, 1985, **83**, 6201.
40. K. Müller-Dethlefs and P. Hobza, *Chem. Rev.*, 2000, **100**, 143.
41. P. R. Bunker and P. Jensen, *Molecular Symmetry and Spectroscopy*, 2nd edn NRC, Ottawa, 1998.
42. G. Herzberg, *Molecular Spectra and Molecular Structure, Vol. III*, 2nd edn Krieger Publishing Company, 1966.
43. E. H. T. Olthof, A. van der Avoird and P. E. S. Wormer, *J. Mol. Struct. Theochem*, 1994, **113**, 201.
44. F. Franks, *Water: A Comprehensive Treatise*, Plenum, New York, 1972.
45. S. S. Xantheas and T. H. Dunning, *Advances in Molecular Vibrations, Collision Dynamics*, JAI Press, Stanford, 1998, pp. 281.
46. R. C. Cohen and R. J. Saykally, *Annu. Rev. Phys. Chem.*, 1991, **42**, 369.
47. Z. S. Huang and R. E. Miller, *J. Chem. Phys.*, 1988, **88**, 8008.
48. N. Goldman, C. LeForestier and R. J. Saykally, *Philos. Trans. Royal Soc. London Series A-Math. Phys. Eng. Sci.*, 2005, **363**, 493.
49. F. N. Keutsch, J. D. Cruzan and R. J. Saykally, *Chem. Rev.*, 2003, **103**, 2533.
50. J. D. Cruzan, L. B. Braly, K. Liu, M. G. Brown, J. G. Loeser and R. J. Saykally, *Science*, 1996, **271**, 59.
51. D. J. Wales, *Energy Landscapes*, Cambridge University Press, Cambridge, 2003.
52. B. J. Mhin, S. J. Lee and K. S. Kim, *Phys. Rev. A.*, 1993, **48**, 3764.
53. H. M. Lee, S. B. Suh and K. S. Kim, *J. Chem. Phys.*, 2000, **112**, 9659.
54. K. Kim and K. D. Jordan, *J. Phys. Chem.*, 1994, **98**, 10089.
55. G. S. Fanourgakis, E. Apra and S. S. Xantheas, *J. Chem. Phys.*, 2004, **121**, 2655.
56. R. S. Fellers, L. B. Braly, R. J. Saykally and C. LeForestier, *J. Chem. Phys.*, 1999, **110**, 6306.
57. J. K. Gregory and D. C. Clary, *Chem. Phys. Lett.*, 1994, **228**, 547.
58. J. K. Gregory and D. C. Clary, *J. Chem. Phys.*, 1995, **102**, 7817.
59. N. Goldman, C. LeForestier and R. J. Saykally, *Philos. Trans. Royal Soc. London Series A-Math. Phys. Eng. Sci.*, 2005, **363**, 493.

60. (a) C. J. Burnham and S. S. Xantheas, *J. Chem. Phys.*, 2002, **116**, 1500; (b) S. S. Xantheas, C. J. Burnham and R. J. Harrison, *J. Chem. Phys.*, 2002, **116**, 1493.

61. N. Goldman, R. S. Fellers, M. G. Brown, L. B. Braly, C. J. Keoshian, C. LeForestier and R. J. Saykally, *J. Chem. Phys.*, 2002, **116**, 10148.

62. K. Liu, M. G. Brown, C. Carter, R. J. Saykally, J. K. Gregory and D. C. Clary, *Nature*, 1996, **381**, 501.

63. R. Bukowski, K. Szalewicz, G. C. Groenenboom and A. van der Avoird, *Science*, 2007, **315**, 1249.

64. G. C. Groenenboom, E. M. Mas, R. Bukowski, K. Szalewicz, P. E. S. Wormer and A. van der Avoird, *Phys. Rev. Lett.*, 2000, **84**, 4072.

65. K. Szalewicz, G. Murdachaew, R. Bukowski, O. Akin-Ojo, C. LeForestier, in *Lecture Series on Computer, Computational Science: ICCMSE 2006*, G. Maroulis, T. Simos, ed. (Brill Academic, Leiden, Netherlands, 2006), vol. 6, pp. 482.

66. W. Cencek, K. Szalewicz, C. LeForestier, R. v. Harrevelt and A. van der Avoird, *Phys. Chem. Chem. Phys.*, 2008, **10**, 4716.

67. R. Bukowski, K. Szalewicz, G. C. Groenenboom and A. van der Avoird, *J. Chem. Phys.*, 2008, **128**, 094313.

68. R. Bukowski, K. Szalewicz, G. C. Groenenboom and A. van der Avoird, *J. Chem. Phys.*, 2008, **128**, 094314.

69. R. Carr and M. Parrinello, *Phys. Rev. Lett.*, 1985, **55**, 2471.

70. M. Matsumoto, S. Saito and I. Ohmine, *Nature*, 2002, **416**, 409.

71. H. Nada and Y. Furukawa, *J. Cryst. Growth*, 2005, **283**, 242.

72. P. M. Felker and A. H. Zewail, *J. Chem. Phys.*, 1987, **86**, 2460.

73. J. S. Baskin, P. M. Felker and A. H. Zewail, *J. Chem. Phys.*, 1987, **86**, 2483.

74. W. Jarzeba, V. V. Matylitsky, C. Riehn and B. Brutschy, *Chem. Phys. Lett.*, 2003, **368**, 680.

75. W. Jarzeba, V. V. Matylitsky, A. Weichert and C. Riehn, *Phys. Chem. Chem. Phys.*, 2002, **4**, 451.

76. J. Gauss and J. F. Stanton, *J. Chem. Phys.*, 2002, **116**, 1773.

77. P. Jurečka, J. Šponer, J. Černý and P. Hobza, *Phys. Chem. Chem. Phys.*, 2006, **8**, 1985.

78. A. Weichert, C. Riehn, A. Weichert and B. Brutschy, *J. Phys. Chem. A*, 2001, **105**, 5679.

79. M. Schmitt, M. Boehm, C. Ratzer, D. Kruegler, K. Kleinermanns, I. Kalman, G. Berden and W. L. Meerts, *Chem. Phys. Chem.*, 2006, **7**, 1241.

80. P. Hobza, Ch. Riehn, A. Weichert and B. Brutschy, *Chem. Phys.*, 2002, **283**, 331.

81. M. Kolář and P. Hobza, *J. Phys. Chem. A*, 2007, **111**, 5851.

82. P. Jurečka, P. Nachtigall and P. Hobza, *Phys. Chem. Chem. Phys.*, 2001, **3**, 4578.

83. M. Feyereisen, G. Fitzgerald and A. Komornicki, *Chem. Phys. Lett.*, 1993, **208**, 359.

84. O. Vahtras, J. Almlöf and M. Feyereisen, *Chem. Phys. Lett.*, 1993, **213**, 514.
85. D. E. Bernholdt and R. J. Harrison, *Chem. Phys. Lett.*, 1996, **250**, 470.
86. R. Ahlrichs, M. Bär, M. Häser, H. Horn and C. Kölmel, *Chem. Phys. Lett.*, 1989, **162**, 165.
87. S. F. Boys and F. Bernardi, *Mol. Phys.*, 1970, **19**, 553.
88. S. Simon, M. Duran and J. J. Dannenberg, *J. Chem. Phys.*, 1996, **105**, 11024.
89. P. Hobza and Z. Havlas, *Theor. Chem. Acc.*, 1998, **99**, 372.
90. I. K. Yanson, A. B. Teplitsky and L. F. Sukhodub, *Biopolymers*, 1979, **18**, 1149.
91. P. R. Kemper, P. Weis and M. T. Bowers, *Int. J. Mass Spectrom. Ion Processes*, 1997, **160**, 17.
92. S. S. Hunnicutt, T. M. Branch, J. B. Everhart and D. S. Dudis, *J. Phys. Chem.*, 1996, **100**, 2083.
93. L. Oudejans and R. E. Miller, *J. Phys. Chem. A*, 1997, **101**, 7582.
94. H. J. Neusser and H. Krause, *Chem. Rev.*, 1994, **94**, 1829.
95. H. J. Neusser and H. Krause, *Int. J. Mass Spectrom. Ion Proc.*, 1994, **131**, 211.
96. C. Lifshitz and N. Ohmichi, *J. Phys. Chem.*, 1989, **93**, 6329.
97. J. R. Grover, E. A. Walters, J. K. Newman and M. G. White, *J. Am. Chem. Soc.*, 1985, **107**, 7329.
98. R. Knochenmuss and S. J. Leutweiler, *J. Chem. Phys.*, 1987, **91**, 1268.
99. S. K. Kim and E. R. Bernstein, *J. Chem. Phys.*, 1991, **95**, 3119.
100. S. K. Kim, J. J. Breen, D. M. Willberg, L. W. Peng, A. Heikal, J. A. Syage and A. H. Zewail, *J. Phys. Chem.*, 1995, **99**, 7421.
101. K. Watanabe, *J. Chem. Phys.*, 1954, **22**, 1564.
102. M. Riese, A. Gaber and J. Grotemeyer, *Z. Phys. Chem.*, 2007, **221**, 663.
103. B. Ernstberger, H. Krause, A. Kiermeier and H. J. Neusser, *J. Chem. Phys.*, 1990, **92**, 5285.
104. N. Ohmichi, Y. Malinovich, J. P. Ziesel and C. Lifshitz, *J. Phys. Chem.*, 1989, **93**, 2491.
105. B. Ernstberger, H. Krause and H. J. Neusser, *Z. Phys. D*, 1991, **20**, 189.
106. H. Krause, B. Ernstberger and H. J. Neusser, *Chem. Phys. Lett.*, 1991, **184**, 411.
107. K. Müller-Dethlefs, M. Sander and E. W. Schlag, *Z. Naturforsch. A*, 1984, **39a**, 1089.
108. K. Müller-Dethlefs, M. Sander and E. W. Schlag, *Chem. Phys. Lett.*, 1984, **112**, 291.
109. K. Müller-Dethlefs, M. Sander and L. A. Chewter, in *Laser Spectroscopy VII*, ed. T.W. Hänsch, Y.R. Shen, Springer Verlag, Berlin, New York, 1985, p. 118.
110. G. Reiser, W. Habenicht, K. Müller-Dethlefs and E. W. Schlag, *Chem. Phys. Lett.*, 1988, **152**, 119.
111. W. Habenicht, G. Reiser and K. Müller-Dethlefs, *J. Chem. Phys.*, 1991, **95**, 4809.

112. K. Müller-Dethlefs, *J. Chem. Phys.*, 1991, **95**, 4822.
113. C. E. H. Dessent, S. R. Haines and K. Müller-Dethlefs, *Chem. Phys. Lett.*, 1999, **315**, 103.
114. C. E. H. Dessent and K. Müller-Dethlefs, *Chem. Rev.*, 2000, **100**, 3999.
115. K. Müller-Dethlefs and E. W. Schlag, *Annu. Rev. Phys. Chem.*, 1991, **42**, 109.
116. E. R. Grant and M. G. White, *Nature*, 1991, **354**, 249.
117. F. Merkt and T. P. Softley, *Int. Rev. Phys. Chem.*, 1993, **12**, 205.
118. K. Müller-Dethlefs, O. Dopfer and T. G. Wright, *Chem. Rev.*, 1994, **94**, 1845.
119. I. Fischer, R. Lindner and K. Müller-Dethlefs, *J. Chem. Soc. Faraday Trans.*, 1994, **90**, 2425.
120. K. Müller-Dethlefs, *Chapter II*, in: *High Resolution Laser Photoionization, Photoelectron Studies*, ed. I. Powis, T. Baer, C. Y. Ng, John Wiley & Sons Ltd, Chichester, 1995, pp. 22.
121. K. Müller-Dethlefs, E.W. Schlag, E.R. Grant, K. Wang and B.V. McKoy, *Adv. Chem. Phys.* ed. I. Prigogine, S. A. Rice, Wiley, Chichester, 1995, **90**, 1.
122. K. Müller-Dethlefs, *J. Electron Spectrosc. Relat. Phenom.*, 1995, **75**, 35.
123. K. Müller-Dethlefs and M. C. R. Cockett, in *Nonlinear Spectroscopy for Molecular Structure Determination*, ed. R.W. Field, E. Hirota, J.P. Maier, S. Tsuchiya, IUPAC, Blackwell Science, Oxford, 1998, p. 164.
124. K. Müller-Dethlefs and E. W. Schlag, *Angew. Chem. Int. Ed.*, 1998, **37**, 1347.
125. M. C. R. Cockett, K. Müller-Dethlefs and T. G. Wright, *Annu. Rep. Prog. Chem. Sect. C*, 1998, **94**, 327.
126. F. Merkt, *Annu. Rev. Phys. Chem.*, 1997, **48**, 675.
127. K. S. Wang and V. McKoy, *Ann. Rev. Phys. Chem.*, 1995, **46**, 275.
128. S. Ullrich, G. Tarczay, X. Tong, C. E. H. Dessent and K. Müller-Dethlefs, *Angew. Chem.-Int. Ed.*, 2002, **41**, 166.
129. S. Ullrich, G. Tarczay, X. Tong, M. S. Ford, C. E. H. Dessent and K. Müller-Dethlefs, *Chem. Phys. Lett.*, 2002, **351**, 121.
130. S. Ullrich, G. Tarczay and K. Müller-Dethlefs, *J. Phys. Chem. A*, 2002, **106**, 1496.
131. S. Ullrich, X. Tong, G. Tarczay, C. E. H. Dessent and K. Müller-Dethlefs, *Phys. Chem. Chem. Phys.*, 2002, **4**, 2897.
132. W. D. Geppert, C. E. H. Dessent, M. C. R. Cockett and K. Müller-Dethlefs, *Chem. Phys. Lett.*, 1999, **303**, 194.
133. W. D. Geppert, C. E. H. Dessent, S. Ullrich and K. Müller-Dethlefs, *J. Phys. Chem. A*, 1999, **103**, 7186.
134. R. Lindner, H.-J. Dietrich and K. Müller-Dethlefs, *Chem. Phys. Lett.*, 1994, **228**, 417.
135. W. A. Chupka, *J. Chem. Phys.*, 1993, **98**, 4520.
136. L. C. Zhu and P. Johnson, *J. Chem. Phys.*, 1991, **94**, 5769.
137. X. Zhang, J. M. Smith and J. L. Knee, *J. Chem. Phys.*, 1993, **99**, 3133.

138. H. J. Dietrich, R. Lindner and K. Müller-Dethlefs, *J. Chem. Phys.*, 1994, **101**, 3399.

139. D. M. Chapman, K. Müller-Dethlefs and J. B. Peel, *J. Chem. Phys.*, 1999, **111**, 1955.

140. S. R. Haines, C. E. H. Dessent and K. Müller-Dethlefs, *J. Chem. Phys.*, 1999, **111**, 1947.

141. S. R. Haines, W. D. Geppert, D. M. Chapman, M. J. Watkins, C. E. H. Dessent, M. C. R. Cockett and K. Müller-Dethlefs, *J. Chem. Phys.*, 1998, **109**, 9244.

142. J. E. Braun, T. Mehnert and H. J. Neusser, *Int. J. Mass Spectrom.*, 2000, **203**, 1.

143. J. E. Braun, H. J. Neusser, P. Harter and M. Stockl, *J. Phys. Chem. A*, 2000, **104**, 2013.

144. S. Chervenkov, P. Q. Wang, J. E. Braun and H. J. Neusser, *J. Chem. Phys.*, 2004, **121**, 7169.

145. S. Georgiev, T. Chakraborty and H. J. Neusser, *J. Phys. Chem. A*, 2004, **108**, 3304.

146. S. Georgiev and H. J. Neusser, *Chem. Phys. Lett.*, 2004, **389**, 24.

147. H. J. Neusser and K. Siglow, *Chem. Rev.*, 2000, **100**, 3921.

148. H.-J. Dietrich, K. Müller-Dethlefs and L. Y. Baranov, *Phys. Rev. Lett.*, 1996, **76**, 3530.

149. Th. L. Grebner, P. von Unold and H. J. Neusser, *J. Phys. Chem. A.*, 1997, **101**, 158.

150. H. Krause and H. J. Neusser, *J. Chem. Phys.*, 1993, **99**, 6278.

151. B. Jeziorski, R. Moszynski and K. Szalewicz, *Chem. Rev.*, 1994, **94**, 1887.

152. A. Hesselmann and G. Jansen, *Phys. Chem. Chem. Phys.*, 2003, **5**, 5010.

153. A. J. Misquitta and K. Szalewicz, *Chem. Phys. Lett.*, 2002, **357**, 301.

154. H. L. Williams and C. F. Chabalowski, *J. Phys. Chem. A*, 2001, **105**, 646.

155. R. Podeszwa, R. Bukowski and K. Szalewicz, *J. Phys. Chem. A*, 2006, **110**, 10345.

156. A. Hesselmann, G. Jansen and M. Schutz, *J. Am. Chem. Soc.*, 2006, **128**, 11730.

157. A. Hesselmann and G. Jansen, *Chem. Phys. Lett.*, 2002, **357**, 464.

158. A. Tekin and G. Jansen, *Phys. Chem. Chem. Phys.*, 2007, **9**, 1680.

159. K. Szalewicz, K. Patkowski and B. Jeziorski, *Struct. Bond.*, 2005, **116**, 43.

160. J. Pittner and P. Hobza, *Chem. Phys. Lett.*, 2004, **390**, 496.

161. M. Pitoňák, P. Neogrády, J. Řezáč, P. Jurečka, M. Urban and P. Hobza, *J. Chem. Theory Comput.*, 2008, **4**, 1829.

162. S. Grimme, *J. Chem. Phys.*, 2003, **118**, 9095.

163. J. G. Hill, J. A. Platts and H. J. Werner, *Phys. Chem. Chem. Phys.*, 2006, **8**, 4072.

164. M. Piacenza and S. Grimme, *J. Am. Chem. Soc.*, 2005, **127**, 14841.

165. M. Piacenza and S. Grimme, *ChemPhysChem*, 2005, **6**, 1554.

166. J. G. Hill and J. A. Platts, *J. Chem. Theory Comput.*, 2007, **3**, 80.
167. J. G. Hill and J. A. Platts, *Phys. Chem. Chem. Phys.*, 2008, **10**, 2785.
168. Y. S. Jung, R. C. Lochan, A. D. Dutoi and M. Head-Gordon, *J. Chem. Phys.*, 2004, **121**, 9793.
169. R. A. Distasio and M. Head-Gordon, *Mol. Phys.*, 2007, **105**, 1073.
170. T. Takatani, E. G. Hohenstein and D. C. Sherrill, *J. Chem. Phys.*, 2008, **128**, 124111.
171. F. Wennmohs and F. Neese, *Chem. Phys.*, 2008, **343**, 217.
172. S. Tsuzuki, K. Honda, T. Uchimaru, M. Mikami and K. Tanabe, *J. Am. Chem. Soc.*, 2002, **124**, 104.
173. M. Pitoňák, P. Neogrády, J. Černý, S. Grimme and P. Hobza, *ChemPhysChem*, 2009, **10**, 282.
174. G. Chalasinski, M. M. Szczesniak and R. A. Kendall, *J. Chem. Phys.*, 1994, **101**, 8860.
175. A. Halkier, T. Helgaker, P. Jørgensen, W. Klopper, H. Koch, J. Olsen and A. K. Wilson, *Chem. Phys. Lett.*, 1998, **286**, 243.
176. D. G. Truhlar, *Chem. Phys. Lett.*, 1998, **294**, 45.
177. M. P. Lara-Castells, R. V. Krems, A. A. Buchachenko, G. Delgado-Barrio and P. Villarreal, *J. Chem. Phys.*, 2001, **115**, 10438.
178. E. C. Lee, D. Kim, P. Jurečka, P. Tarakeshwar, P. Hobza and K. S. Kim, *J. Phys. Chem. A*, 2007, **111**, 3446.
179. M.-S. Liao, Y. Lu, V. D. Parker and S. Scheiner, *J. Phys. Chem. A*, 2003, **107**, 8939.
180. C. Desplanches, E. Ruiz, A. Rodríguez-Fortea and S. Alvarez, *J. Am. Chem. Soc.*, 2002, **124**, 5197.
181. S. D. Zaric, *Eur. J. Inorg. Chem.*, 2003, **2197**.
182. S. Kristyan and P. Pulay, *Chem. Phys. Lett.*, 1994, **229**, 175.
183. P. Hobza, J. Šponer and T. Reschel, *J. Comput. Chem.*, 1995, **16**, 1315.
184. F. Sim, A. Stamant, I. Papai and D. R. Salahub, *J. Am. Chem. Soc.*, 1992, **114**, 4391.
185. S. Sirois, E. I. Proynov, D. T. Nguyen and D. R. Salahub, *J. Chem. Phys.*, 1997, **107**, 6770.
186. Y. K. Zhang, W. Pan and W. T. Yang, *J. Chem. Phys.*, 1997, **107**, 7921.
187. H. Y. Liu, M. Elstner, E. Kaxiras, T. Frauenheim, J. Hermans and W. T. Yang, *Proteins: Struct. Funct. Genet.*, 2001, **44**, 484.
188. D. Řeha, H. Valdes, J. Vondrášek, P. Hobza, A. Abu-Riziq, B. Crews and M. S. de Vries, *Chem. Eur. J.*, 2005, **11**, 6803.
189. M. Elstner, P. Hobza, T. Frauenheim, S. Suhai and E. Kaxiras, *J. Chem. Phys.*, 2001, **114**, 5149.
190. S. Grimme, *J. Comput. Chem.*, 2004, **25**, 1463.
191. P. Jurečka, J. Černý, P. Hobza and D. R. Salahub, *J. Comput. Chem.*, 2007, **28**, 555.
192. P. Jurečka and P. Hobza, *J. Am. Chem. Soc.*, 2003, **125**, 15608.
193. J. Šponer, P. Jurečka and P. Hobza, *J. Am. Chem. Soc.*, 2004, **126**, 10142.

194. P. Jurečka, J. Šponer and P. Hobza, *J. Phys. Chem. B*, 2004, **108**, 5466.

195. J. Šponer, P. Jurečka, I. Marchan, F. J. Luque, M. Orozco and P. Hobza, *Chem. Eur. J.*, 2006, **12**, 2854.

196. A. Hesselmann, G. Jansen and M. Schutz, *J. Chem. Phys.*, 2005, **122**, 014103.

197. A. J. Misquitta, B. Jeziorski and K. Szalewicz, *Phys. Rev. Lett.*, 2003, **91**, 033201.

198. A. J. Misquitta, R. Podeszwa, B. Jeziorski and K. Szalewicz, *J. Chem. Phys.*, 2005, **123**, 214103.

199. V. P. Osinga, S. J. A. van Gisbergen, J. G. Snijders and E. J. Baerends, *J. Chem. Phys.*, 1997, **106**, 5091.

200. I. Adamovic and M. S. Gordon, *Mol. Phys.*, 2005, **103**, 379.

201. T. Sato, T. Tsuneda and K. Hirao, *J. Chem. Phys.*, 2005, **123**, 104307.

202. Y. Andersson, D. C. Langreth and B. I. Lundqvist, *Phys. Rev. Lett.*, 1996, **76**, 102.

203. M. Dion, H. Rydberg, E. Schrõder, D. C. Langreth and B. I. Lundqvist, *Phys. Rev. Lett.*, 2004, **92**, 246401.

204. T. Thonhauser, V. R. Cooper, S. Li, A. Puzder, P. Hyldgaard and D. C. Langreth, *Phys. Rev. Lett.*, 2007, **76**, 125112.

205. J. F. Dobson, *Int. J. Quantum Chem.*, 1998, **69**, 615.

206. K. Rapcewicz and N. W. Ashcroft, *Phys. Rev. B*, 1991, **44**, 4032.

207. T. A. Wesolowski and F. Tran, *J. Chem. Phys.*, 2003, **118**, 2072.

208. A. D. Becke and E. R. Johnson, *J. Chem. Phys.*, 2005, **122**, 154104.

209. T. R. Walsh, *Phys. Chem. Chem. Phys.*, 2005, **7**, 443.

210. W. Kohn, Y. Meir and D. E. Makarov, *Phys. Rev. Lett.*, 1998, **80**, 4153.

211. M. Dion, H. Rydberg, E. Schroder, D. C. Langreth and B. I. Lundqvist, *Phys. Rev. Lett.*, 2004, **92**, 246401.

212. D. J. Lacks and R. G. Gordon, *Phys. Rev. A*, 1993, **47**, 4681.

213. C. Adamo and V. Barone, *J. Chem. Phys.*, 1998, **108**, 664.

214. X. Xu and W. A. Goddard, *Proc. Natl. Acad. Sci. USA*, 2004, **101**, 2673.

215. N. Kurita, H. Inoue and H. Sekino, *Chem. Phys. Lett.*, 2003, **370**, 161.

216. J. Černý and P. Hobza, *Phys. Chem. Chem. Phys.*, 2005, **7**, 1624.

217. J. Antony and S. Grimme, *Phys. Chem. Chem. Phys.*, 2006, **8**, 5287.

218. S. Grimme, *J. Comput. Chem.*, 2006, **27**, 1787.

219. S. Grimme, *J. Chem. Phys.*, 2006, **124**, 034108.

220. Y. Zhao and D. G. Truhlar, *Phys. Chem. Chem. Phys.*, 2005, **7**, 2701.

221. D. G. Truhlar, Y. Zhao and N. E. Schultz, *Abstr. Pap. Am. Chem. Soc.*, 2006, **231**, 47–COMP.

222. Y. Zhao, N. E. Schultz and D. G. Truhlar, *J. Chem. Theory Comput.*, 2006, **2**, 364.

223. Y. Zhao and D. G. Truhlar, *J. Phys. Chem. A*, 2006, **110**, 13126.

224. Y. Zhao and D. G. Truhlar, *Org. Lett.*, 2006, **8**, 5753.

225. Y. Zhao and D. G. Truhlar, *J. Chem. Phys.*, 2006, **125**.

226. Y. Zhao and D. G. Truhlar, *J. Chem. Theory Comput.*, 2007, **3**, 289.
227. J. J. Zheng, Y. Zhao and D. G. Truhlar, *J. Chem. Theory Comput.*, 2007, **3**, 569.
228. M. P. Waller, A. Robertazzi, J. A. Platts, D. E. Hibbs and P. A. Williams, *J. Comput. Chem.*, 2006, **27**, 491.
229. M. Gutowski, K. D. Jordan and P. Skurski, *J. Phys. Chem. A*, 1998, **102**, 2624.
230. D. Svozil, T. Frigato, Z. Havlas and P. Jungwirth, *Phys. Chem. Chem. Phys.*, 2005, **7**, 840.
231. J. Hepburn, G. Scoles and R. Penco, *Chem. Phys. Lett.*, 1975, **36**, 451.
232. R. Ahlrichs, R. Penco and G. Scoles, *Chem. Phys.*, 1977, **19**, 119.
233. P. Hobza and C. Sandorfy, *J. Am. Chem. Soc.*, 1987, **109**, 1302.
234. P. Hobza, F. Mulder and C. Sandorfy, *J. Am. Chem. Soc.*, 1981, **103**, 1360.
235. E. J. Meijer and M. Sprik, *J. Chem. Phys.*, 1996, **105**, 8684.
236. W. T. M. Mooij, F. B. van Duijneveldt, J.G.C.M. Van Duijneveldt-van de Rijdt and B. P. van Eijck, *J. Phys. Chem. A*, 1999, **103**, 9872.
237. W. T. M. Mooij, B. P. van Eijck and J. Kroon, *J. Phys. Chem. A*, 1999, **103**, 9883.
238. X. Wu, M. C. Vargas, S. Nayak, V. Lotrich and G. Scoles, *J. Chem. Phys.*, 2001, **115**, 8748.
239. L. Zhechkov, T. Heine, S. Patchkovskii, G. Seifert and H. A. Duarte, *J. Chem. Theory Comput.*, 2005, **1**, 841.
240. T. Kubař, P. Jurečka, J. Černý, J. Řezáč, M. Otyepka, H. Valdes and P. Hobza, *J. Phys. Chem. A*, 2007, **111**, 5642.
241. Q. Wu and W. T. Yang, *J. Chem. Phys.*, 2002, **116**, 515.
242. U. Zimmerli, M. Parrinello and P. Koumoutsakos, *J. Chem. Phys.*, 2004, **120**, 2693.
243. C. Morgado, M. A. Vincent, I. H. Hillier and X. Shan, *Phys. Chem. Chem. Phys.*, 2007, **9**, 448.
244. O. Bludský, M. Rubeš, P. Soldán and P. Nachtigall, *J. Chem. Phys.*, 2008, **128**, 114102.
245. M. Rubeš, O. Bludský and P. Nachtigall, *ChemPhysChem.*, 2008, **9**, 1702.
246. B. Martin and T. Clark, *Int. J. Quantum. Chem.*, 2006, **106**, 1208.
247. J. P. McNamara and I. H. Hillier, *Phys. Chem. Chem. Phys.*, 2007, **9**, 2362.
248. W. Weber and W. Thiel, *Theor. Chim. Acta*, 2000, **103**, 495.
249. T. Tuttle and W. Thiel, *Phys. Chem. Chem. Phys.*, 2008, **10**, 2159.
250. B. Civalleri, C. M. Zicovich-Wilson, L. Valenzano and P. Ugliengo, *Cryst. Eng. Comm.*, 2008, **10**, 405.
251. J. Černý, P. Jurečka, P. Hobza and H. Valdes, *J. Phys. Chem. A*, 2007, **111**, 1146.
252. H. Valdes, D. Řeha and P. Hobza, *J. Phys. Chem. B*, 2006, **110**, 6385.

253. M. D. Wodrich, D. F. Jana, P. von R. Schleyer and C. Corminboeuf, *J. Phys. Chem. A*, 2008, **112**, 11495.

254. O. A. von Lilienfeld, I. Tavernelli, U. Rothlisberger and D. Sebastiani, *Phys. Rev. Lett.*, 2004, **93**, 153004.

255. M. D. Towler, *Phys. Status Solidi B*, 2006, **243**, 2573.

256. K. Raghavachari and J. B. Anderson, *J. Phys. Chem.*, 1996, **100**, 12960.

257. C. Diedrich, A. Luchow and S. Grimme, *J. Chem. Phys.*, 2005, **123**, 184106.

258. M. R. Aliev and J. K. G. Watson, in *Molecular Spectroscopy: Modern Research*, ed. K. Nahari Rao, Vol. III., Academic Press, New York, 1985.

259. I. M. Mills, in *Molecular Spectroscopy Modern Research*, ed. K. Nahari Rao, C.W. Mathews, Vol. I. Academic Press, New York, 1972, p.115.

260. P. E. Maslen, N. C. Handy, R. D. Amos and D. Jayatilaka, *J. Chem. Phys.*, 1992, **97**, 4233.

261. N. C. Handy, P. E. Masle, R. D. Amos, J. S. Andrews, C. C. W. Murray and G. J. Laming, *Chem. Phys. Lett.*, 1992, **197**, 506.

262. O. Bludský, V. Špirko, R. Kobayashi and P. Jorgensen, *Chem Phys. Lett.*, 1994, **228**, 568.

263. V. Špirko, J. Šponer and P. Hobza, *J. Chem. Phys.*, 1997, **106**, 1472.

264. P. Hobza, O. Bludský and S. Suhai, *Phys. Chem. Chem. Phys.*, 1999, **1**, 3073.

265. R. S. Fellers, L. B. Braly, R. J. Saykally and C. LeForestier, *J. Chem. Phys.*, 1999, **110**, 6306.

266. M. J. Elrod and R. J. Saykally, *J. Chem. Phys.*, 1995, **103**, 933.

267. R. E. Miller, T. G. A. Heijmen, P. E. S. Wormer, A. van der Avoird and R. Moszynski, *J. Chem. Phys.*, 1999, **110**, 5651.

268. G. C. M. van der Sanden, P. E. S. Wormer, A. van der Aoird, C. A. Schuttenmaer and R. Saykally, *J. Chem. Phys. Lett.*, 1994, **226**, 22.

269. R. S. Cohen and R. J. Saykally, *J. Chem. Phys.*, 1991, **95**, 7891.

270. B. Brauer, R. B. Gerber, M. Kabeláč, P. Hobza, J. M. Bakker, A. G. A. Riziq and M. S. de Vries, *J. Phys. Chem. A*, 2005, **109**, 6974.

271. A. P. Scott and L. Radom, *J. Phys. Chem. A*, 1996, **100**, 16502.

272. P. Sinha, S. E. Boesch, C. Gu, R. A. Wheeler and A. K. Wilson, *J. Phys. Chem. A*, 2004, **108**, 9213.

273. K. Giese, M. Petkovic, H. Naundorf and O. Kuhn, *Phys. Rep.*, 2006, **430**, 211.

274. J. M. Bakker, C. Plutzer, I. Hünig, T. Huber, I. Compagnon, G. von Helden and G. Meijer, *Chem. Phys. Chem.*, 2005, **6**, 120.

275. W. P. Kraemer, V. Špirko and O. Bludský, *J. Mol. Spectrosc.*, 1994, **164**, 500.

276. J. T. Hougen, P. R. Bunker and J. W. C. Johns, *J. Mol. Spectrosc.*, 1970, **34**, 136.

277. P. R. Bunker and B. M. Landsberg, *J. Mol. Spectrosc.*, 1977, **67**, 374.

278. D. Nachtigallová, P. Hobza and V. Špirko, *J. Phys. Chem. A*, 2008, **112**, 1854.
279. M. Mons, I. Dimicoli, F. Piuzzi, B. Tardivel and M. Elhanine, *J. Phys. Chem. A*, 2002, **106**, 5088.
280. E. Nir, C. Janzen, P. Imhof, K. Kleinermanns and M. S. de Vries, *J Chem. Phys.*, 2001, **115**, 4604.
281. E. Nir, C. Plutzer, K. Kleinermanns and M. S. de Vries, *Eur. Phys. J. D*, 2002, **20**, 317.
282. M. Y. Choi and R. E. Miller, *J. Am. Chem. Soc.*, 2006, **128**, 7320.
283. M. Mons, F. Piuzzi, I. Dimicoli, L. Gorb and J. Leszczynski, *J. Phys. Chem. A*, 2006, **110**, 10921.
284. H. Chen and S. Li, *J. Phys. Chem. A*, 2006, **110**, 12360.
285. C. E. Dateo, T. J. Lee and D. W. Schwenke, *J. Chem. Phys.*, 1994, **101**, 5853.
286. J. Čížek, V. Špirko and O. Bludský, *J. Chem. Phys.*, 1993, **99**, 7331.
287. H.-C. Chang and W. Klemperer, *J. Chem. Phys.*, 1993, **98**, 9266.
288. H.-C. Chang and W. Klemperer, *J. Chem. Phys.*, 1994, **100**, 1.
289. W. Klopper, M. Quack and M. A. Suhm, *J. Chem. Phys.*, 1998, **108**, 10096.
290. M. Quack and M. A. Suhm, *Chem. Phys. Lett.*, 1990, **171**, 517.
291. S. Davis, D. T. Anderson and D. J. Nesbitt, *J. Chem. Phys.*, 1996, **105**, 6645.
292. P. Salvador and M. M. Szczêæniak, *J. Chem. Phys.*, 2003, **118**, 537.
293. J. T. Farrell Jr, O. Sneh and D. J. Nesbitt, *J. Phys. Chem.*, 1994, **98**, 6068.
294. S. N. Tsang, H.-C. Chang and W. Klemperer, *J. Phys. Chem.*, 1994, **98**, 7313.
295. R. Prosmiti, C. Cunha, A. A. Buchachenko, G. Delgado-Barrio and P. Villarreal, *J. Chem. Phys.*, 2002, **117**, 10019.
296. V. Aquilanti, M. Bartolomei, D. Cappelletti, E. Carmona-Novillo and F. Pirani, *J. Chem. Phys.*, 2002, **117**, 615.
297. P. Çarçabal, N. Seurre, M. Chevalier, M. Broquier and V. Brenner, *J. Chem. Phys.*, 2002, **117**, 1522.
298. M. J. Watkins, D. Belcher and M. C. R. Cockett, *J. Chem. Phys.*, 2002, **116**, 7868.
299. M. D. Marshall, B. V. Pond and M. I. Lester, *J. Chem. Phys.*, 2003, **118**, 1196.
300. N. L. Pivonka, C. Kaposta, G. von Helden, G. Meijer, L. Wöste, D. M. Neumark and K. R. Asmis, *J. Chem. Phys.*, 2002, **117**, 6493.
301. N. Solcà and O. Dopfer, *Chem. Eur. J.*, 2003, **9**, 3154.

CHAPTER 3

Potential-Energy and Free-Energy Surfaces

3.1 Benzene Dimer

The benzene dimer is one of the most frequently studied non-covalent complexes and this reflects its importance in science. First, it is the prototype of complexes of aromatic systems or, in a broader sense of the π–π interactions. Such π–π interactions involving aromatic rings are important in molecular/biomolecular assembly and engineering. These complexes play a key role in determining the structure of biomacromolecules like DNA and RNA and also of proteins (only Trp, His, Tyr, and Phe of the twenty amino acids are aromatic but they play a decisive role). The aromatic–aromatic interactions also play a key role in designing functional nanomaterials. Second, the potential-energy surface is characteristic of non-covalent interactions in the sense that energy barriers between energy minima are very small, which makes the determination of the equilibrium geometry doubtful. Finally, it makes a borderline between medium and extended complexes.

The historic development of investigations of the benzene dimer is interesting. Unlike H-bonded systems, where dipolar forces are dominant in deciding the ultimate geometry of the complex, the benzene dipole moment is vanishing for symmetry reasons. Nevertheless, the benzene dimer geometry is determined by electrostatic forces: the quadrupole–quadrupole interaction (benzene quadrupole moment: $-28.3\pm1.2\times10^{-40}$ Cm^2).[1] The importance of the quadrupole moment for the stabilisation of the benzene dimer was already shown in 1983 by Karlstrom et al.[2] Being ultimately weaker than dipole–dipole interactions, the balance of the electrostatic quadrupole–quadrupole interaction (which can be attractive or repulsive) and attractive dispersion interaction becomes much more important and, consequently, presents a difficult challenge to experimental and theoretical interpretation in terms of structure and energetics determination.

RSC Theoretical and Computational Chemistry Series No. 2
Non-covalent Interactions: Theory and Experiment
By Pavel Hobza and Klaus Müller-Dethlefs
Published by the Royal Society of Chemistry, www.rsc.org

The decisive role of dispersion energy for a π–π interaction was not recognised for a long time. Early interpretations of the nature of π–π interactions[3,4] were largely attributed to be electrostatic in nature. Later, however, the important role of dispersion energy was highlighted.[5]

Indeed, in both experimental and computational studies published, various proposals for the expected global minimum structure on the benzene dimer PES exist. If, for a moment, we neglect the existence of the benzene quadrupole, then we may only conclude that the dimer structure will most likely take a completely overlapping "sandwich" structure (S, see Figure 3.1) to allow for the maximum dispersive interaction between the two rings. Experiments performed about 15 years ago initially supported this conclusion,[6,7] but were soon contradicted by hole-burning[8] and microwave[9] studies. Both these studies supported a T-shaped (T and TT, see Figure 3.1) global minimum structure and, in Ref. 8, showed evidence of three isomeric forms.

Consideration of the quadrupole-quadrupole interaction is important to account for the observation of all possible dimer species; this has been amply demonstrated by qualitative estimates[10,11] of this contribution and,

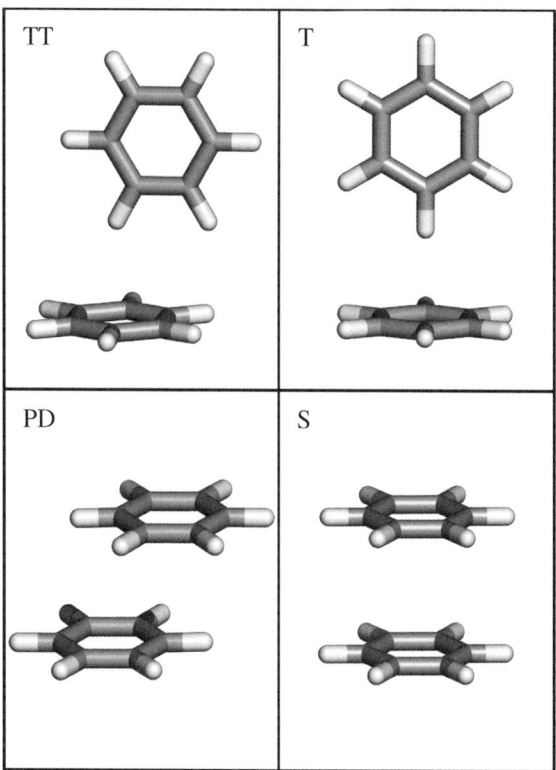

Figure 3.1 Structures of the benzene dimer: T-shaped tilted C_S (TT), T-shaped C_{2v} (T), parallel-displaced C_{2h} (PD) and sandwich D_{6h} (S).

subsequently, by high-level correlated *ab initio* calculations.[12] In the MP2 and CCSD(T) computational studies, a T-shaped structure was also found to be the global minimum in accord with experiment. Similar studies conducted at the same time showed that the Perdew–Wang functional describes the benzene dimer PES relatively correctly while the BLYP and B3LYP functionals fail.[13] With respect to Refs. 11 to 13 we may conclude that whilst a consideration of the quadrupole–quadrupole interaction is certainly important, the failure of the BLYP and B3LYP models – and the relatively good performance of correlated *ab initio* calculations – suggests that the dispersive contribution to binding is significant to the extent that both electrostatics and dispersion must be treated in an even-handed manner. Further computational studies have examined this requirement in detail. Whilst MP2/aug-cc-pVDZ has been found to be sufficient for structural predictions, much higher basis sets and highly correlated methods of the CCSD(T) type – that is, the higher angular momentum functions, diffuse functions and methodologies important for calculating dispersion contributions – were required for accurate prediction of binding energies,[14–16] suggesting that dispersion energy is the dominant energy contribution.

The relative stability between the parallel-displaced (face-to-face) and the T-shaped (edge-to-face) aromatic–aromatic interactions has been the focus of several studies. In the gas-phase studies of the benzene dimer that were pioneered by Klemperer and coworkers,[17,18] followed by Schlag and coworkers,[8,19–22] and Felker and coworkers,[23,25] the T-shaped conformers seem to be more favored.[20–26] However, in some studies the displaced-stacked conformers were also present.[8] On the other hand, in crystals, aromatic rings exhibit displaced-stacked conformers more frequently than T-shaped conformers.[27–33] To explain these experimental observations, one needs to understand the energetic basis of T-shaped *vs.* displaced stacked conformational stability. In this regard, a number of theoretical calculations have been carried out.[34–63]

As mentioned above the PES of the dimer is very flat (what is characteristic of any non-covalent complex) which makes special requirements for structure and geometry determination. In this case geometry should be determined as accurately as stabilisation energy and we will first mention procedure and results from Ref. 64. One important finding resulted from this study (and also from Ref. 65) namely that the T-shaped C_{2v} structure (T in Figure 3.1) does not correspond to the energy minimum but only to the first-order saddle point. The minimum corresponds to the tilted T-shaped structure (TT in Figure 3.1 having the C_s symmetry) and the C_{2v} structure separates two minima. The energy barrier separating these two minima is, however, very small (see later).

The geometries and the associated binding energies of four different energy minima (see Figure 3.1) were evaluated at the complete basis set (CBS) limit of RI-MP2 and coupled cluster theory with singles, doubles, and perturbative triples excitations [CCSD(T)].[64] Contrary to the Helgaker-type extrapolation, in the present case the CBS value is extrapolated with only two energies corresponding to the aug-cc-pVDZ and aug-cc-pVTZ basis sets (BSSE corrected and BSSE-uncorrected energies were utilised). This type of intrapolation is based on the fact that both energy curves cross in one point. The advantage of this extrapolation (in comparison with the Helgaker type) is the fact that it does not require any parameter.

At the RI-MP2/CBS level, the PD is a global minimum with binding energy 5.03 kcal/mol and tilted T-shaped structure is a local minimum (3.84 kcal/mol). Thus, the displaced-stacked conformers are lower in energy by 1.2 kcal/mol than the T-shaped conformers. On the other hand, CCSD(T)/CBS predicts that tilted T-shaped structures (2.84 kcal/mol) are 0.1 kcal/mol lower than the PD ones (2.73 kcal/mol). This clearly demonstrates that the MP2/CBS binding energy of PD structure is highly overestimated and further discussion will be based on the CCSD(T) results. Then, tilted T-shaped structures would be slightly more stable, or at least both TT and PD structures would be nearly equally stable. The potential surface has two flat minima composed of iso-energetic configurations, and the barrier between the two minima is very small (0.1 kcal/mol). This results in an extremely floppy structure that encompasses diverse configurations with quantum-statistical distributions.

The ZPE-corrected binding energies of both the TT and PD conformers of the benzene dimer are estimated to be 2.59 kcal/mol at the BSSE-corrected geometry-optimised CCSD(T)/CBS level. Quadruple excitations in the coupled cluster theory are likely to destabilise the T-shaped and PD structures by 0.13/0.24 kcal/mol (these conclusions were made from calculations on model complexes and should thus be considered with care; more accurate numbers derived for the benzene dimer will be discussed later). It means that consideration of quadruply excitations favoured the T-shaped structures by about 0.1 kcal/mol. Therefore, the binding energy for TT and PD conformers would be 2.46/2.35 kcal/mol, which is in excellent agreement with experiment. At very low temperatures, both nearly isoenergetic T-shaped and displaced-stacked conformers exhibit quantum-statistical distributions for various configurations around their extremely shallow minima, with only a small barrier separating them. The ZPE correction does not significantly alter the energy difference between both basic conformers. However, the T-shaped conformer is more flexible (with both rotation and twisting of the axial benzene) and therefore would be more stabilised than the stacked conformer at nonzero temperatures due to the entropic effect (see Section 5.5).

The C–H distance of the axial benzene in the tilted T-shaped structure is slightly shortened by 0.025 Å, showing the blueshifted C–H frequency by 14 cm^{-1} (based on BSSE-corrected MP2/aug-cc-pVDZ). This result was previously discussed as the blueshift of the improper H-bond (see Section 4.1). It is interesting to note that a recent study based on DFT-SAPT calculations with optimised functionals by Podeszwa *et al.*[65] yields energies that are in good agreement with the CCSD(T)/CBS results. This seems to indicate that the DFT-SAPT employing optimised functionals would be a promising approach to yield reliable interaction energies for a π-electron-containing systems. The study is also important since it presents the analysis of single energy terms. The decomposition confirmed what was known before (see above) that stabilisation in the dimer is mainly due to dispersion energy while the structure of the dimer is due to electrostatic forces with some role of the balance between dispersion and exchange terms.

All structures of the dimer were in all studies discussed so far obtained by step-by-step optimisation. The other strategy was applied in a recent study[66] where the geometries of the benzene dimer structures were obtained *via* full coordinate gradient geometry optimisation using the DFT-D/BLYP method, covering the empirical dispersion-correction fitted exclusively for this system. The fit was carried out against two estimated CCSD(T)/CBS potential-energy curves corresponding to the distance variation between two benzene rings for the parallel-displaced (PD) and T-shaped (T) structures.

The CCSD(T)/CBS results[66] based on high-level OVOS (optimised virtual orbital space; this is a technique that allows the virtual space to be considerably reduced without losing the accuracy) CCSD(T) interaction energy calculations (up to the aug-cc-pVQZ basis set) and various extrapolations towards the CBS limit confirmed the previous results and show, within the error bars of the applied methodology, that the energetically lowest-lying structure is the tilted T structure, which is nearly 0.1 kcal/mol more stable than the almost isoenergetic PD and T structures. Evidently, to obtain reliable relative energy values all subtle effects should be taken into consideration. The effect of the connected quadruple excitations on the interaction energy was in the paper mentioned estimated (for the benzene dimer) using the CCSD(TQf) method in a 6-31G*(0.25) basis set. It was shown that it destabilises the T and T-shaped tilted structures by –0.02 kcal/mol and the PD structure by –0.04 kcal/mol. It is thus possible to conclude that these effects can be safely neglected.

A very important feature of the benzene dimer (characteristic also for other non-covalently bound molecular clusters) is the fact that energy barriers separating the TS and PD minima, as well as two TS minima, are very low, of the order of 0.1 kcal/mol. These extremely low barriers indicate that the concept of equilibrium structure is misleading and should be replaced by a dynamical averaged structure. This is even more true for non-zero temperatures, where besides the enthalpy term the entropy term becomes important as well. Since the vast majority of experiments are performed at non-zero temperatures passing from the PES to the free-energy surface (FES) is imperative. Determination of the dynamic structure of the benzene dimer is topical also from the point of explanation of the nature of C–H stretching mode shift observed in the infrared spectrum upon dimerisation (for detailed discussion see Section 5.5).

The quadrupole–quadrupole interaction discussed in the benzene dimer plays an important role not only in non-covalent complexes but is also important for the structure of proteins. The structure of crystalline phenyl-alanine is to a large extent determined by the interaction of benzene rings. In an investigation of the crystal structures of phenylalanine, a high occur-rence of T-shaped and parallel-displaced structures of the benzene rings was found.[67] The quadrupole–quadrupole interaction is attractive only for these two structures; all other structures show repulsive quadrupole–quadrupole interaction.

3.1.1 Benzene-Containing Complexes

As mentioned in the previous section, the potential-energy surface of the benzene dimer possesses two energy minima, *i.e.* T-shaped and parallel-displaced ones, having very similar stabilisation energies. The stacked structure, originally believed to be the global minimum, does not represent an energy minimum but corresponds to the saddle point. The explanation is simple; the quadrupole–quadrupole interaction (quadrupole being the first nonzero multipole moment) is repulsive for the parallel structure, while it is attractive for the T and PD structures. In the stacked structure, the dispersion energy is largest among all structures (maximal overlap of both subsystems), but electrostatic repulsion is also large here. Electrostatic and dispersion energy contributions in T-shaped and parallel-displaced structures are attractive but are, due to the large distance between the centres of mass of both subsystems, rather small. Consequently, the total stabilisation energy of the dimer is also rather small. The parallel structure of the dimer is interesting due to its large dispersion energy; providing that we can change the repulsive quadrupole–quadrupole interaction for an attractive interaction, a dramatic increase of the stabilisation energy should result. This is certainly impractical for the dimers made by identical subsystems but in the case of different subsystems it is viable. The simplest case is the benzene . . . hexafluorobenzene complex. The quadrupole moment of hexafluorobenzene is in absolute value similar to that of benzene but of the opposite sign. This means that the stacked structure of the heterodimer possesses an attractive quadrupole–quadrupole electrostatic term, whereas in the case of the T-shaped and parallel-displaced ones this term is repulsive. The idea is not new and Williams[68] already in the 1990s pointed out a higher melting point of the cocrystal of the two compared with either pure compound. The first theoretical studies on heterodimers also appeared in the 1990s,[69] with the authors reporting "fairly strong stabilisation" of about 3.7 kcal mol^{-1} (MP2/6-31G** level). It was concluded that an important part of the overall stabilisation originates in the London dispersion energy. Evidently, this stabilisation energy is underestimated and evaluating the accurate numbers is important since these interactions and structures can play a role in supramolecular construction.

Stabilisation energies of stacked structures of $C_6H_6 \ldots C_6X_6$ (X=F, Cl, Br, CN) complexes were studied[70] at the CCSD(T) complete basis-set (CBS) limit level. The respective energies were constructed from MP2/CBS stabilisation energies and a CCSD(T) correction term determined with a medium basis set (6-31G**). The former energies were extrapolated using the Helgaker two-point formula from aug-cc-pVDZ and aug-cc-pVTZ Hartree–Fock energies and MP2 correlation energies. The CCSD(T) correction term is as in any other stacked structure systematically repulsive and reaches substantial values. The largest one was found for the hexabromobenzene complex where it is as large as 5.7 kcal/mol. Let us recall here that among stacked DNA base pairs the largest CCSD(T) correction term was determined for methyl adenine . . . methyl thymine and reached "only" about 4 kcal/mol. The final CCSD(T)/CBS stabilisation energies are substantial, *i.e.* considerably larger than previously calculated and increase in the series as follows: hexafluorobenzene (6.3 kcal/mol), hexachlorobenzene (8.8 kcal/mol), hexabromobenzene (8.1 kcal/mol), and hexacyanobenzene (11.0 kcal/mol). MP2/SDD** relativistic calculations performed for all complexes mentioned and also for benzene . . . hexaiodobenzene have clearly shown that due to relativistic effects the stabilisation energy of the hexaiodobenzene complex is lower than that of hexabromobenzene complex. The decomposition of the total interaction energy to physically defined energy components was made by using the symmetry-adapted perturbation treatment (DFT-SAPT). The main stabilisation contribution for all complexes investigated is due to the London dispersion energy, with the induction term being smaller. Electrostatic and induction terms that are attractive are compensated by their exchange counterparts.

The stabilisation energies of the stacked complexes are substantial, which suggests that this motif and these constructing blocks may be considered as a powerful tool in supramolecular construction requiring the stable orientation of molecular subsystems. The $C_6H_6 \ldots C_6X_6$ (X=F, Cl, Br, I, CN) recognition motif shows a significant stabilisation well comparable to, or even stronger than hydrogen bonding. The stacking interactions thus exhibit comparable supramolecular activity as hydrogen bonding, which was believed to be the only recognition factor.

The previous part of the section was devoted to complexes between benzene and fully substituted benzene. The role of substituent effects on benzene in substituent benzene dimers in general was investigated in detail by Sherrill *et al.*[71,72] and Kim *et al.*[64,73,74] Regardless of the character of substituent (in the sense of electron donating or electron withdrawing) the aromatic–aromatic interactions were enlarged. It was concluded that the substituent effect is stronger in parallel-displaced structures; consequently, displaced structures were more stabilised than the T-shaped structures. This is the explanation for the significant prevalence of displaced structures (over T-shaped ones) in organic crystals. Dispersion energy was dominated even in substituted benzene dimers but substituent and conformational effects were correlated with the electrostatic interaction.

3.2   Nucleic Acid–Base Pairs

The structure and dynamics of nucleic acid molecules are influenced by a variety of contributions. Among those, the interactions occurring between the nucleic acid base heterocycles are of particular importance. In DNA, the bases are involved in two qualitatively different mutual interaction types: H-bonding and aromatic stacking. The H-bonded base-pair geometries observed in crystal structures of DNA fragments correspond to the minima on potential-energy surfaces of isolated DNA base pairs. In contrast, stacked configurations present in crystals of DNA fragments are rather variable and in many cases do not correspond to energetically optimal stacked arrangements. In order to understand the nature of binding in these pairs, it is necessary to study isolated base pairs, *i.e.* base pairs *in vacuo*. Let us mention here that DNA does not contain any water inside the molecule and that the real situation is better described by gas-phase interactions between DNA bases than by interactions modified by an environment.

It is difficult to treat the PES of as extended complexes as DNA base pairs containing large numbers of transition structures and energy minima separated by low-energy barriers.[75]

Let us note here that the gradient-optimisation techniques localise the nearest energy minimum and then stop the calculation. On encountering such a stationary point (for which the gradient is zero by definition), which can be a minimum or, in fact, a saddle point, it is then necessary to restart the optimisation from a different geometry and to hope that, after several trials, the whole PES will be sampled. It is true though that neither chemical intuition nor experience will necessarily be of any considerable help in elucidating a potential surface, especially in more complex systems. The use of an efficient sampling technique is thus inevitable and computer experiments offer an ideal solution. The aim is not only to localise the global minimum but also to identify all other energy minima, since a full description of the PES is of key importance for subsequent comparison of theory with experimental results. For this reason, the use of methods like simulated annealing, which aims to find global minima only, is of limited use. Instead, techniques like molecular-dynamics (MD) simulations in combination with quenching techniques[76] (MD/Q) are more useful. Variation of the potential energy (it should be higher than the energy of the highest transition structure) and the length of a quench ensure the proper sampling of the whole surface. In performing the longer MD simulations, one can obtain information about the population of various energy minima, which corresponds to the Gibbs energy change. Simulations can be performed in the *NVT* canonical or *NVE* microcanonical ensemble (*N*, *V*, *E* and *T* refer to the number of molecules in a system, its volume, energy and temperature, respectively). In the *NVT* canonical ensemble, the cluster is in thermal equilibrium with the surroundings and, accordingly, the *NVT* ensemble yields information about the behaviour of the cluster when it is interacting with the surroundings. In the *NVE* microcanonical ensemble, all systems have the same energy and each system is individually isolated. Performing simulations in either ensemble (depending on the type of experiment) allows passing from the potential-energy surface to the Gibbs-energy surface. It is not surprising that entropy plays a different role for different types of molecular clusters and, thus, the potential-energy surface and Gibbs-energy surface can differ. Indeed, for the most part the two surfaces are significantly different. Very frequently, the structure of a global minimum at the potential-energy surface differs from that at the Gibbs-energy surface.

The MD/Q technique allows sampling the surface, but the final description of the surfaces strongly depends also on the quality of the potential used. Despite the enormous progress of *ab initio* MD simulations, they are still limited to rather small systems; the size of DNA or RNA base pairs is still unattainable, being completely intractable for such procedures. The only chance for elucidation of such large systems is thus the use of empirical potentials. We must repeat that the quality of MD/Q results is strongly affected by the quality of the potential and the use of an empirical potential that does not correctly describe the structure of a complex only leads to wrong and therefore misleading results. Thus, care should be taken in the choice of the potential, since not every potential used for simulations of DNA and RNA is also suitable for the description of base pairs. We accumulated extensive evidence that the Cornell *et al.* potential,[77] prepared and parameterised in the Kollman laboratory, is well suited for these purposes. This evidence is based on a comparison of structures and stabilisation energies of a large number of DNA and RNA base pairs evaluated by this potential and by correlated nonempirical *ab initio* calculations.[78] Frequently, we were surprised how well this potential (which is in fact rather simple and does not contain any features of the advanced potentials of the last generation, such as polarisation terms or inclusion of higher than harmonic terms) describes various structural types of nucleic acid–base pairs. Evidently, Kollman and his team were fortunate in the parameterisation of this potential, which, due to compensation of many errors, describes nucleic acid bases and their complexes so accurately.

The MD/Q calculations were used intensively for the study of potential-energy and Gibbs-energy surfaces of nucleic acid–base pairs. First, individual base pairs were studied (uracil dimer,[79] the adenine...2,4-difluorotoluene pair,[80] methyluracil dimers,[81] the adenine...thymine pair[82] and the methyladenine...methylthymine pair[83]), and, later, all the DNA base pairs and methylated DNA base pairs were also considered.[84]

The complexity of the problem will be demonstrated on the above-mentioned case of the adenine...thymine (nonmethylated) base pairs.[82] MD/Q calculations revealed twenty-seven energy minima, of which nine were H-bonded, eight T-shaped and ten stacked (see Figure 3.2). The H-bonded structures were the most stable ($\sim 12\,\text{kcal/mol}$), with stacked and T-shaped structures found to be less stable by at least $4\,\text{kcal/mol}$. The global minimum and first two local minima surprisingly correspond neither to Watson–Crick nor to Hoogsteen structural types; the bonding is realised through N_9–H and N_3 functional groups of adenine (and not through the N_6 amino group and the ring N_1 and N_7 adenine positions like in the Watson–Crick and Hoogsteen structures). We are aware of the fact that these structures cannot occur in nucleic acids since the N_9 position is blocked by the attached sugar ring. These results are, however, significant for gas-phase molecular beam experiments where the knowledge of the structure of the global minimum (or mostly populated minimum) is of key importance for interpretation of measured IR spectra. The Hoogsteen and Watson–Crick types of complexes represent the third and fourth local minima and are less stable by about $3\,\text{kcal/mol}$ than the global minimum. The surprising energy preference of the global and the first two local energy minima was confirmed by correlated MP2 *ab initio* calculations using 6-31G** and 6-311G(2d,p) basis sets (structural types from MD/Q

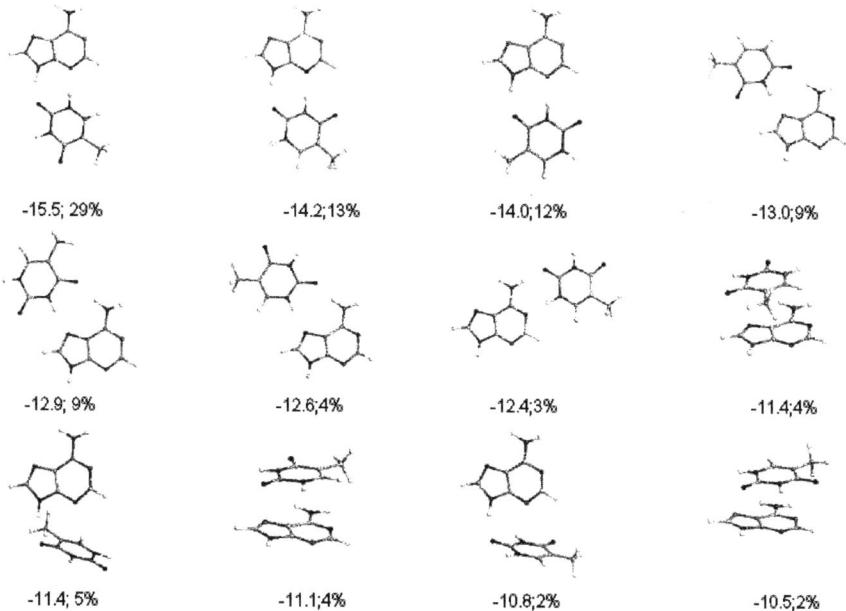

-15.5; 29%   -14.2;13%   -14.0;12%   -13.0;9%

-12.9; 9%   -12.6;4%   -12.4;3%   -11.4;4%

-11.4; 5%   -11.1;4%   -10.8;2%   -10.5;2%

Figure 3.2   Structures of the adenine...thymine (nonmethylated) base pairs. Numbers refer to empirical interaction energy and relative population (in kcal/mol and %).

calculations were fully reoptimised at the *ab initio* level). For the sake of comparison with experiment performed at nonzero temperatures, the relevant data are obtained from analysis of the Gibbs-energy surface and not of the potential-energy surface. The relative population of various structures (a quantity proportional to ΔG of base-pair formation) was determined by MD simulations in the NVE microcanonical ensemble. Although the stability order of the global and first two local minima is unaffected by including the entropy contribution, the stability order of the remaining structures is altered rather significantly in favour of stacked and T-shaped structures. The simulations further show that the population of the global minimum is about 29%, meaning that experimental gas-phase studies are likely to detect a large number of mutually coexisting structures.

The potential-energy and Gibbs-energy surfaces of all ten canonical and methylated nucleic acid–base pairs were studied[84] and the results can be summarised as follows. More than a dozen energy minima were located on the PES of each base pair. The global and first local minima of the nonmethylated base pairs do systematically exhibit a planar H-bonded structure, while T-shaped and stacked structures are less stable. Entropy does not play an important role and, therefore, the relative order of individual structures on the PES and FES does not differ to a large extent. However, methylation at purine N_9 and pyrimidine N_1 (positions where a sugar unit is attached) causes dramatic changes of the PESs and FESs. The main observation is that the most stable and most populated H-bonded structures found for the nonmethylated pairs are eliminated. For the methylated base pairs, entropy plays an important role and the structure of the global minimum does not usually correspond to the most populated structure. Frequently, it is a stacked structure that is the most populated one with entropy favouring stacking over H-bonding. Calculations reveal that the PESs and FESs of most base pairs are very complex and are characterised by the coexistence of several structures, which makes assignment of various experimental characteristics difficult.

The weak point of all studies discussed above is the use of empirical potential in the MD simulations. More reliable treatment will be based on on-the-fly MD simulations similarly as is done for the benzene dimer (see later).

3.2.1 Accurate Stabilisation Energies of H-Bonded and Stacked Nucleic Acid–Base Pairs

Besides their role in stabilising DNA and RNA there is another reason to investigate base pairs in detail: the existence of experimental stabilisation enthalpies for dimers of methylated bases, specifically for mG...mC, mA...mT, mU...mU and mT...mT.[85]

In this section we will discuss the theoretical characteristics of base pairs; their comparison with experiment will be examined in detail in Section 5.7. Planar H-bonded and stacked structures of the 9-methylguanine...1-methylcytosine and 9-methyladenine...1-methylthymine (as well as of nonmethylated species) were optimized[86] at the RI-MP2 level using the TZVPP (5s3p2d1f/ 3s2p1d) basis set and optimised structures are shown in Figure 3.3. The planar H-bonded structure of the mG...mC corresponds to the Watson–Crick (WC) arrangement, whereas the mA...mT possess the Hoogsteen (H) structure. The MP2/CBS interaction energies as well as the CCSD(T) correction terms are summarised in Table 3.1. Stabilisation energies for all structures were determined as the sum of the complete basis-set limit of MP2 energies and the CCSD(T) ($\Delta E^{CCSD(T)} - \Delta E^{MP2}$) correction term evaluated with the 6-31G** (0.25, 0.15) basis set. The complete basis-set limit was determined by a two-point extrapolation using the aug-cc-pVXZ basis sets for X=D and T, and T and Q, respectively. The convergence of the MP2 interaction energy for the studied complexes is rather slow and it is thus inevitable to include the extrapolation to the complete basis-set limit. The MP2/aug-cc-pVQZ stabilisation energies for all complexes were already very close to the complete basis-set limit. Much cheaper D→T extrapolation provided a complete basis-set limit that is very close (by less than 0.7 kcal/mol) to the accurate T→Q term and can be recommended for evaluation of complete basis-set limits of more extended complexes (e.g. larger motifs of DNA). The convergence of the ($\Delta E^{CCSD(T)} - \Delta E^{MP2}$) term is known to be faster than that of the MP2 or CCSD(T) correlation energies themselves, and the 6-31G** (0.25, 0.15) basis set provides reasonable values for planar H-bonded as well as stacked structures.[87]

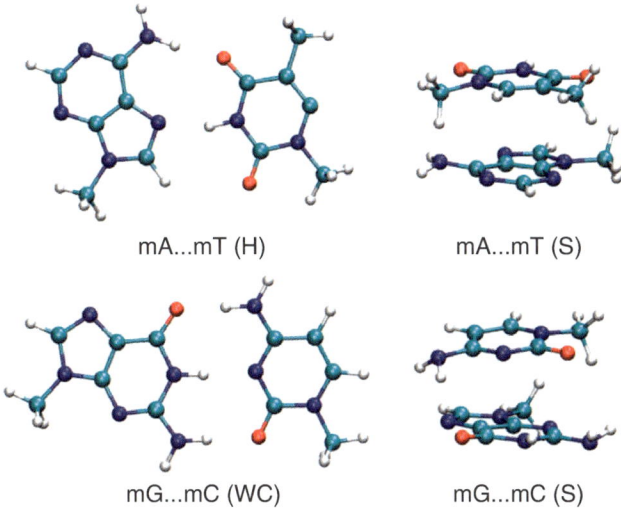

mA...mT (H)   mA...mT (S)

mG...mC (WC)   mG...mC (S)

Figure 3.3 Structures of methyl adenine...methyl thymine (Hoogsteen (H) and stacked (S)) and methyl guanin...methyl cytosine (Watson–Crick (WC) and stacked (S)).

Table 3.1 Interaction energies (in kcal/mol) of methylated and nonmethylated A...T and G...C DNA base pairs determined with different basis sets. The aug-cc-pVXZ basis sets (X=D,T,Q) were abbreviated as aXZ; CBS means complete basis-set limit. ΔCCSD(T) is the CCSD(T) correction term determined as a difference between CCSD(T) and MP2 interaction energies. WC: Watson–Crick. H: Hoogsteen. S: stacked.

Method	AT WC	mAmT H	AT S	mAmT S	GC WC	mGmC WC	GC S	mGmC S
MP2/TZVPP	−14.3	−14.8	−12.1	−14.4	−25.8	−25.6	−16.3	−17.7
MP2/aDZ	−13.8	−15.2	−12.8	−14.9	−25.6	−25.4	−16.9	−18.3
MP2/aTZ	−14.7	−15.9	−13.8	−16.2	−27.0	−26.8	−18.1	−19.6
MP2/aQZ	−15.1	−16.2	−14.1	−16.4	−27.7	−27.5	−18.5	−20.2
CBS(D→T)	−15.0	−16.2	−14.3	−16.8	−27.5	−27.4	−18.6	−20.1
CBS(T→Q)	−15.4	−16.4	−14.4	−16.6	−28.2	−27.9	−18.8	−20.5
ΔCCSD(T)	0.0	0.1	2.8	3.5	−0.6	−0.6	1.9	2.5
ΔE	−15.4	−16.3	−11.6	−13.1	−28.8	−28.5	−16.9	−18.0

Inclusion of CCSD(T) correlation corrections is inevitable for obtaining reliable relative values between planar H-bonding and stacking interactions; their neglect results in large errors of 2.5–3.5 kcal/ mol (in relative energies).

The CCSD(T) correction terms are negligible for H-bonded complexes, while they are large and positive (*i.e.* repulsive) in the case of stacking interactions. It should be added that the same conclusions have been drawn for about 200 H-bonded and stacked DNA base pairs studied up to now,[88] which is of key importance as the very expensive CCSD(T) calculations can be omitted for

H-bonding but never for stacking. However, (nearly planar) H-bonded and vertical stacked interactions can be clearly separated only in the case of DNA base pairs. The situation in proteins is much more complicated (see below) as it is difficult to find such clearly defined geometry motifs. Final stabilisation energies for the base pairs studied were very large (see Table 3.1), much larger than the values previously published, which is especially true for stacked pairs. From Table 3.1 it follows that a stacked mGmC pair without any H-bond is more stable than an mAmT pair having two strong H-bonds. This conclusion has important consequences on the nature of stability in DNA (see later).

Data discussed in the previous section were based on optimised (*i.e.* gas-phase) geometries. In the following section we refer to the crystal structures.[89] In the case of H-bonded pair structures there exists a deep similarity between crystal and gas-phase geometries. The situation is, however, entirely different for stacked pairs and Table 3.2 contains interaction energies of *interstrand* and *intrastrand* base pairs determined at various computational levels. These energies were determined for ten unique B-DNA base-pair steps (a base step consists of two intrastrand and two interstrand contributions; *e.g.* the GC step means stacked G...C and C...G base pairs). The idealised geometries were used since the individual X-ray geometries in medium-resolution structures are often affected by data and refinement inaccuracies and are thus not suitable for accurate QM reference calculations. Let us only add here that selection of suitable "experimental" geometries is in fact a large problem for many QM studies.

The stacking energies of base steps are large and the GC (and CG) steps are the most stable ones. The stability of the step depends on the sequence (GC and CG, as well as AT and TA possess different stabilisation energies). The individual pair *intrastrand* stacking contributions (not presented in Table 3.2) are in the range of −10.8 to −1.6 kcal/mol, while the individual *interstrand* terms lie between −4.8 and +3.1 kcal/mol. The ΔCCSD(T) correction terms are in the range of −0.1 to +2.5 kcal/mol and, evidently, its neglect can change the stability order of individual steps. The very expensive CCSD(T)/CBS calculations

Table 3.2 Stacking energies (kcal/mol) in B-DNA base-pair steps determined at *ab initio* CCSD(T)/CBS, MP2/6-316*(0.25) and empirical Amber levels.

B-DNA step	*CCSD(T)/CBS* intrastrand	interstrand	total	*MP2/6-31G* (0.25)* total	*Amber* total
GC	−21.6	5.1	−16.6	−14.1	−15.6
CG	−15.8	−2.7	−18.4	−13.8	−16.3
GG	−5.2	−8.5	−13.7	−11.5	−13.8
GA	−13.8	0.3	−13.6	−12.1	−13.7
AG	−13.6	−0.7	−14.3	−12.2	−14.9
TG	−10.6	−5.4	−16.0	−12.5	−15.7
GT	−10.4	−3.8	−14.2	−12.3	−14.6
AT	−13.3	0.0	−13.3	−11.6	−15.6
TA	−12.1	−0.9	−13.0	−11.2	−14.2
AA	−16.1	3.0	−13.1	−12.0	−14.7

show only a modest increase in the stacking stabilisation compared to the MP2/6-31G*(0.25) data. A very big surprise comes from the force-field results. The absolute values are in better agreement with the reference CCSD(T)/CBS data, while the relative discrepancies between force-field and QM reference values are modestly enlarged (in comparison to the MP2 level). These data suggest that absolute stabilisation energies reflected by the AMBER force field should be correct. Further, the MP2/6-31G*(0.25) method, which was used for the first time more than a decade ago as a reference method[90] proves to be qualitatively correct. The difference between the cheap MP2/6-31G*(0.25) and the extremely expensive CCSD(T)/CBS levels is found to be below 2 kcal/mol. Going to stacked bases it was found that 80% of the CCSD(T)/CBS stacking energy is recovered at the MP2/6-31G*(0.25) level. Even better correlation was recently found[91] for complexes of adenine with aromatic amino acids where the MP2/6-31G*(0.25) calculations account for 91–105% of the CCSD(T)/CBS interaction energies for the T-shaped clusters. Despite the fact that the data presented in Table 3.2 are highly accurate we have no solid clues how to relate these base–base forces to the folding and stability of DNA. We still do not know exactly how stacking contributes to the variability of nucleic acids.

3.2.2 Verification of Accurate Stabilisation Energies

Comparison with experiment. Comparison of theoretical and experimental data is in this case not straightforward. The reason is that experimental stabilisation enthalpies do not correspond to one structural type but rather to an average value. Further, no experimental evidence exists about dimer structures. The situation is further complicated by the fact that experiments were performed at rather high temperatures, which means that it is necessary to pass from the PES to the FES. For a more detailed discussion see Chapter 5.

Comparison with the DFT-SAPT and QMC method. Let us repeat that interaction energies discussed in the previous section were derived from variation calculations. The present values were obtained by a completely different technique, a perturbation method. This means the total interaction energy was not constructed as the difference between energies of a supersystem and subsystems (as in the variation method) but as a sum of various energy terms. Besides this, a perturbation method allows decomposition of an interaction energy into physically well-defined energy terms. With the introduction of the DFT-SAPT procedure, it becomes possible to determine the energy components even for such large complexes as DNA base pairs. Hesselmann et al.[92] have investigated the optimised DNA base pairs and shown that the largest attractive contribution for H-bonded pairs is the electrostatic energy; second-order induction and dispersion contributions are, however, substantial as well. The situation is entirely different for stacking, where the dominant contribution stems from dispersion energy. A very important conclusion concerns the total interaction energy determined as a sum of first- and second-order perturbative contributions. These energies for both H-bonded and stacked DNA base pairs agree very well with the interaction energies presented in Table 3.1 (geometries were in both cases identical). This finding is important as almost identical stabilisation energies are yielded by quite different procedures (variation – supermolecular and perturbative) which gives confidence to both procedures.

The further verification comes from the fixed-node diffusion Monte Carlo (FNDMC) method. This quantum Monte Carlo method differs completely from the WFT and DFT procedures discussed above. The FNDMC was used for H-bonded and stacked structures of the UU, AT, and GC dimers.[93] The agreement with our CCSD(T)CBS results is reasonably good. For the H-bonded uracil dimer, the difference between FNQMC and CCSD(T)/CBS is –0.95 kcal/mol. The differences are 0.49, –0.60 and 1.34 kcal/mol for the adenine-thymine Watson–Crick base pair, stacked uracil dimer, and stacked adenine-uracil dimer (respectively), in all cases utilising identical geometries. Therefore, the CCSDC(T)/CBS and FNQMC methods seem to provide, on average, very similar absolute values of stabilisations while there are modest random differences for the individual systems. To understand the origin of these differences further investigation will be required.

We can conclude this section by stating that interaction of DNA bases is very strong, which concerns not only H-bonded pairs (where it was expected) but also stacked pairs. Surprisingly large stabilisation energies calculated at the CCSD(T)/CBS level were confirmed independently by being compared with theoretical values taken from perturbative SAPT and QMC methods. All three theoretical procedures are based on entirely different grounds. Finally, the theoretical stabilisation energies were compared with the experimental values and the agreement was also satisfactory (for a more detailed analysis see Section 5.7). It is, however, clear that agreement obtained with either of these methods cannot justify the chemical accuracy of the procedure described, *i.e.* accuracy better than ± 1 kcal/mol. Such a verification can thus only be obtained from even higher-level quantum-chemical calculations.

Highly accurate CCSD(T) stabilisation energies of the uracil dimer. These calculations were performed[94] for the smallest nucleic acid–base pair, the uracil dimer. The C_{2h} and C_S structures (see Figure 3.4) of the H-bonded and stacked uracil dimers were considered. Geometries of both isomers were taken from the S22 set[88] and were determined by counterpoise-corrected gradient optimisation

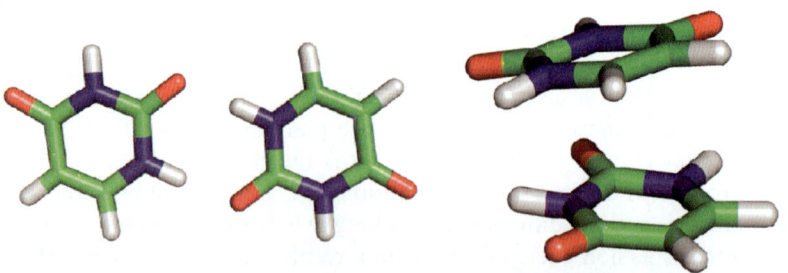

Figure 3.4   H-bonded and stacked structures of the uracil dimer.

at the MP2/cc-pVTZ level. Table 3.3 summarises the interaction energies for both structures determined with various methods and basis sets. Considering the MP2 results it can be seen that, when passing from the aug-cc-pVDZ to the larger aug-cc-pVTZ and aug-cc-pVQZ basis sets, there is a significant stabilisation energy increase. In the case of the H-bonded structure the respective increases amount to 1.19 and 0.47 kcal/mol. Stacked stabilisation energies are smaller than H-bonded ones and passing to a larger basis set is connected with

Table 3.3 MP2, SCS-MP2, MP3, CCSD(T), DFT/M06 and DFT-D/TPSS interaction energies calculated for H-bonded and stacked energy minima of the uracil dimer with various basis sets (energies in kcal/mol). BF means bond functions. The aug-cc-pVXZ basis sets (X = D, T, Q) were abbreviated as a-pVXZ; CBS means complete basis set limit.

Method	H-bond	Stacked
MP2/a-pVDZ	−18.41	−9.80
MP2/a-pVTZ	−19.60	−10.63
MP2/a-pVQZ	−20.07	−10.90
MP2/CBS	−20.37	−11.08
CCSD(T)/a-pVDZ	−18.43	−8.54
CCSD(T)/a-pVTZ (OVOS)	−19.81	−9.33
CCSD(T)/CBS (A)	−20.40	−9.67
CCSD(T)/CBS (B)	−20.64	−9.77
CCSD(T)/CBS (C)	−20.50	−9.68
CCSD(T)/da-pVDZ	−18.54	−8.66
CCSD(T)/a-pVDZ + BF	–	−8.83
SCS-MP2/pVTZ	−16.91	−6.60
SCS-MP2/pVQZ	−17.85	−7.60
SCS-MP2/pV(DT)Z	−17.82	−7.57
SCS-MP2/pV(TQ)Z	−18.45	−8.30
SCS(MI)-MP2/pVTZ	−20.87	−9.38
SCS(MI)-MP2/pVQZ	−20.83	−9.51
SCS(MI)-MP2/pV(DT)Z	−20.85	−9.42
SCS(MI)-MP2/pV(TQ)Z	−20.79	−9.60
MP3/a-pVDZ	−18.51	−6.58
MP3/a-pVTZ	−19.88	−7.37
MP3/a-pVQZ	−20.46	−7.70
MP3/CBS	−20.84	−7.92
DFT-SAPT/a-pVDZ	−17.94	−8.51
DFT-SAPT/a-pVTZ	−19.04	−9.26
DFT-SAPT/CBS	−19.51	−9.58
M06-2X/6-311 + G(2df,f2p)	−18.97	−10.00
DFT-D/TPSS/6-311 + + G(3df,3pd)	−20.45	−9.92

slightly smaller stabilisation energy increases (0.83 and 0.27 kcal/mol, respectively). These data agree with the slow convergence mentioned above for methylated and nonmethylated AT and GC pairs.

Three different extrapolation types were investigated. The extrapolation based solely on the CCSD(T) correlation contribution is labelled in the following text as "A", while a separate extrapolation of the MP2 and ΔCCSD(T) is labelled as "B". The "interpolation-like" procedure, labelled as "C", relies on the fact that BSSE corrected and uncorrected quantities converge to the same value in the CBS limit. The A-type extrapolation was possible by using the OVOS technique which enables one to perform the CCSD(T) calculation with aug-cc-pVDZ and aug-cc-pVTZ basis sets. Interaction energies obtained at the CCSD(T) level using such relatively large basis sets (for systems of this size) are shown in Table 3.3. CCSD(T) values for the H-bonded complex differ from the MP2 values only marginally, –0.02 kcal/mol for aug-cc-pVDZ, –0.2 kcal/mol for aug-cc-pVTZ and at most –0.27 kcal/mol for the CBS value (extrapolation "B"). The difference between MP2 and CCSD(T) is more pronounced for the stacked complex, where the stabilisation energy is overestimated by the MP2 method, by 1.26 kcal/mol for aug-cc-pVDZ, 1.30 kcal/mol for aug-cc-pVTZ basis set and (at most) by 1.41 kcal/mol at the CBS level (extrapolation "A"). To investigate the importance of the very diffuse basis functions, calculations in Dunning's doubly augmented correlation consistent basis set, daug-cc-pVDZ, were performed. The results, especially for the stacked structure, were rather surprising. Less than 0.1 kcal/mol of additional stabilisation is achieved for both structures compared to the aug-cc-pVDZ basis set. A slightly larger effect on the stabilisation of the stacked structure ($\sim$0.3 kcal/mol compared to the aug-cc-pVDZ basis set) is obtained from augmentation of the aug-cc-pVDZ basis set with bond functions with three s- (coefficients: 0.9, 0.3 and 0.1), three p- (same coefficients as s-functions) and two d-functions (coefficients: 0.6 and 0.2) placed at the centre of mass of the complex. However, inclusion of basis functions with higher angular momenta in the aug-cc-pVTZ basis set is obviously of more importance and leads to an increase of the stabilisation energy by $\sim$1.4 kcal/mol for the H-bonded complex and $\sim$0.8 kcal/mol for the stacked complex.

More detailed insight into basis-set effects can be obtained from Table 3.4. CCSD(T) interaction energies as well as ΔCCSD(T), ΔCCSD, and (T) terms were separately analysed for various basis sets. The H-bonded structure will be discussed first. The CCSD(T) stabilisation energy depends strongly on the quality of the basis set and standard basis sets not containing diffuse polarisation functions significantly underestimate the stabilisation energies. The difference between aug-cc-pVDZ and aug-cc-pVTZ CCSD(T) values are also remarkably large, suggesting that the aug-cc-pVDZ basis set is not extended enough to describe this system properly. This is certainly "bad" news since such calculations are extremely demanding and are almost impractical with today's computers for complexes larger than the studied pair. Among the split valence basis sets the largest stabilisation energies are produced by the modified 6-31G** basis set (denoted as a-pVDZ diffuse; polarisation functions are taken

Table 3.4 CCSD(T) interaction energies and ΔCCSD(T) term (difference between CCSD(T) and MP2 interaction energies) determined for energy minima of the uracil dimer with various basis sets (energies in kcal/mol). BF means bond functions.

	H-bonded structure				Stacked structure			
	ΔE	$\Delta CCSD(T)$	$\Delta CCSD$	$\Delta(T)$	ΔE	$\Delta CCSD(T)$	$\Delta CCSD$	$\Delta(T)$
6–31G*(0.25)	−17.08	−0.37	0.35	−0.72	−7.59	0.91	2.30	−1.29
6–31G**(0.25,0.15)	−17.40	−0.43	0.34	−0.77	−7.75	0.81	2.28	−1.47
6–31G(a–pVDZ diffuse)	−17.73	−0.28	0.50	−0.78	−7.90	0.61	2.11	−1.50
6–31G**	−17.07	0.12	0.61	−0.48	−4.57	1.12	2.04	−0.92
6–31 + G**	−17.26	−0.03	0.56	−0.60	−6.26	1.05	2.31	−1.26
cc–pVDZ	−16.10	0.03	0.50	−0.47	−4.77	1.16	2.14	−0.99
daug–cc–pVDZ	−18.54	−0.08	0.77	−0.85	−8.66	1.23	2.92	−1.68
aug–cc–pVDZ + BF	–	–	–	–	−8.83	1.24	2.95	−1.71
aug–cc–pVDZ	−18.43	−0.02	0.83	−0.85	−8.54	1.26	2.92	−1.66
aug–cc–pVTZ (OVOS)	−19.81	−0.21	0.73	−0.94	−9.33	1.29	3.11	−1.81
CBS (B)	−20.64	−0.27	0.71	−0.98	−9.77	1.31	3.19	−1.88

from the most diffuse set of the aug-cc-pVDZ basis set). It should be noted that there is a difference of more than 1.5 kcal/mol with respect to the cc-pVDZ result. Concerning the ΔCCSD(T) correction term we found that it depends dramatically on the quality of the basis set. However, its absolute value is smaller, by an order of magnitude, than the MP2 correlation correction, thus affecting the total interaction energy only marginally. The most reliable ΔCCSD value, obtained with the aug-cc-pVTZ basis set, is –0.21 kcal/mol and all values determined with considerably smaller basis sets differ in absolute value by less than ±0.4 kcal/mol. Quite surprising is the performance of the aug-cc-pVDZ basis set: the correction term, –0.02 kcal/mol, is one of the poorest estimates of the ΔCCSD(T) term. However, this is a consequence of a compensation of the ΔCCSD and (T) terms, which both individually are described within 0.1 kcal/mol accuracy. Among split valence basis sets, the 6-31G**(with polarisation functions taken from the aug-cc-pVDZ basis set, abbreviated here a-pVDZ diffuse) basis set provides the closest agreement. The CCSD(T) stabilisation energies for the stacked structures are smaller but still remarkably large. The underestimation of these energies when using bases not containing diffuse basis sets is even more pronounced than in the case of the H-bonded structures. Typically, aug-cc-pVDZ stabilisation energies are 80% larger than those determined by the respective nonaugmented basis set. Similarly, as in the previous case, the a-pVDZ diffuse basis set provides the largest energies, which compare well with energies determined using the much larger aug-cc-pVDZ basis set. The performance of the basis sets on the ΔCCSD(T) correction term is more balanced compared to the H-bonded structure. The ΔCCSD(T) term is already almost converged to the CBS value at the aug-cc-pVDZ level, with an error of ~0.05 kcal/mol. Generally, results in different basis sets fall into slightly larger error bars of ±0.7 kcal/mol, as for the H-bonded complex, but in this case the ΔCCSD(T) term is repulsive and its absolute value is almost five times larger, ∼1.3 kcal/mol. It is interesting to look at the decomposition of the ΔCCSD(T) term into the contributions from the ΔCCSD and (T) terms. In analogy with the H-bonded structure, these contributions are directed opposite to each other, but for the stacked structure they are (especially the ΔCCSD term) much larger. Neglecting the (T) contribution would lead to a strong underestimation of the stabilisation energy, by almost 2 kcal/mol. At first glance, the best-performing split valence basis set is 6-31G**, but this is an artefact of underestimation of the effect of the perturbative triples correction. To obtain reasonably accurate results for the right reasons, at least the 6-31 + G** basis set should be used.

We can conclude this section by stating that presently the most accurate CCSD(T)/CBS results (Schemes "A" and "C") agree very well with CCSD(T)/CBS estimates (scheme "B") shown in the previous paragraph for methylated and nonmethylated A...T and G...C base pairs. This confirms that these results were computed with chemical accuracy (better than 1 kcal/mol). Contrary to previous expectations the previously published stacked stabilisation energies of the uracil dimer are larger than the present ones. This means that CCSD(T)/CBS stabilisation energies determined for the S22 set[88] represent rather the upper (and not as expected the lower) limit of the true values. These conclusions are based on the highly accurate CCSD(T) calculations (levels A and C) performed for the uracil dimer. The CCSD(T) correction term for this complex is, however, rather small (about 1 kcal/mol) and this affects the conclusions obtained. To prove the validity of the conclusions mentioned above highly accurate CCSD(T) calculations should be performed for other stacked dimers having a much higher CCSD(T) correction term (*e.g.* mA...mT or benzene...hexacyanobenzene).

The presence of highly accurate CCSD(T) results allows us to test the performance also for other computationally less demanding methods like SCS-MP2, SCS(MI)-MP2 and MP3 (see Table 3.3). Introducing scaling coefficients for the parallel and antiparallel MP2 spin contribution in SCS-MP2 significantly reduces the overestimation of the interaction energy in the stacked complex (by ∼2.30 kcal/mol). However, compared to the CCSD(T)/CBS results, the SCS-MP2 interaction becomes significantly underestimated by 1.4 kcal/mol. This is, in its absolute value, an error even larger than that of MP2 itself (∼1.3 kcal/mol). For the H-bonded complex, the performance of SCS-MP2 compared to MP2 is (as expected) even poorer; the interaction energy is underestimated by more than 2 kcal/mol, whereas MP2, as mentioned above, describes the interaction within an accuracy of 0.3 kcal/mol. When scaling coefficients optimised for molecular interactions calculations are used, *i.e.* SCS(MI)-MP2, the conclusions change dramatically. The error of SCS(MI)-MP2 extrapolated from cc-pVTZ and cc-pVQZ basis sets is as small as 0.15 kcal/mol for the H-bonded structure and 0.17 kcal/mol for the stacked structure. This unbelievable agreement with the CCSD(T)/CBS values is perhaps a consequence of the fact that both of the uracil dimer structures studied in this work were included in the training set (S22) for obtaining the optimised spin scaling coefficients. Since there are no SCS(MI)-MP2 data published yet for systems out of the S22 training set, it is impossible to make any generally valid statements. However, if the performance of the method on other systems is comparable with its performance on the uracil dimer, it would be a computationally inexpensive method for highly accurate calculations of molecular interactions. The error of the MP3 method extrapolated from aug-cc-pVTZ and aug-cc-pVQZ for H-bonded structure is $\sim 0.2\,$kcal/mol, which is smaller than that of MP2. On the other hand, the performance for the stacked structure mimics the performance of CCSD, meaning that the stabilisation energy is underestimated by $\sim 1.85\,$kcal/mol due to neglecting the effect of perturbative triples. Let us, however, add here that no additional parameters are used in MP3, contrary to SCS-MP2 or SCS (MI)-MP2 procedures discussed above. Among lower-level methods, the best results, not only in terms of the energy minima of H-bonded and stacked structures, but also for the entire potential-energy curves, were exhibited by the SCS (MI)-MP2 technique. The original SCS-MP2 methods works well for stacking but underestimates H-bonding. Let us, however, stress again that both SCS procedures were parametrised.

3.2.3 Decomposition of Stabilisation Energy Using the Perturbation Calculation

The energy decomposition using the DFT-SAPT method was first performed for the gas-phase optimised structures of the adenine ... thymine and guanine ... cytosine and in both cases the planar H-bonded Watson–Crick and stacked structures were investigated (see above). The complete basis-set limit of DFT-SAPT was generated by performing all calculations at the cc-pVQZ level with the exception of the dispersion energy, which was extrapolated from aug-cc-pVTZ and aug-cc-pVQZ values. It was shown that H-bonded structures were mainly stabilised by the electrostatic term but these energies are over-compensated by the exchange-repulsion term. In the case of stacked structures the dispersion energy was dominant.

More detailed analysis was performed[95] for crystal geometries of both H-bonded and stacked pairs. The energy components as well as the CCSD(T)/CBS stabilisation energies from dataset S22 are summarised in Table 3.5. The DFT-SAPT stabilisation energies are systematically smaller than the reference CCSD(T)/CBS values, and this finding is in full agreement with the conclusion drawn in Ref. 92.

Table 3.5 includes 6 planar H-bonded complexes, among which are 3 GC WC, 2 AT WC and one GA miss-pair. The GC WC structures are considerably more stable than the AT WC pairs, with the GA miss-pair being the least stable. Dominant attraction for all H-bonded structures originates in the $E_{el}^{(1)}$ electrostatic term. Due to much larger dipole moments of G and C (in comparison with A and T), the electrostatic $E_{el}^{(1)}$ energy of the GC pairs is much more attractive than that of AT pairs and also the GA pair. The attractive $E_{el}^{(1)}$ term

Table 3.5 First- and second-order perturbative DFT-SAPT energies and the reference CCSD(T)/CBS interaction energies (energies in kcal/mol) for planar H-bonded DNA base pairs (Watson Crick, WC) (1–6), intrastrand stacked pairs (S) (7–14) and interstrand pairs (IS) (15–27); all geometries taken from crystal data. Second-order energies include the respective exchange terms; the exchange-dispersion energy is presented in parentheses.

	Structure[a]	$E^{(1)}_{el}$	$E^{(1)}_{Ex}$	$E^{(1)}$	$E^{(2)}_i$	$E^{(2)}_D$	$E^{(2)}$	$\delta(HF)$	E	ΔE^{b}	E_{disp}^{c}
1	mAmT WC	-26.71	33.29	6.58	-6.13	-9.45(2.43)	-15.58	-5.13	-14.13	-16.40	-9.01
2	mCmG WC	-48.28	52.51	4.22	-12.39	-13.17(3.75)	-25.56	-9.42	-30.76	-35.80	-13.38
3	mAmT WC	-27.36	32.29	4.92	-6.21	-9.30(2.40)	-15.51	-5.23	-15.82	-18.40	-8.74
4	GA	-24.96	34.77	9.81	-5.74	-9.88(2.49)	-15.62	-5.03	-10.84	-11.30	-9.79
5	GC WC	-48.77	56.73	7.96	-12.75	-13.60(3.92)	-26.35	-9.94	-28.33	-30.70	-14.18
6	GC WC	-47.96	55.39	7.44	-12.55	-13.44(3.84)	-25.99	-9.70	-28.25	-31.40	-13.96
7	AT S	-3.30	9.83	6.53	-0.62	-11.84(1.53)	-12.46	-0.46	-6.39	-8.10	-10.31
8	mCmG S	-2.98	3.08	0.10	-1.10	-5.84(0.58)	-6.94	-0.20	-7.04	-7.90	-4.57
9	mAmC S	-1.02	7.95	6.93	-1.12	-10.55(1.39)	-11.67	-0.31	-5.05	-6.70	-8.70
10	mTmG S	-0.53	7.11	6.58	-0.77	-10.21(1.10)	-10.98	-0.28	-4.68	-6.20	-9.13
11	CG S	-5.93	11.54	5.61	-0.91	-10.07(1.67)	-10.98	-0.69	-6.06	-7.70	-8.54
12	AG S	-4.39	12.78	8.39	-0.80	-11.91(1.87)	-12.71	-0.60	-4.92	-6.50	-9.97
13	GC S	-6.97	9.78	2.81	-1.39	-10.48(1.56)	-11.87	-0.54	-9.60	-12.40	-8.36
14	GC S	-7.68	9.51	1.83	-1.10	-9.89(1.49)	-10.99	-0.50	-9.66	-11.60	-8.05
15	GC IS	-1.98	0.95	-1.03	-0.57	-1.93(0.16)	-2.50	-0.08	-3.61	-3.68	-1.53
16	GG IS	1.24	8.54	9.78	-0.93	-10.48(1.39)	-11.41	-0.19	-1.82	-4.82	-8.90
17	AT IS	-1.99	2.77	0.78	-0.34	-2.21(0.32)	-2.55	-0.22	-1.99	-2.34	-1.67
18	TT IS	-1.31	8.69	7.38	-0.59	-9.28(1.19)	-9.87	-0.33	-2.82	-2.16	-8.22
19	GG IS	-2.04	7.27	5.23	-0.72	-7.07(1.03)	-7.79	-.33	-2.89	1.24	-6.23
20	AG IS	-2.75	5.70	2.95	-0.48	-5.76(0.84)	-6.24	-0.27	-3.56	-4.22	-4.82
21	TC IS	-0.60	0.10	-0.50	-0.12	-0.55(0.02)	-0.67	-0.01	-1.18	-1.15	-0.45
22	AG IS	-2.71	3.47	0.76	-0.76	-3.43(0.49)	-4.19	0.25	-3.18	-4.06	-2.56
23	AT IS	-0.57	0.36	-0.21	-0.09	-1.34(0.06)	-1.43	-0.02	-1.66	-1.71	-1.09
24	mGmG IS	-1.86	6.70	4.84	-0.50	-7.57(1.02)	-8.07	-0.30	-3.53	-4.50	-6.48
25	mAmG IS	-2.81	2.51	-0.30	-0.88	-3.13(0.38)	-4.01	-0.19	-4.50	-4.80	-2.31
26	CA IS	-1.93	4.32	2.39	-0.83	-3.15(0.54)	-3.98	-0.36	-1.95	-3.00	-2.35
27	GG IS	-4.06	1.71	-2.35	-0.61	-2.10(0.32)	-2.71	-0.16	-5.22	-5.20	-1.62

[a] See Ref. 88
[b] CCSD(T)/CBS
[c] non-damped empirical dispersion energy as in Ref. 96

is systematically overcompensated by the exchange-repulsion term, thus the first-order $E^{(1)}$ energy is systematically repulsive and this repulsion is significant (more than 4 kcal/mol).

The second-order induction energy is large and again for the GC pairs it is much larger than for the other pairs, the explanation for which should be again sought in the varying values of electric dipole moments of various bases. The exchange component is now smaller (not shown), and the resulting effective second-order induction energy is systematically attractive. In the case of GC pairs, this attraction is approximately twice as big as that of the other pairs.

The δ(HF) terms are rather large (more than half of the dispersion energy) and attractive for planar H-bonded pairs. For the GC pairs, they are about twice as large as for other pairs. Evidently, the value of δ(HF) term correlates with the value of induction energy.

In conclusion, we ascertained that the stabilisations in induction and dispersion terms are comparable, whereas the δ(HF) term is smaller but definitely not negligible. The electrostatic $E^{(1)}{}_{el}$ term is significantly more attractive than the previous terms, but the absolute values of the exchange-repulsion term are even higher. The $E^{(1)}$ energies are repulsive, while the $E^{(2)}$ and δ(HF) energies are attractive. All these findings basically agree with the conclusions drawn by Hesselmann *et al.*[92] The situation found for stacked complexes was different, where intrasystem stacking will be investigated first. As in the case of H-bonded pairs, the DFT-SAPT stabilisation energies are systematically underestimated when compared with the accurate values. The $E_{el}{}^{(1)}$ electrostatic energy is systematically attractive and is the largest for GC pairs and also for the AG pair. The electrostatic attraction is roughly compensated by exchange-repulsion terms (as in the case of H-bonding). The range of $E^{(1)}$ energy is larger here (from 0.1 to 8.4 kcal/mol) than in the case of H-bonding (from 4.2 to 9.8 kcal/mol), and GC stacked pairs exhibit more favourable (less repulsive) $E^{(1)}$ energies. The $E^{(1)}$ energies are, however, surprisingly comparable for planar H-bonded and stacked base pairs. Attractive and repulsive components of induction energy are virtually compensated (not shown), and the resulting effective induction energy is negligible. This is an important difference in comparison with H-bonded systems, where the effective induction energy is very large. Unambiguously, dominant attraction originating in all the cases is effective dispersion energy, containing only a small (repulsive) exchange part. Dispersion energy is systematically more attractive than the electrostatic term, and the ratio ranges from 10 (mAmC and mTmG) to approximately 1.5 (GC pairs). It is to be mentioned that even with two GC stacked pairs, the dispersion energy is about twice as large as the electrostatic term. Yet having compared the dispersion energies for stacked and H-bonded complexes, we reached a similar conclusion as with the $E^{(1)}$ energies, which indicates that they are more or less comparable! This finding is surprising as it had been expected that the dispersion energy should be larger for stacked pairs and it deserves a more detailed discussion, which is presented in the following section. The δ(HF) term is negligible for all stacked pairs, which makes the stacked pairs significantly different from H-bonded systems.

Intuitively, the similarity of magnitudes of the dispersion energy in stacked and H-bonded complexes is indeed puzzling. In the stacked complexes the distance of centres of masses is considerably smaller than in the H-bonded ones, and, what is more important, both monomers seem to have larger geometrical overlap in the stacking arrangement. Consequently, one would anticipate larger dispersion energy. An explanation can be found – at least in part – in comparing the SAPT dispersion energies with the empirical dispersion energies as calculated by the well-known the C_6/r^6 formula. Table 3.5 shows the SAPT values along with the calculated empirical dispersion energy contributions (for details on the method and C_6 coefficients used see Ref. 96). Interestingly, the nondamped values are in very good agreement with the SAPT values for both stacked and H-bonded complexes, being on average 0.6 kcal/mol weaker and with standard deviation 0.6 kcal/mol. Obviously our intuitive perception of the dispersion interaction is not accurate and very close contacts in the H-bonded complexes (X and Y atoms in X–H . . . Y H-bond are separated by less than 3 Å and the H . . . Y distance is closer to 2 Å) can bring as much dispersion stabilisation as is found in the stacked molecules. The remaining 0.6 kcal/mol missing in the empirical description (and likely more – note that our SAPT values are underestimated in the aug-cc-pVDZ basis) can be explained by overlap effects, which result in a small, but exponentially growing dispersion contribution at very short distances. We would like to point out that the present comparison is illustrative only, because the C_6/r^6 term is unphysically divergent at short distances. However, very similar magnitudes of the reference SAPT calculations indicate that the short-range effects (damping) are still not so profound even at the short H-bonding distances, so the results drawn are likely not affected.

Interstrand stacking is characterised by smaller total interaction energies, which are attractive and repulsive. A total of 32 interstrand pairs have been investigated, of which only such pairs are shown in Table 3.5, which had DFT-SAPT stabilisation energy larger than 1 kcal/mol. Six pairs (of 27 presented in Table 3.5) possess stabilisation energy larger than 3 kcal/mol, including only one GC pair. The largest interstrand stacking was found for the GG pair. Three pairs had $E^{(1)}$ energy attractive (due to attractive electrostatic energy and the rather small exchange-repulsion term). Notice that this energy was systematically repulsive for all H-bonded and stacked pairs. The δ(HF) term is mostly negligible and never exceeds 0.5 kcal/mol.

Summarising all data we can state that, in the DFT-SAPT type of analysis, the $E^{(1)}$ energies for H-bonded and intrastrand stacked pairs (in crystal geometries) are similar. The $E^{(2)}$ energies are much larger for planar H-bonded pairs, which is caused by $E^{(2)}_i$ induction energy. The dispersion energy is surprisingly similar for both structural types. The $E^{(1)}$ and $E^{(2)}_D$ energies thus do not show a preference for either of the two motifs.

A similar study for crystal structures of the stacked complementary base pairs was performed by Fiethen et al.[97] In this case the authors used geometries taken as average values from high-resolution crystal structures. Such an averaging is, however, not justified and these geometries are far from the real values.

Table 3.6 DFT-SAPT interaction energy decomposition results for the hydrogen-bonded and stacked structures of the uracil dimer. (HB and ST refer to the H-bonded and stacked base pair configurations respectively, all energies are given in kcal/mol.)

	a-pVDZ		a-pVTZ		CBS	
	HB	ST	HB	ST	HB	ST
E(elec.)	−29.62	−8.78	−29.66	−8.71	−29.67	−8.68
E(ind.)	−7.40	−0.91	−7.47	−0.93	−7.50	−0.93
E(disp.)	−8.14	−10.15	−9.04	−10.87	−9.42	−11.17
E(exch.)	32.55	11.87	32.52	11.80	32.51	11.77
$\delta(HF)$	−5.34	−0.54	−5.40	−0.56	−5.43	−0.57
ΔE_{int}^{SAPT}	−17.94	−8.51	−19.04	−9.26	−19.51	−9.58

A rather detailed study using the DFT-SAPT decomposition was published for H-bonded and stacked structures of the uracil dimer.[94] An important advantage of this study is the fact that for this dimer there exist very accurate CCSD(T)/CBS data, allowing comparison between CBS DFT-SAPT and CCSD(T)/CBS values. Table 3.6 gives the DFT-SAPT interaction energy decomposition results for the hydrogen-bonded and stacked uracil dimer with the aug-cc-pVDZ and aug-cc-pVTZ basis sets and at the extrapolated complete basis-set limit. One of the most prominent aspects of these data is the fact that, as expected, the electrostatic interaction is dominant in stabilising the hydrogen-bonded dimer (−29.67 kcal/mol) while the stacked system is bound chiefly by dispersion (−11.17 kcal/mol). It should, however, be noted that dispersion (−9.42 kcal/mol) (and induction (−7.50 kcal/mol)) play a non-negligible role in stabilising the hydrogen-bonding complex and that there is also an appreciable contribution from the electrostatic interaction component for the stacked structure (−8.68 kcal/mol). These results are generally in good agreement with those discussed above for the DNA base pairs either in the gas phase or crystal geometry.

The DFT-SAPT binding energies for the hydrogen-bonded uracil dimer complex are generally too low (underbound) by ∼0.50–1.00 kcal/mol compared with the CCSD(T) results for a given basis set. Comparing the DFT-SAPT (−19.51 kcal/mol) and CCSD(T) (−20.64 kcal/mol) CBS (CCSD(T)/CBS "A") stabilisation energies, it can be seen that DFT-SAPT underestimates this value by 1.13 kcal/mol. It should be kept in mind that, considering the large absolute values of these binding energies, the DFT-SAPT/CBS result is within 6% of the CCSD(T)/CBS one. For the stacked uracil dimer the DFT-SAPT method provides results that are in much better agreement with those of CCSD(T). Considering the CBS binding energies, it can be seen that the DFT-SAPT value (−9.58 kcal/mol) is within ∼0.2 kcal/mol of the CCSD(T) value (−9.77 kcal/mol). It is interesting to note that, in their work on AT and GC pairs, Hasselman and coworkers[92] found that the DFT-SAPT method underestimated all binding energies (hydrogen bonded and stacked) compared to estimated CCSD(T)/CBS results.

In the past several years, as more data concerning binding energies for stacked nucleobase structures have become available, many questions have arisen about the strengths of these types of interactions. The relatively strong stacking interaction found in the benzene dimer, which has been studied extensively, is known to be attributable principally to dispersion interactions. However, the sandwich configuration of the benzene dimer has been estimated to have a (CCSD(T)/CBS) binding energy of 1.81 kcal/mol. How is it that stacked nucleobases, such as the stacked uracil dimer, have binding energies that are so much more favorable than those of the benzene dimer? In order to gain some insights that may be useful in answering this question we have computed the DFT-SAPT interaction energy components for a benzene dimer restricted in such a way that the relative orientation of the monomers is the same as in our stacked uracil dimer, see Figure 3.4.

Table 3.7 gives the DFT-SAPT/aug-cc-pVDZ interaction energy results for the stacked uracil dimer and the restricted geometry benzene dimer. The exchange repulsion term for the benzene dimer is much larger than that of the uracil dimer, this seems reasonable because the optimum separation distance for a stacked benzene dimer (~ 3.8 Å) is much larger than that for the uracil dimer. One of the most prominent features of these data is the fact that, for both dimers, the dispersion term has approximately the same value (~ 10 kcal/mol). The stacking interaction involving uracil is more electrostatic in character by about 3 kcal/mol. Based on the behavior of these interactions, in terms of the dispersion and electrostatic components, it seems likely that the key difference between the binding modes of the benzene and uracil dimers involves forces that are electrostatic in nature.

The H-bonded and stacked structures of the uracil dimer were also studied[98] by using DFT-SAPT and SAPT methods and the energy components as well as total perturbation energies differ. In the case of stacked structures the dispersion energy (including the exchange component) is for the latter method more attractive by 1.4 kcal/mol, while for the H-bonded structure the opposite is true and the difference is smaller (0.3 kcal/mol). The situation with respect to the first-order electrostatic energy is different. Here, both methods provide similar

Table 3.7 DFT-SAPT interaction energy decomposition results for the stacked configurations of the benzene dimer (BD) and uracil dimer (UD) as computed using the aug-cc-pVDZ basis set (the benzene dimer is restrained to have the same geometry as the stacked uracil dimer, all energies are in kcal/mol).

	BD	*UD*
E(elec.)	−5.84	−8.78
E(ind.)	−0.25	−0.91
E(disp.)	−9.45	−10.15
E(exch.)	19.12	11.87
$\delta(HF)$	−1.02	−0.54
$\Delta E_{\text{int}}^{\text{SAPT}}$	2.56	−8.51

values for stacked structure (SAPT being more attractive by 0.6 kcal/mol) but different values for the H-bonded one (the SAPT value is more attractive by 2.1 kcal/mol). The authors explained this by the fact that in the DFT-SAPT method the intramonomer correlation energy is integrated within the monomer's electron density, while in the second case SAPT is performed with the HF intramolecular wavefunctions (which do not cover any correlation energy).

3.2.4 Microhydrated and Microsolvated Nucleic Acid Bases and Base Pairs

It is well known that bulk water dramatically changes the structure, properties and reactivity of systems and this is particularly true for nucleic acid–base pairs. Most of the base pairs *in vacuo* possess a planar H-bonded structure but after placing them into water, the structure is changed to the stacked one. It is assumed that this is due to the entropy of bulk water (hydrophobic effect). However, the question arises whether this is true and whether this pronounced effect is not just due to the action of a few water molecules. Microhydration (addition of a few water molecules) has become popular among theoreticians as well as experimentalists and this process is under study in many laboratories. But even in microhydrated environment (in which the temperature is very low), the static approach is no longer adequate and dynamic calculations are required. The dynamic structure of all ten possible nucleic acid and methylated nucleic acid–base pairs hydrated by a small number of water molecules (from 1 to 16) was determined[99,100] using molecular-dynamics simulations in the *NVE* microcanonical ensemble with the Cornell *et al.* force field.[77] Figure 3.5 shows as an example the microhydration for adenine...thymine and 9-methyladenine...1-methylthymine base pairs. The presence of one water molecule does not affect the structure of any H-bonded base pair but a higher number (mostly just two) of water molecules does. An equal population of the H-bonded and stacked structures of adenine...adenine, adenine...guanine and adenine ...thymine pairs is reached if as few as two water molecules are present, while obtaining an equal population of these structures in the case of adenine... cytosine, cytosine...thymine, guanine...guanine and guanine...thymine pairs required the presence of four water molecules, and in the case of a guanine...cytosine pair as many as six water molecules. A comparable population of H-bonded and stacked structures for cytosine...cytosine and thymine... thymine base pairs was only obtained if at least eight water molecules hydrated the nucleobase dimer.

Methylation of bases changes the situation dramatically and the stacked structures were favoured over the H-bonded ones even in the absence of water molecules in the majority of cases (this is, of course, also partially due to the fact that the hydrogen atom, most suitable for the formation of H-bonds was replaced by the methyl group). The data supply evidence that the preferred stacked structure of DNA base pairs in a water solution might be due to the hydrophilic interaction of a small number of water molecules and not only due

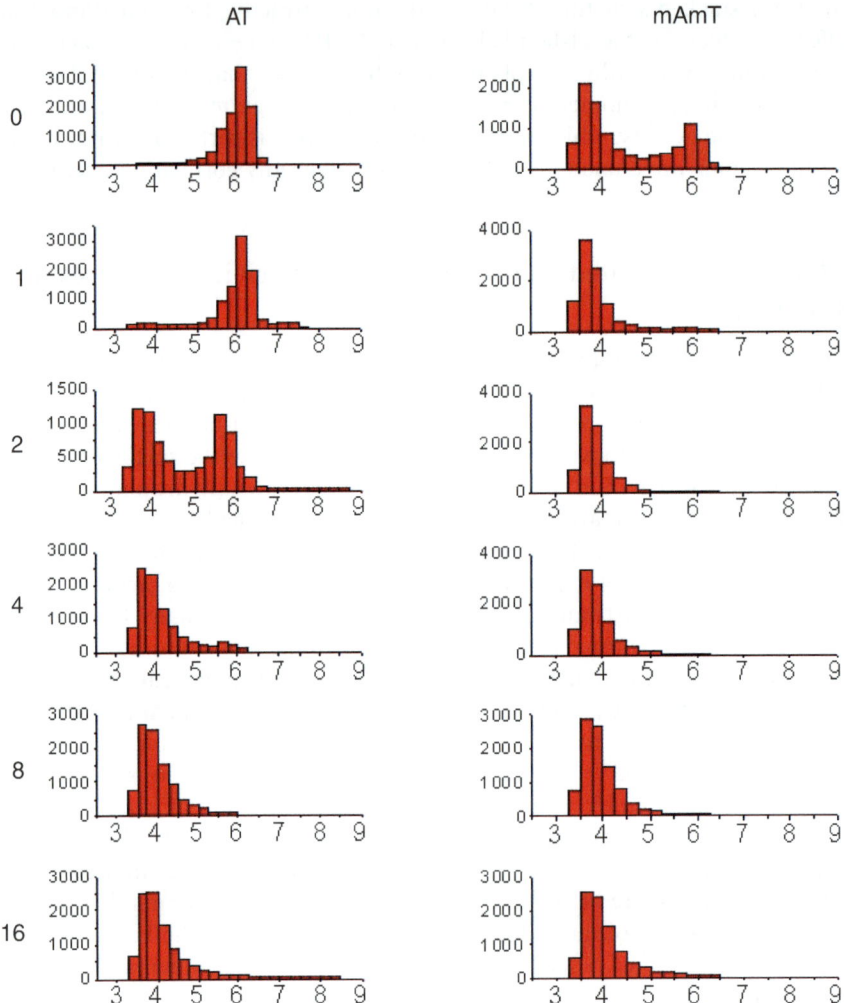

Figure 3.5 Histograms for microhydration (0 to 16 water molecules) of adenine...
thymine and 9-methyladenine...1-methylthymine base pairs. X and Y
axes describe distance (in Å) and population.

to hydrophobic effect of bulk water. The main conclusion from the study
described, however, concerns the very pronounced role of microhydration,
which is able to dramatically change the structure and thus also the properties
of the cluster studied. Mono- and dihydration of the adenine...thymine pair
was investigated[100] using MD simulations and correlated *ab initio* optimisations. The latter study fully confirmed results from the previous study[99] that
were based on structures optimised with an empirical potential.

A more detailed study of hydration of DNA base pairs as well as of
their tautomers were performed recently[101] and the main conclusions are

as follows: (i) the higher the hydration number of the base pairs, the lower the difference between the stability of HB and stacked (S) complexes. This means that S structures hydrate better than the HB ones, which is caused by the presence of a denser network of H-bonds than in the case of HB structures, as we convincingly demonstrated by using MD, MD/Q and correlated *ab initio* calculations. Already, the occurrence of two water molecules carries an advantage of 6 kcal/mol to the stability of S structures in comparison with the HB base pair, regardless of which bases are present in the base pair. This amount of energy is sufficient for the promotion of the S structures to become global minima or leads at least to a comparable stability of HB and S structures at the PES of dihydrated base pairs. (ii) Methylation of bases leads to an even stronger preference for stacked structures, which are favoured over the H-bonded ones even in the absence of water molecules in the majority of cases (this is also due to the fact that two hydrogen atoms suitable for the formation of H-bonds have been replaced by the methyl group). We can conclude that the preference for stacked structures in the DNA base pairs in the water solution might arise from the hydrophilic interaction of a small number of water molecules and not only, as expected, from the hydrophobic effect of bulk water. (iii) The strong influence of a solvent on the tautomeric equilibrium between the tautomers of bases and on the spatial arrangement of the bases in a base pair was demonstrated. The results provide clear evidence that the prevalence of either the stacked or hydrogen-bonded structures of the base pairs in the solvent is not determined only by its bulk properties, but rather by specific hydrophilic interactions of the base pair with a small number of solvent molecules.

The tautomeric equilibria of nucleic acid bases were studied in the gas phase, in a microhydrated environment and in aqueous solution.[102–105] It was shown that bulk water can change the relative stability of base tautomers significantly, for which the tautomerisation of guanine serves as an excellent example. The canonical form of guanine (the form which exists in DNA) and the 7-tautomer (hydrogen is placed at N_7 instead of at N_9) are the most stable forms in the gas phase, while the 7,9-tautomer (having hydrogens both at N_7 and N_9) is strongly destabilised (relative Gibbs energy amounts to about 20 kcal/mol). The Gibbs energy of hydration of this tautomer is extremely high (about –31 kcal/mol) and as a result makes the relative Gibbs energy of this tautomer in an aqueous environment very favourable (–11 kcal/mol). A reason for this huge stabilisation is the very large dipole moment of the 7,9-tautomer (9.1 D; the canonical form has a dipole moment of 6.3 D). A very large dipole moment should, however, also manifest itself by the large stabilisation energy of this tautomer complexed with water molecules; without water, the bare 7,9-tautomer is about 20 kcal/mol less stable than the canonical form. This very large energy difference is dramatically reduced upon complexation with one and two water molecules and becomes equal to about 13 and 9 kcal/mol, respectively. A single water or two water molecules are not able to change the tautomeric equilibrium of isolated guanine and we expect that this is also true when a higher number of water molecules is added. Let us add that similar results were also obtained for the microhydration of adenine and thymine. All these results indicate that microhydration plays a very significant role and must be properly considered.

The role of hydration in tautomeric equilibria was reinvestigated[101] at higher and more sophisticated theoretical levels and the following conclusions were made. The canonical form of cytosine, which is the global minimum in the gas phase, becomes clearly favored already in the presence of two water molecules and also has the most favourable free energies of hydration. For guanine, rare tautomers with very large dipole moments, albeit extremely disfavoured energetically in the gas phase (by about 20 kcal/mol), are stabilised by water. The hydration of adenine also reduced the difference between the stability of the canonical form (global minimum) and the first two local minima, which resulted in the coexistence of these three forms. The canonical form of thymine and uracil is undoubtedly favoured both in the gas phase and in the water environment. These trends were confirmed by calculations of the hydration free energies. We found only marginal differences between the calculated hydration energies of tautomers of bases using a broad palette of approaches covering C-PCM, MD-TI and hybrid methods. The largest discrepancy between the hydration energies was found for tautomers with high dipole moments, for which the results should be interpreted with care.

It was demonstrated that the water environment changes the structure of nucleic acid bases as well as base pairs dramatically. The role of other solvents is less clear and prompted a detailed theoretical study[106] where the dynamic structure and potential-energy surface of adenine...thymine and guanine... cytosine base pairs and their methylated analogues interacting with a small number (from 1 to 16 molecules) of organic solvents (methanol, dimethylsulfoxide and chloroform) were investigated. Various theoretical approaches, starting from the simple empirical methods employing the Cornell *et al.* force field[77] to highly accurate *ab initio* quantum-chemical calculations (MP2 and particularly CCSD(T) methods) were applied. Following the simple molecular-dynamics simulation, the molecular dynamics in combination with quenching technique was also used.

The molecular-dynamics simulations have confirmed previous experimental and theoretical results from the bulk solvents showing that whereas in chloroform the base pairs create hydrogen-bonded structures, in methanol, stacked structures are preferred. While methanol (like water) can stabilise the stacked structures of the base pairs by a higher number of hydrogen bonds than is possible in hydrogen-bonded pairs, the chloroform molecule lacks such a property, and the hydrogen-bonded structures are preferred in this solvent. The large volume of the dimethylsulfoxide molecule is an obstacle for the creation of very stable hydrogen-bonded and stacked systems and a preference for T-shaped structures, especially for complexes of methylated adenine... thymine base pairs, was observed.

These results provide clear evidence that the preference of either the stacked or hydrogen-bonded structures of the base pairs in the solvent is not determined only by bulk properties or the solvent polarity, but rather by specific interactions of the base pair with a small number of solvent molecules.

These conclusions obtained at the empirical level were verified also by high-level *ab initio* correlated calculations.

3.2.5 On the Role of Dispersion Energy on Stabilisation of DNA Double Helix

The double-helical structure of DNA, responsible for storing and transferring genetic information, is determined by a subtle balance of non-covalent interactions among DNA building blocks (see above). The most prominent role is played by interactions between DNA bases and two binding motifs can be recognised: planar hydrogen bonding and vertical stacking. In DNA, contrary to RNA, exclusively only two types of H-bonded arrangement exist, the Watson–Crick (WC) one of guanine (G) and cytosine (C), and adenine (A) and thymine (T), while there are ten different stacked arrangements of four bases. It was believed that the H-bonding is more important than stacking and the stability of DNA is thus governed by the presence of H-bonds. The role of stacking has not been known for a long time, which was partially due to fact that the absolute as well as relative stability of both binding motifs in the crystal geometry can hardly be determined experimentally. Further, theoretical description of stacking of DNA bases, where the London dispersion energy play a key role, is more involved than that of H-bonding.

The transfer of genetic information is triggered by opening (unwinding) of DNA followed by synthesis of two daughter strands. The unwinding rate of DNA is proportional to the stability of DNA and several empirical models were suggested to deduce the stability from the sequence of DNA bases.[107] It was shown that the unwinding free energy can be correlated with the H-bonded and stacked interaction energies and since the process occurs in the water environment, the solvation free energy should be also considered.[108] The main (and surprising) conclusion of the study was that H-bonding contributes less to the stability of DNA than stacking since the H-bonding (contrary to stacking) is penalised by a large desolvation energy.

There exists a completely independent way to prove the role of stacking – namely to perform geometry optimisation or run the molecular-dynamics simulations with modified energy functions. When the dispersion energy (which is responsible for stacking stabilisation) term is reduced to zero (or, in the opposite way when it is enlarged) then the structure of DNA should be affected.

The stability of DNA in the water environment was investigated[109] along two directions. First, the quantum-mechanical (QM) optimisation of small fragment of DNA in water environment (and for the sake of comparison, also in vacuum) was performed. A similar study with limited information was already published[110] before, showing that when the dispersion energy is not covered the structure of DNA is modified. These authors mention that "it is not surprising that fragments of DNA are structurally unstable...when computed with self consistent charge density fitting tight binding (SCC-DFTB)" (*i.e.* with a method that does not cover the dispersion energy). In the second step various molecular dynamics (MD) simulations explicitly based on an empirical potential in water were performed. Each type of MD simulation is performed first with standard (unmodified) energy functions and, then, with modified energy functions. In the case of dispersion energy, its weakening as well as strengthening was considered, while in the case of electrostatic energy only weakening was taken into account.

The MD simulations were carried out for the 12-mer of B-DNA with the sequence: 5'-CGCGAATTCGCG-3' given in a box containing 3373 TIP3P water molecules and 22 sodium cations to reach system electroneutrality. The molecular-dynamics simulations were performed with the AMBER empirical force field and the following modifications in the nonbonded part were introduced: The nonbonded part describing the potential energy of the system (see eqn (3.1)) is divided into electrostatic and Lennard-Jones terms; the electrostatic term is modelled by the Couloumb interaction of atomic point-charges and the Lennard-Jones term describes the exchange-repulsion and dispersion energies.

$$V(r) = \frac{q_i q_j}{4\pi\varepsilon_0 r_{ij}} + 4\varepsilon \left[\left(\frac{\sigma}{r_{ij}}\right)^{12} - \left(\frac{\sigma}{r_{ij}}\right)^6 \right] \tag{3.1}$$

Modifications of the dispersion energy were introduced by scaling the parameter ε. The following values of the ε were considered: 0.01, 0.5, 1.0, and 2.0; *i.e.* the first value almost completely removes the dispersion energy, the second yields its 50% reduction and the last one magnifies the dispersion energy by a factor 2. All modifications were performed for all atoms of DNA, keeping the original parameters for the water molecules (modification of the water parameters resulted in significant artifacts in the solvent structure).

The role of neglecting the dispersion energy was first investigated by performing the minimisation of the 5'-TATA-3' tetramer using the RI-DFT-D method. It was shown that the final structure differs only slightly from the

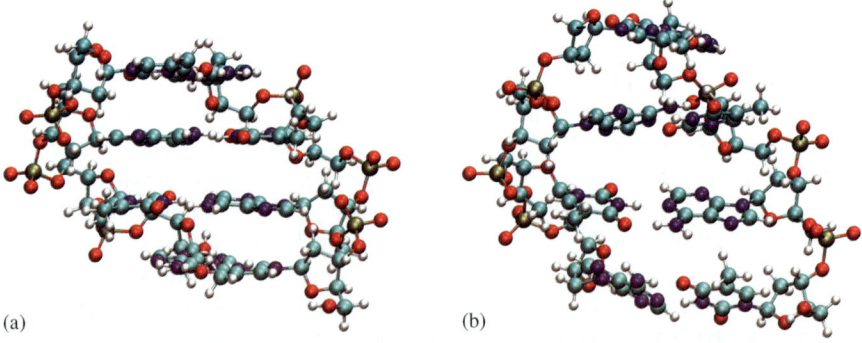

(a)   (b)

Figure 3.6   Optimised structures of the 5'-TATA-3' tetramer obtained using the RI-DFT-D method including the empirical dispersion term (a) and using the RI-DFT method (*i.e.* the dispersion energy was not taken into account) (b).

initial one (see Figure 3.6(a)). On the other hand, when optimisations were performed using the RI-DFT method (*i.e.* the dispersion energy was not taken into account) significantly distorted structure resulted (see Figure 3.6(b)). This structure shows the huge enlargement of vertical distance of bases. Further, the central bases are enormously twisted trying to create the unnatural H-bonded contact with the peripheral bases.

MD simulations performed for the 12-mer of B-DNA confirmed these results. By varying the parameters that influence the dispersion energy we modify mainly the stacking interactions, the H-bonding is affected considerably less (see later). The Figure 3.7 shows the mean distance between nitrogen atoms connected to the sugar moiety in all stacked base pairs in the DNA investigated. The blue curve corresponds to the simulation with the standard (unmodified) potential and the mean distance between nitrogens is about 4.5 Å. This distance is practically unchanged during simulation. Only between 20 and 25 ns of simulation time is a small enlargement of this distance apparent. Reducing the dispersion energy to 50% (green curve) we obtained a less-folded double-helical structure where the nitrogens distance increased by about 1 Å. Reducing the dispersion energy to almost zero (red curve) leads to a rather dramatic change of the double-helical

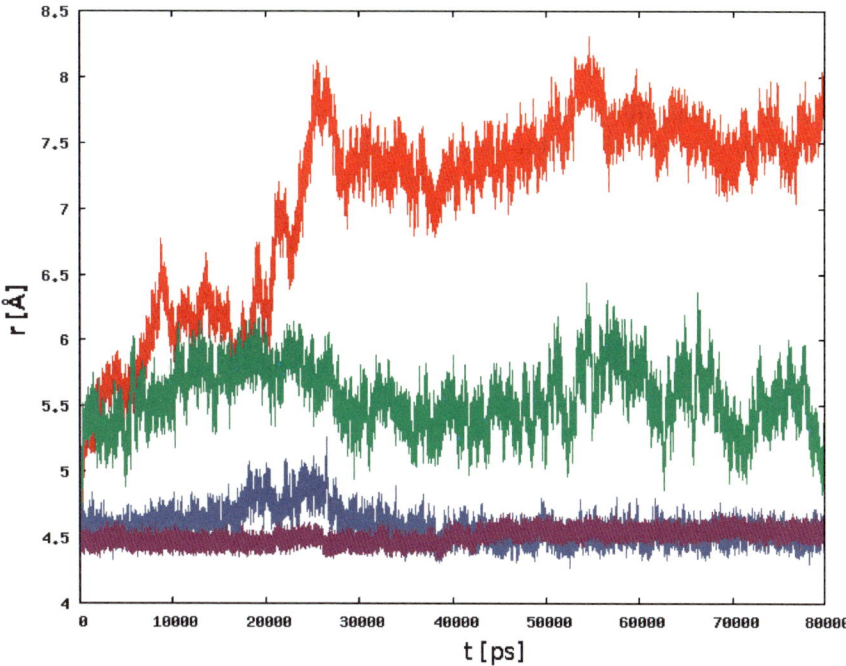

Figure 3.7 The mean distance (Å) between nitrogen atoms (connected to the sugar moiety) of consequent NA bases along the DNA strands (Y axis) against time (in ps). Results involving the ε parameter scaled by 0.01, 0.5, 1.0, and 2.0 (red, green, blue, and violet lines, respectively) are presented for the whole 80 ns force-field simulation.

structure. The folded structure is almost fully transferred to the ladder-like structure characterised by considerably larger N . . . N distance (about 7.5 Å). The final N . . . N distance of about 7.5 Å is reached within less than 20 ns (at this time of the simulation water molecules are accessing the space between the stacked bases). When, on the other hand, the strength of dispersion is increased (by a factor of 2, violet curve in Figure 3.7) the folded structure is more rigid, however, the average distance differs only negligibly from that obtained from the simulation with original parameters. Qualitatively very similar results can be obtained when considering for example the distance of the centre of mass of the respective DNA bases or the distance between the C1 carbon atoms of the deoxyribose moieties. Figure 3.8 shows the final ladder-like (unfolded) structure, which is virtually two-dimensional and, for the sake of comparison also the initial folded structure. The fundamental difference is clearly apparent.

The results obtained by MD/empirical potential simulations were fully confirmed by more sophisticated QM/MM MD simulations where four central DNA base pairs were described by the SCC-DF-TB-D method.

The electrostatic energy modifications affect mainly the planar H-bonding. When the atomic charges in eqn (3.1) are reduced to 10% (the electrostatic energy is reduced to 1%) the distances between two strands rapidly increased. This provided evidence of a complete breaking of all H-bonds, yielding two separate strands. Since stacking is present (dispersion energy is not modified) in each strand, some degree of folding (helicity) is retained.

It can be concluded that electrostatic and dispersion forces are essential for the structure and full biological functionality of DNA. The role of the electrostatic force is obvious – the lack of Coulomb attraction would lead to the separation of two strands with a partial conservation of the helical structure in each strand. The importance of the dispersion forces for the stability has been rather unclear until now. The MD simulations showed that the lack of dispersion leads quickly to the complete unwinding of the strands from the folded double-helical structure to a practically two-dimensional ladder-like arrangement. The vertical distance of DNA bases rose from 3.5 Å to about 7 Å. Such drastic geometrical changes have fundamental biological consequences – large base separation in one strand leads to a loss of replication and transcription activity, proteins like transcription factors cannot bind to specific sequences of the bases exposed in the major groove, water can easily penetrate between the bases, *etc.* Each of these above-mentioned points would surely be lethal for the cell/living organism.

3.3 Amino Acid Pairs

Building blocks of DNA are nucleic acid bases and we have shown that interaction between these blocks determine the structure and thus also the function of DNA. In the case of proteins the situation is more complicated since we have 20 amino acids, which form a side chain of proteins. Only 4 of 20 amino acids are aromatic but these 4 amino acids play a much more important

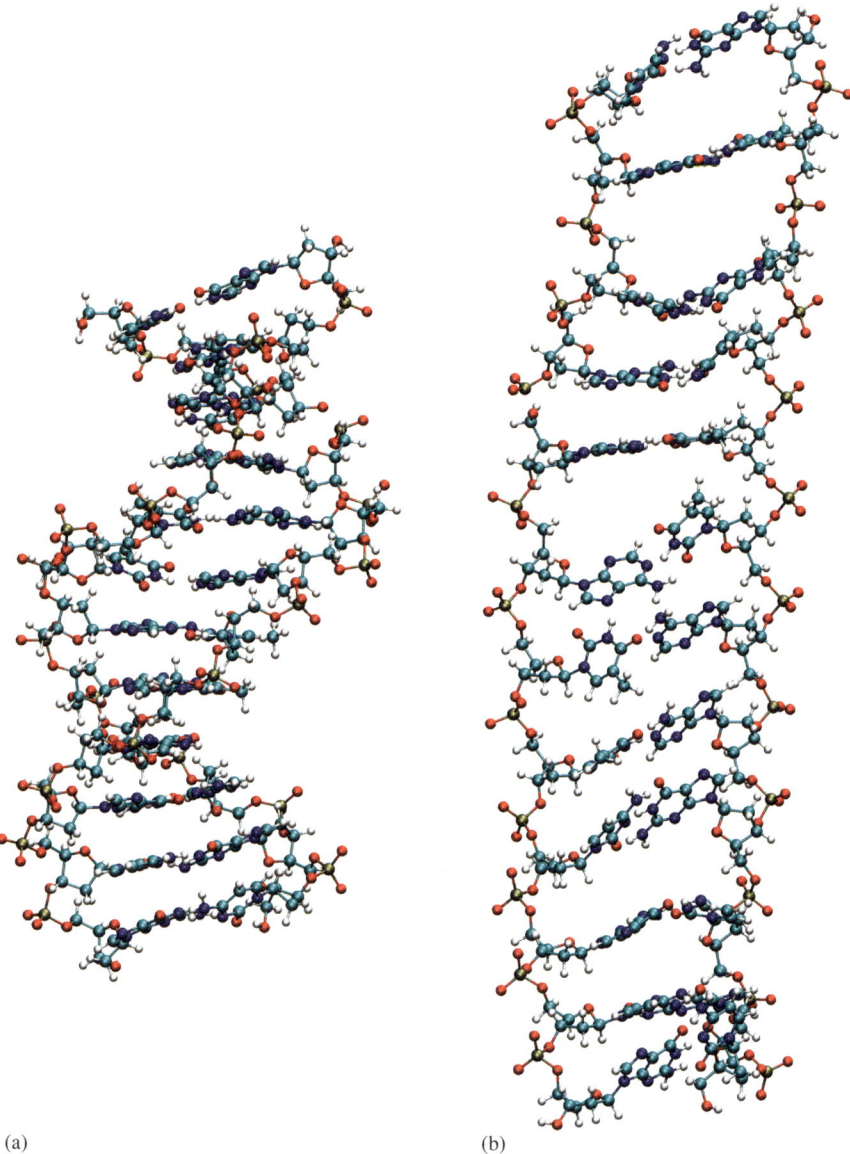

(a)   (b)

Figure 3.8 Snapshot figure of the initial crystal structure (a) and final ladder-like
structure (b) of DNA corresponding to the force field simulation involving
the ε parameter scaled by 0.01.

role than expected from their occurrence simply since their interaction energies
are much larger than those of any other pairs.

After the synthesis of a protein it should fold and this process belongs to one
of the critical steps in cell life. Protein folding involves two critical elements,

stability and specificity. The native structure of a typical protein is only 5–15 kcal/mol more stable than the unfolded state.[111]

Hence, small differences in energy between multitudes of possible non-covalent interactions are summed up to provide the properly folded structure. To gain control of the protein secondary and tertiary structure requires an understanding of how these non-covalent interactions provide both stabilisation and specificity.[112]

Every globular and water-soluble protein has a hydrophobic core. The core is an arrangement of hydrophobic residues buried in the protein interior. The formation of a hydrophobic core, which is the driving process of protein folding in terms of energy, is connected with the existence of a folding nucleus,[113,114] a conserved region of protein that initiates the folding.[115,116]

Evidence for a nucleation condensation mechanism can be found in the work of Itzhaki *et al.*,[117] which can be taken as one of the most important works in the field. Core formation is believed to be the consequence of exterior hydrophobic forces of entropic nature,[118,119] an example of the classical hydrophobic effect[120] characterised by a small contribution (repulsive or attractive) of complexation enthalpy. This, together with the low occurrence of hydrogen bonds in the protein core, leads to the assumption that the energy (enthalpy) contribution of the core formation to protein folding is small or negligible. Theoretical and experimental investigations of various types of non-covalent interactions have shown[121] that a rather large attraction could be gained not only from hydrogen bonding but also from other types of non-covalent interactions. Thus, the question arises of how strong are the stabilising contributions of amino acids in a hydrophobic core. This question is of key importance for understanding the mechanism of protein folding as well as understanding protein secondary and tertiary structure.

The stabilisation energy of a model hydrophobic core (see Figure 3.9), based on a high-resolution X-ray structure of rubredoxin, a small soluble FeS protein

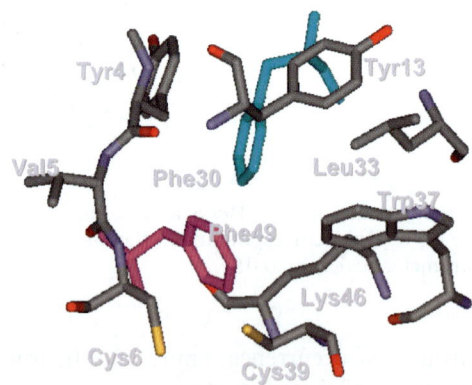

Figure 3.9   Hydrophobic core of rubredoxin with central phenylalanines Phe30 and Phe49.

(PDB code 1RB9), was investigated in detail in Ref. 122. The stabilisation energy of the core was determined using high-level correlated *ab initio* calculations, specifically, as the sum of the complete basis-set (CBS) limit of the MP2 stabilisation energy and the CCSD(T) correction term. The whole cluster was partitioned into two distinct clusters (named after the central residues, F30 and F49) and was further fragmented into well-defined, chemically distinct pairs of neutral amino acids (modelled as methylated aminoacid residues). The central F30 and F49 phenylalanines thus interact with five (F49, K46, L33, Y13, and Y4) and seven (C39, C6, F30, K46, V5, W37, and Y4) amino acids, respectively. There is one H-bond ascribed to the F30 cluster (a classical CO . . . HN H-bond in the F30 . . . L33 pair) and another two H-bonds are ascribed to the F49 cluster (a classical CO . . . HN H-bond in the F49 . . . K46 pair, as well as an unusual CH . . . π interaction between the methyl group of the capped O terminus of V5 and the π system of the phenylalanine in the F49 . . . V5 pair; see Figure 3.9).

The stabilisation energies for all pairs of amino acids in clusters F30 and F49 determined at the RIMP2 level using a complete atomic orbital basis-set limit and, for a few selected pairs, also at the CCSD(T) level, are presented in Tables 3.8 and 3.9. Inspecting these results we found that interaction energies of all 12 pairs are stabilising and the stabilisation energy in 6 pairs was higher than 4.5 kcal/mol. Especially important were F30 . . . Y4 and F49 . . . V5 interactions reaching stabilisations of 7 kcal/mol. Table 3.8 further shows that time-consuming CCSD(T) correction terms are (contrary to, *e.g.* stacking of DNA bases) rather small here. It is worth mentioning the fact that when the DFT/B3LYP procedure was used all interaction energies were repulsive: this clearly indicates that the classic DFT method is useless for describing interaction of amino acids.

The stabilisation energy inside the hydrophobic core of rubredoxin is very high and amounted to about 50 kcal/mol. These results clearly demonstrate the strong attraction inside a hydrophobic core and this finding may lead to changes in the current view of protein folding.

Table 3.8 Pair interaction energies (in kcal/mol) of the selected residues clustered around Phe30. a-pVDZ and a-pVTZ mean aug-cc-pVDZ and aug-cc-pVTZ basis sets, respectively.

	MP2			$\Delta CCSD(T)^a$	$CCSD(T)/CBS$
Residue	a-pVDZ	a-pVTZ	CBS	6-31G*(0.25)	
Phe49	−3.1	−3.3	−3.3	−/0.6	
Lys46	−3.1	−3.3	−3.4	0.3/0.2	−3.10
Leu33	−4.9	−5.3	−5.5	0.5/0.2	−5.00
Tyr13	−4.2	−4.4	−4.5	0.6/0.4	−3.90
Tyr4	−6.5	−6.8	−7.0	−/1.7	
sum	−21.8	−23.2	−23.7		

[a]The first number is the correction for wholly modelled residue; the second number is the correction for the side chain only (side chain modelled from C_α atom).

Table 3.9 Pair interaction energies (in kcal/mol) of the selected residues clustered around Phe49.

| residue | MP2 | | |
	aug-cc-pVDZ	aug-cc-pVTZ	CBS
Cys39	–1.7	–2.0	–2.1
Cys6	–4.4	–4.8	–5.0
Phe30	–3.1	–3.3	–3.3
Lys46	–4.0	–4.6	–4.8
Val5	–5.6	–6.4	–6.7
Trp37	–2.3	–2.4	–2.5
Tyr4	–2.7	–3.0	–3.1
Sum	–23.8	–26.5	–27.5

The role of the stabilisation energy inside a hydrophobic core was further investigated in Ref. 123. The reason is that Rubredoxin from the hyperthermophile Pyrococcus furiosus (Pf Rd) is an extremely thermostable protein (its melting temperature approaches 200 °C), which makes it an attractive subject of protein folding and stability studies. A fundamental question arises as to what the reason for such extreme stability is and how it can be elucidated from a complex set of interatomic interactions. We addressed this issue first theoretically through a computational analysis of the hydrophobic core of the protein and its mutants, including the interactions taking place inside the core. In this study the DFT-D method, which closely mimics the highly accurate CCSD(T)/CBS results, was utilised. Besides the interaction energy we also determined the change of the Gibbs energy accompanying the mutation. It was shown that a single mutation of one of phenylalanine's residues inside the protein's hydrophobic core (phenylalanine was mutated to alanine or glycine) results in a dramatic decrease in its thermal stability. The calculated unfolding Gibbs energy as well as the stabilisation energy differences between a few core residues follows the same trend as the melting temperature of protein variants determined experimentally by microcalorimetry measurements. NMR spectroscopy experiments have shown that the only part of the protein affected by mutation is the rearranged hydrophobic core. It is hence concluded that stabilisation energies, which are dominated by London dispersion, represent the main source of stability of this protein.

3.3.1 On the Role of Dispersion and Electrostatic Energy on Stabilisation and Folding of Proteins

In Section 3.2.5 we discussed the role of dispersion and electrostatic energy contributions for the stabilisation of DNA. It was demonstrated that the loss of the dispersion-energy term induces the transition from a three-dimensional double-helical structure to the quasi-two-dimensional ladder-like structure.[109]

What is the situation in the case of proteins? How important is the dispersion and electrostatic term for a reasonable description of protein structure and stability or even for protein folding? Such a question is fully justified considering the common belief of the biochemical community that the dispersion interaction between amino acids of the hydrophobic core or between a protein and a nonpolar ligand is roughly of the same magnitude as for interactions of these molecules with water and thus cannot be the driving force of the processes. It is known[122,124] that the role of dispersion is significant for the stabilisation of the hydrophobic core or for protein–ligand complexes. However, this description lacks the important features of the process – its dynamics and the presence of the environment.

The role of two different types of interactions – dispersion and electrostatic or H-bonding – on the protein stability and structure in the water environment was investigated (similarly as previously in the case of DNA) by two different methods.[125] The MD simulations were performed with standard empirical energy functions and, subsequently, with modified empirical energy functions (weakening and strengthening of dispersion energy and weakening of the electrostatic energy; see eqn (3.1)). In the second step, the MD simulations were performed with the more reliable QM/MM potential where the QM part was described by the self-consistent charges density-fitting tight binding method with or without an empirical dispersion-energy term[126] (SCC-DF-TB-D) and the MM part was described by a standard empirical potential. Calculations were performed for the Trp-cage protein (Figure 3.10), which is an artificial small protein designed by utilisation of a known protein scaffold and its role for theoretical as well as experimental purposes is enormous. We have shown

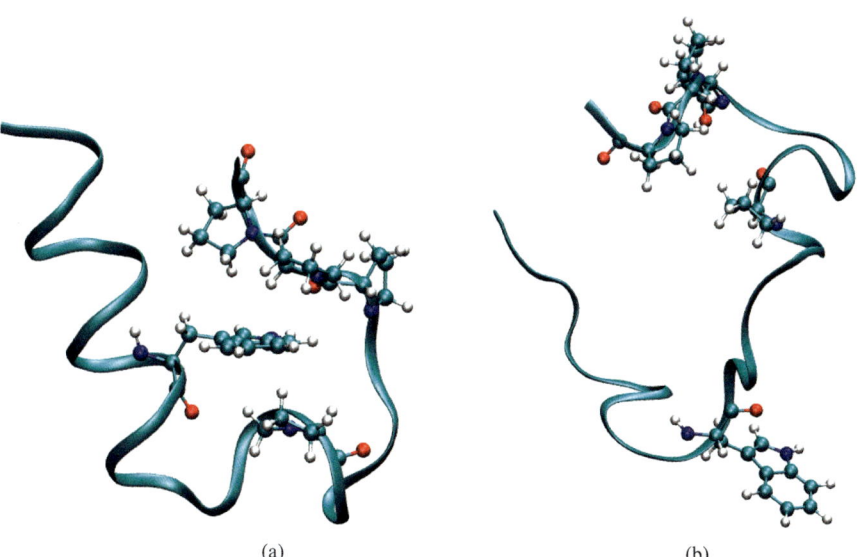

(a)   (b)

Figure 3.10   The initial structure (a) of the protein and the structure obtained by performing the 100-ns MD simulation with the ε scaled by 0.01 (b).

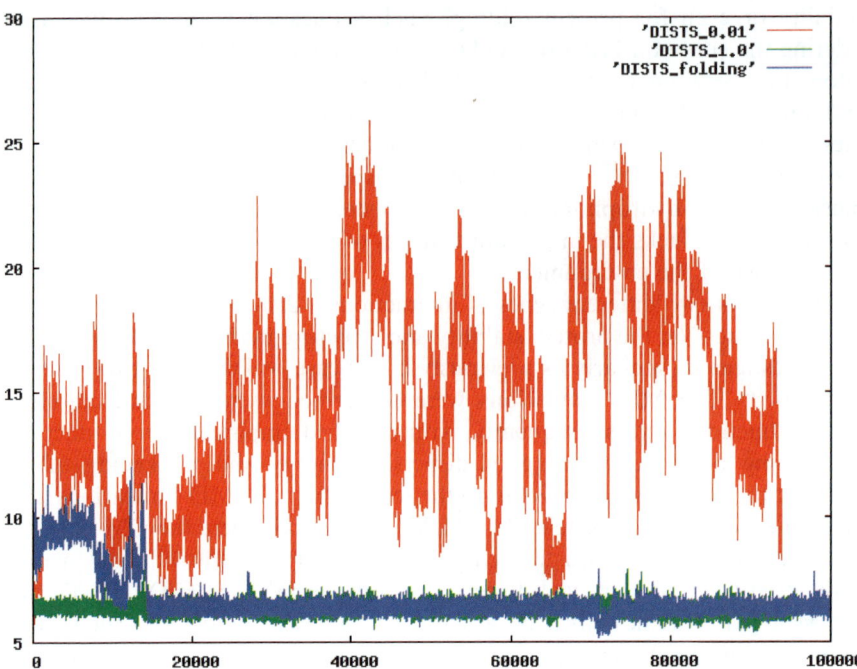

Figure 3.11   The mean distance of the centres of mass of Trp and Pro in Å (Y axis)
against time (in ps). The green curve corresponds to the simulations with
standard (unmodified) potential and the mean distance is about 6 Å. A
reduction of the dispersion energy almost to zero, ε scaled by 0.01 (red
curve), leads to a dramatic change of the protein structure characterised
by considerably larger mean distances between Trp and Pro (see also
Figure 3.10(b)). When, however, the dispersion energy is fully recovered,
the unfolded structure (found after a 100-ns simulation without dispersion energy) is reversibly folded (blue curve).

recently[127] the substantial stabilisation between tryptophane and prolines in
this protein as well as the key role played by dispersion energy.

Figure 3.11 shows the mean distance of the centres of mass for Tryptophan
(Trp) and Proline (Pro) residues in the structure of Trp cage. The green curve
corresponds to the simulations with standard (unmodified) potential and the
mean distance is about 6 Å. Neglect of the dispersion energy (red curve) leads to
a dramatic change of the Trp-cage structure characterised by considerably
larger mean distances between Trp and Pro. Evidently, the binding between
Trp and Pro amino acids vanished and the Trp residue moved out from the
protein interior and it is fully exposed to the solvent. Figure 3.10(b) shows the
final structure. Besides the changed position of Trp it also demonstrates that
reduction of the dispersion energy also caused damage of the α helix.

Loss of the dispersion energy leads to damage of the structure so the question emerges what is going to happen when the dispersion-energy term is fully
recovered. The blue curve in Figure 3.11 shows the trajectory for this case. We

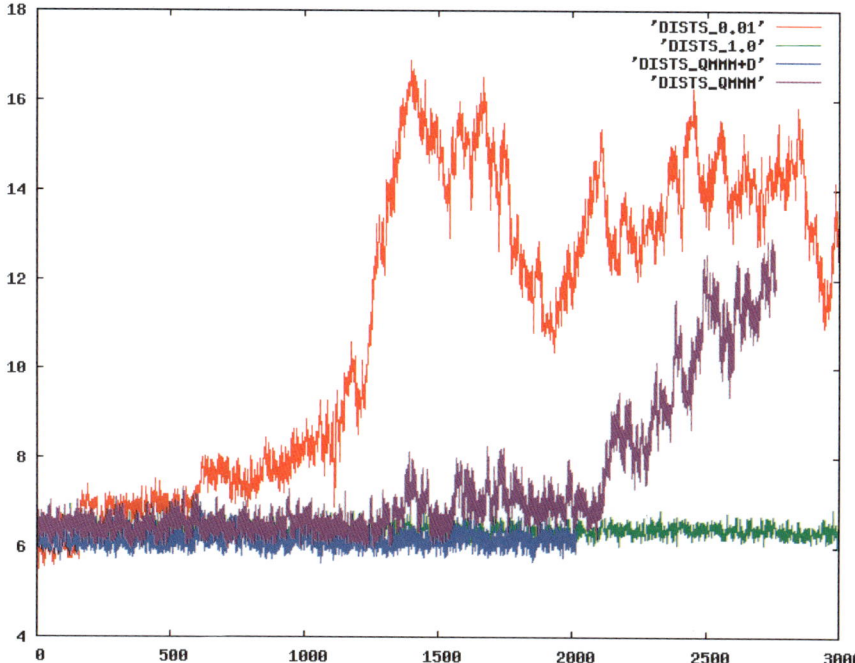

Figure 3.12 The mean Pro Trp distance in Å (Y axis) against time (in ps) for an unmodified MD simulation (green line) and for a MD simulation with ε scaled by 0.01 (red line), compared to the results of the QM/MM MD simulations. The blue line represents the simulation where the SCC-DFTB-D method was used as the QM part, while the violet line corresponds to the SCC-DFTB (without the empirical dispersion correction term).

started from the damaged structure and then the 100-ns MD simulation was performed. Evidently, after a short time (≈ 10 ns) the structure is fully recovered to the native state and the blue line fully coincides with the green one.

The blue and violet curves in Figure 3.12 correspond to QM/MM MD simulations where the dispersion energy is fully covered (DFT augmented by dispersion term) or completely missing. Results obtained by force-field methods (green and red curves) are shown for comparison. Evidently, the structural change in the case of QM/MM simulations occurs later than in MM simulations (2 and 1 ns, respectively) and this can be due to a different model of dispersion-energy scaling.

The fact that the QM/MM model fully agreed with the simpler MM one is significant. The reduction of the dispersion energy in the latter model is accompanied by a reduction of the repulsion energy and the overall interaction energy is thus described only by the charge–charge electrostatic term. In the QM/MM model this is not true and it is only the dispersion energy that is either covered or deleted. When the dispersion energy is not covered the remaining SCC-DF-TB interaction energy determined quantum mechanically fully covers

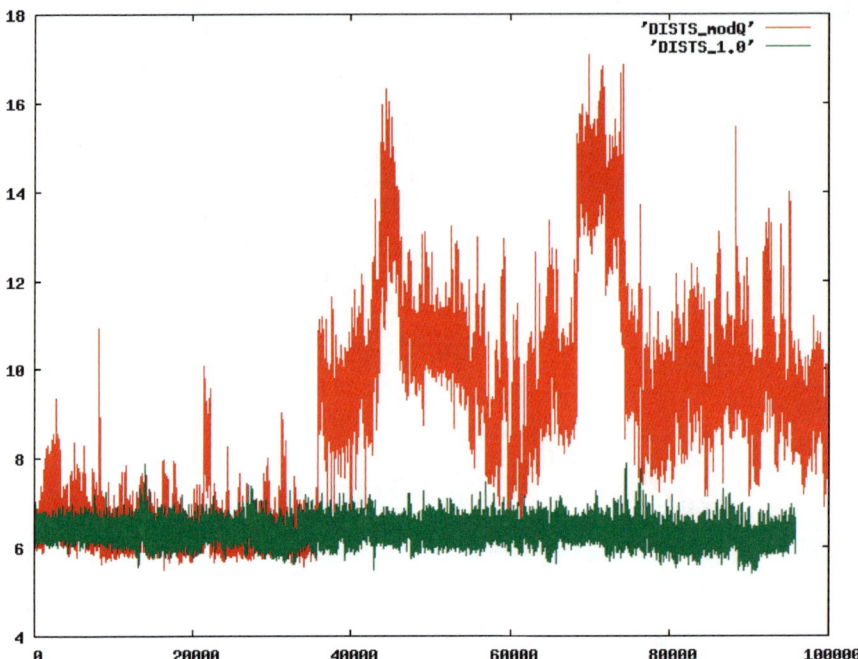

Figure 3.13   The mean Pro Trp distance in Å (Y axis) against time (in ps) for a
unmodified MD simulation (green line) and for a simulation with a
modified electrostatic term, where the atomic point charges of the protein
atoms were scaled by 0.1 (red line).

all other interaction energy terms. The agreement between QM/MM and MM
simulations thus indicates that the deep changes in the structure of the protein
can be assigned solely to the missing dispersion energy and not to the physically
oversimplified energy term in the MM approach.

The reduction of the atomic charges decreases the electrostatic energy and
correspondingly the H-bonding. The green curve in Figure 3.13 indicates the
mean distance between Trp and Pro systems obtained from simulations with
the standard (unmodified) empirical potential. The distance around 6 Å
remains stable in the 100-ns simulations. When the atomic charges are reduced
to 10%, *i.e.* the electrostatic interaction is reduced to 1% (red curve), the mean
distance between the Trp and Pro systems oscillates heavily between folded and
completely random structures. Figure 3.14 shows snapshots of Trp-cage
structures obtained by the simulations after 20, 40 and 60 ns, respectively.
Figure 3.14 also shows that the α-helical structure, which is supposed to be held
by H-bonds, is surprisingly retained at a noticeable level of stability. It seems
that the important part of α helix stabilisation can also be assigned to the
dispersion energy, which is fully covered in these simulations.

Let us remind ourselves that a different picture resulted upon analogous
reduction of the electrostatic energy in the case of DNA[109] when the DNA
double-helical structure was denatured to two separate strands. This was,

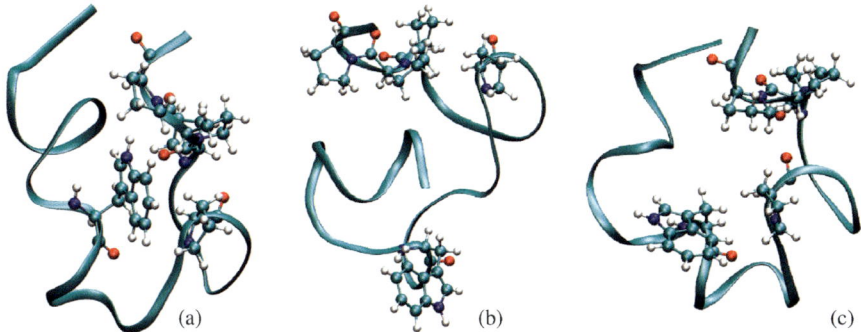

Figure 3.14 The snapshots of protein structures obtained from the simulations with modified electrostatic term (atomic point charges scaled by 0.1) after about 20 (a), 40 (b) and 60 ns (c), respectively.

however, not surprising since H-bonds between DNA bases are much stronger than these in the present α helix. On the other hand, helicity within single DNA strands was retained to a high degree, which supports the above-mentioned conclusion that the structure of the α helix is determined not only by H-bonding but also by dispersion forces.

The situation in proteins is different and apparently more complex. The protein structure deteriorates completely upon reduction of either electrostatic or dispersion energy and does not keep a trace of the former structural arrangement. This is evidently due to the fact that both H-bonding as well as stacking is stronger in DNA than in proteins.

In summary, we have shown the decisive role of dispersion energy for the structure of the Trp-cage protein. Without inclusion of dispersion, the folded structure of the protein was deteriorated neither in the empirical potential function nor in the QM treatment and stayed in its non-native state. When the dispersion-energy term is fully recovered the non-native structure was established in a rather short time. Surprisingly, the absence of dispersion energy also affected the Trp-cage α helix, which seems to indicate that this structure is held not only by H-bonding but also by dispersion forces. The role of the electrostatic forces (and thus H-bonding) was surprisingly less apparent. When electrostatic interactions are virtually eliminated the Trp–Pro distance oscillates between unstructured and native states and the α-helical motif was retained to a surprisingly high degree. All these findings indicate that dispersion energy plays an important role not only in the determination of structure but can also be a crucial driving force during the folding process of the Trp-cage protein.

3.4 Carboxylic Acid Dimers

Formic acid dimers,[128] acetic acid dimers[129] and glycine dimers[130] were studied using the MD/Q simulations in the same way as the above-mentioned nucleic acid–base pairs. The overall situation is similar in both cases and again, the global minimum of the PES sometimes differs from that of the FES. This is, in fact, a surprising result since the cyclic structures of the carboxylic acid dimers possessing two strong C=O...H–O hydrogen bonds (see Figure 3.15) rank among the most stable complexes. The RI-MP2/TZVPP stabilisation energies of formic acid and acetic acid dimers are 14.4 and 14.7 kcal/mol, respectively, while the stabilisation energy of the cyclic structure of the glycine dimer is slightly larger. The other structures of the first two acids are significantly less stable. At low temperatures the cyclic structure of the formic acid dimer remains the global minimum. On increasing the temperature of the system, the population of the other cyclic structure (II on the left of

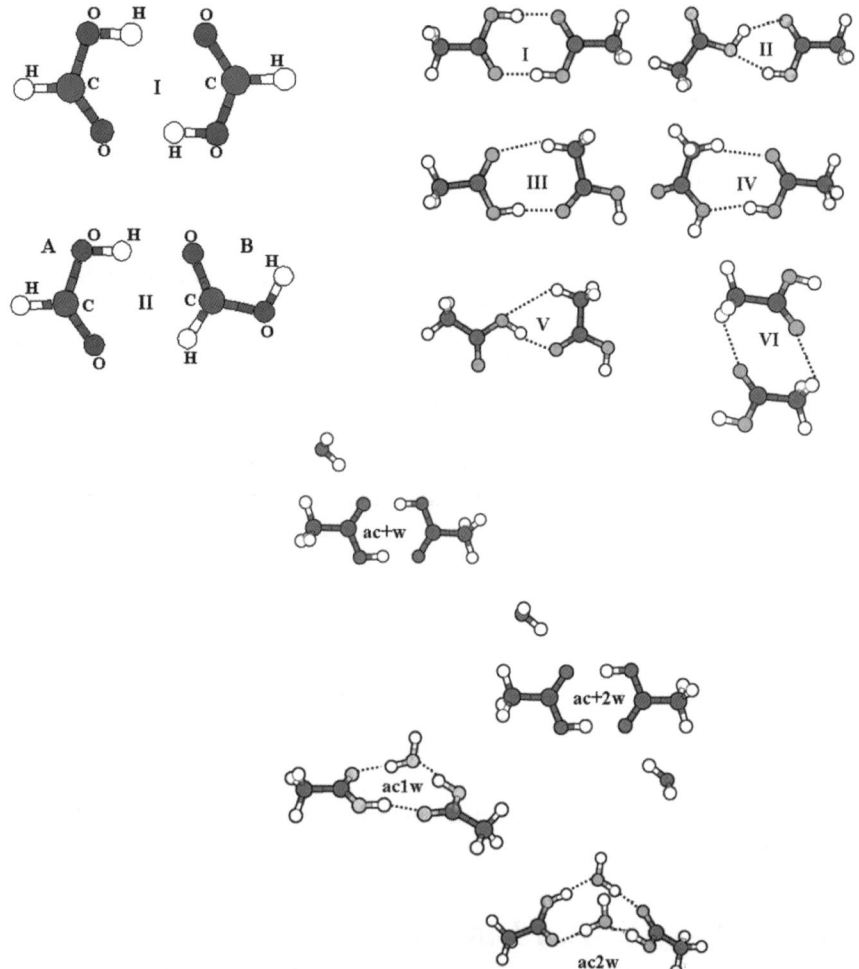

Figure 3.15   Structures of carboxylic acid dimers and microhydrated carboxylic acid dimers. Reproduced from Refs. 128 and 129 with permission.

Figure 3.15) having one strong $C=O\ldots H-O$ and one weak $C=O\ldots H-C$ hydrogen bond becomes comparable; at higher temperatures (300 K) this structure even becomes the global minimum, as entropy considerations lead to stabilisation of this structure. Evidently, entropy disfavours the strong (rather rigid) cyclic structure with two $C=O\ldots H-O$ hydrogen bonds and favours the weaker, more floppy cyclic structure having one strong $C=O\ldots H$ O and one weak $C=O\ldots H-C$ hydrogen bond. This theoretically predicted behaviour was fully confirmed by experiments of Sander *et al.*[131] who measured IR spectra of the formic acid dimer in an argon matrix (though it must be mentioned that gas-phase experiments did not detect any evidence of this structure[132]). When performing their experiment at low temperature, the authors detected only the cyclic structure I (see Figure 3.15) with two $C=O\ldots H-O$ hydrogen bonds. Upon increasing the temperature, the population of the second cyclic dimer II (see Figure 3.15) having one strong $C=O\ldots H-O$ and one weak $C=O\ldots H-C$ H-bond-becomes dominant. Surprisingly, experiments using helium droplets[133] confirmed only the existence of the less-stable cyclic structure. Since this cannot be due to entropy effects (the temperature in He-droplets is extremely low), further explanation of this phenomenon is required (see Section 5.11).

In the case of acetic acid dimer the cyclic structure I, with two $C=O\ldots O-H$ bonds, remains the dominantly populated structure even when the temperature increases. Microhydration of the dimer (studied by MD/Q methods), however, shows changes in the dimer structure with the addition of only a single water moiety, resulting in breaking one of the $O-H\ldots O$ bonds and allowing the water molecule to be incorporated into the structure of the dimer (see Figure 3.15). The second water molecule behaves in a similar way and usually breaks the other $O-H\ldots O$ bond, and similarly is also incorporated into the structure (see Figure 3.15). On addition of further water molecules, different water-separated complexes of the acetic acid dimer are formed; the most frequently appearing complexes correspond to those incorporating one and two water molecules.

Experiments on the glycine dimer performed in He-droplets (see above) gave unambiguous evidence of the presence of a free (unbound) O–H group. Evidently, the dimer cannot have an expected cyclic structure with two $C=O\ldots H-O$ H-bonds (C1 structure in Figure 3.16) since the redshift of the O–H stretching vibration in the dimer is large. The only way of interpreting the experiment is to admit that this cyclic structure is not present and another dimer structure having an unbound O–H group is dominantly populated. We performed the MD/Q simulations[130] using the Cornell *et al.* potential[77] and found several hundred structures, of which 22 with the lowest energy were investigated in detail. To our surprise, the empirical potential favoured the stacked structure (S1 in Figure 3.16) of the dimer over the cyclic C1 structure by about 3 kcal/mol. Both C1 and S1 structures were reoptimised at the *ab initio* correlated level (MP2/6-31G**) for which the results followed their expectations: The C1 structure was found to be the global minimum, whilst the S1 structure was higher in energy by about 2 kcal/mol. On the basis of our experience with stacking structures, however, we realised that the

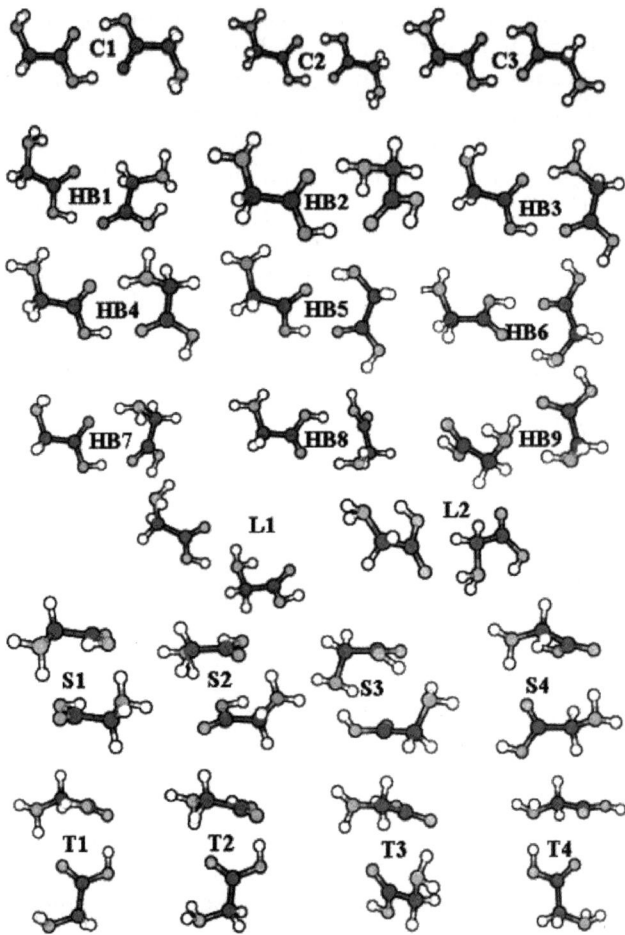

Figure 3.16   Structures of glycine dimer. Reproduced from Ref. 130 with permission.

MP2/6-31G** description is not adequate (it is, on the other hand, quite sufficient for planar H-bonding). Upon increasing the theoretical level (enlarging
the AO basis set and increasing the level of electron correlation), we came to the
conclusion that the stabilisation energy of both structures is comparable, and
from the extrapolation results obtained it may even be expected that the
stacked structure becomes more stable. However, this surprising result did not
solve the experimental finding since the stacked structure investigated also
exhibits a rather large redshift of the O–H stretching frequencies. Evidently,
another explanation, as for the case of experiments on formic acid in He-
droplets, is required. Further, also higher-level theoretical calculations
including the on-the-fly *ab initio* MD simulations and highly accurate CCSD(T)
calculations should be performed.

3.5 Peptides

The formation of protein secondary and tertiary structures depends on the chemical and physical constraints imposed by the individual properties of the protein-building blocks; that is, the 20 DNA-coded amino acids and their sequence in the polypeptide chain. Prediction of side-chain conformations plays an essential role in the modelling of protein structure. A number of prevailing approaches utilise rotamer libraries. The basic idea of this concept, which simplifies the prediction problem, is the grouping of side-chain positions in a relatively small number of statistically likely orientations called rotamers. These rotamers generally correspond to the minimum-energy positions expected for tetrahedral or trigonal carbon atoms.[134]

Tripeptides can represent efficient building blocks for protein-structure prediction. A tripeptide already constitutes a model containing all important forces resulting in distinct conformer states of the participating amino acids. Generally, a tripeptide contains all necessary factors that influence the behaviour of the rotamer, maintain the interaction of side-chain with the main chain, and take into account the φ and χ preferences of allowed regions in protein structures. Recent findings show that there is relative structural rigidity between C_α and C_β atoms in some tripeptides.[135] The carrier of such structural rigidity is often a hydrophobic residue; however, there are also tripeptides with polar chains that form rigid structures.

Determination of the peptide structure is an extremely difficult task since here not only covalent but also non-covalent interactions play a role. Evaluation of non-covalent interactions in such extended systems like peptides is difficult and the main reason is that peptides are very flexible systems showing an extensive conformational landscape. The ideal solution represents non-empirical *ab initio* on-the-fly MD simulations but this procedure is still far from being practical. The philosophy nowadays is to use a MD simulation based on a lower-level potential to generate the conformational landscape and next, select only a few of the most stable conformers for a more accurate quantum-chemical treatment. The application of QM methods is, however, also not unambiguous. A main problem is connected with the basis-set superposition error (see also Section 2.2.5). The BSSE plays an important role with molecular clusters and when working with small or medium basis sets its effect should be eliminated. This is an easy task and function counterpoise procedure provides a satisfactory solution. It was long believed that the basis-set extension effect was characteristic for molecular clusters only and did not affect isolated systems. Only recently has it been pointed out that the same effect, *i.e.* improving the basis set of one part of a system by orbitals localised on the atoms of the other part, affects the calculation of the relative energies of isolated systems, especially in very flexible systems like peptides.[136–138]

In analogy with the BSSE in clusters, this effect is referred to as intramolecular BSSE. What intermolecular BSSE has in common with intramolecular BSSE is that it is also repulsive and unless it is eliminated the stabilisation energy is artificially too high. The definition of intramolecular BSSE is, however, less straightforward and the use of the counterpoise procedure is impractical for isolated systems. The only possibility of eliminating the intramolecular BSSE is to use procedures that are BSSE-free or at least, which have only small values of BSSE. The use of the SAPT procedure that provides the BSSE-free interaction energies is impractical here simply because we do not have two interacting systems. The problem is even more involved since we must use the optimisation technique and, in fact, up to now only two possibilities exist. Either to use a correlated method with an extended basis set, which is close to the CBS limit (and thus not requiring the elimination of the BSSE), or to use a suitable DFT procedure covering the London dispersion energy, which is characterise by a negligible BSSE.

Tripeptides containing at least one aromatic ring are of special interest, because of a possible strong interaction between delocalised π-electrons of the aromatic ring with peptide bonds. It has been recently shown[122,139] that attraction between phenylalanine and a peptide bond modelled by N-methyl formamide is surprisingly large, ranging up to about 10 kcal/mol. A proper description of this attraction originating in London dispersion forces requires the performance of high-level correlated *ab initio* methods. Popular DFT methods are of limited use, since these methods do not cover the dispersion attraction.[122] The smallest tripeptide for which the computational treatment can be performed at the highest theoretical level allowing comparison with experiment is the phenylalanyl-glycyl-glycine (FGG). At the same time, FGG is already a very complex molecule, because of its very extensive conformational landscape. In fact, this molecule can serve as a testing ground for computations of a system of this magnitude. In addition to their biological importance as building blocks of peptides and proteins, these types of molecules are also of interest from a purely chemical point of view, because they form typical multiconformer systems with numerous local minimum structures associated with different conformational arrangements of the backbone and the side chain.

The study of the FGG tripeptide exemplifies some of the methods used.[140] The potential-energy surface of the tripeptide was studied in the gas phase by means of IR/UV double-resonance spectroscopy, and quantum-chemical and statistical-thermodynamic calculations. Experimentally, only four conformational structures were detected and their IR spectra in the spectral region of 3000–4000 cm^{-1} were recorded. The experimental information is clearly limited and without close cooperation with high-level theoretical studies it would be not possible to identify these structures. The PES was studied by a combination of molecular-dynamics/quenching procedures with correlated *ab initio* calculations. The PES is very complex and the quenching procedure localised more than 1000 energy minima. Finally, structures and relative energies of about 15 lowest-energy isomers were determined on the basis of various level calculations up to CCSD(T)/CBS ones. Probably not surprisingly, neither empirical potentials nor various DFT functionals provide satisfactory results. On the other hand, the approximate DFT method covering the dispersion energy yields a reliable set of the most stable structures, which were subsequently investigated with an accurate, correlated *ab initio* treatment. Figure 3.17 shows

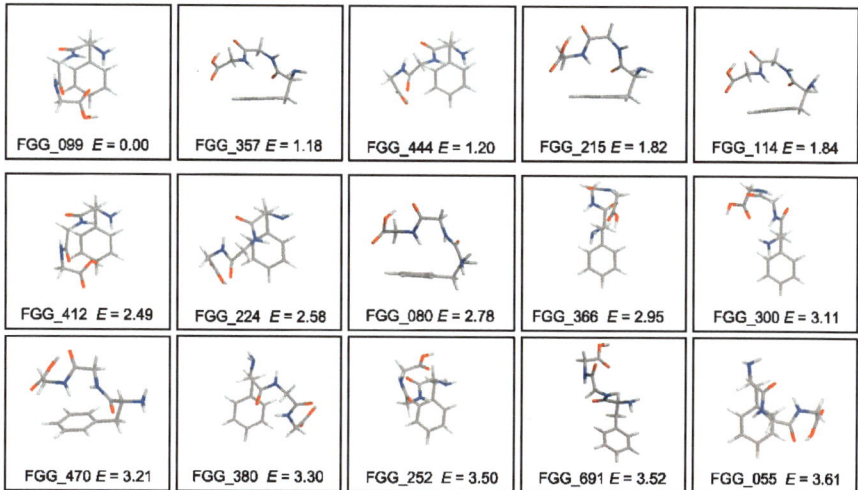

Figure 3.17 MP2/cc-pVTZ geometries and relative energies (in kcal/mol) for the 15 most stable structures of the FGG tripeptide

MP2/cc-pVTZ geometries and relative energies (in kcal/mol) for the 15 most stable structures. The global minimum corresponds to a folded structure and is stabilised by three moderately strong intramolecular hydrogen bonds as well as by London dispersion forces between the phenyl ring, the carboxylic acid group, and various peptide bonds. A proper description of the last type of interaction is difficult and requires accurate correlated *ab initio* calculations.

Since in the beam experiments the conformations are frozen out by cooling from a higher temperature, it is necessary to localise the most stable structures on the free-energy surface rather than on the PES. Two different procedures (rigid rotor/harmonic oscillator/ideal gas approximation based on *ab initio* characteristics and evaluation of relative populations from the molecular dynamic simulations using the AMBER potential) were applied and both yield four structures, the global minimum and three local minima. These four structures were among the 15 most energetically stable structures obtained from accurate *ab initio* optimisation. The calculated IR spectra for these four structures agree well with the experimental frequencies, which validates the localisation procedure.

Evidently, the proper evaluation of the FES is essential and can strongly affect the selection of the most stable structures. Both procedures discussed above are to some extent questionable, the former one due to the harmonic approximation and the latter due to the application of an empirical force field. There exists another possibility how to determine thermodynamic character-istics, namely to scan the FES using metadynamics, a free-energy modelling technique effectively covering anharmonic effects.[141] The method is, in its standard form, combined with an empirical force field but in the recent study[142] on glycyl-phenyalanyl-alanine tripetide it was combined with tight-binding

DFT-D. An important advantage is the fact that tight-binding DFT-D is used systematically in evaluating both the PES and FES. The FES obtained by means of the combination of MD/Q simulations using the tight-binding DFT-D method with high-level correlated *ab initio* quantum-chemical calculations followed by statistical-thermodynamics RR-HO-IG calculations was confirmed by the FES obtained independently with the metadynamics calculations based on the tight-binding DFT-D method. It is needless to say that this proves that both methodologies are suitable for the study of isolated small peptides.

Five peptides (WG, WGG, FGG, GGF and GFA) containing the residues phenylalanyl (F), glycyl (G), tryptophyl (W) and alanyl (A), where F and W are of aromatic character, were carefully investigated and several important conclusions valid for other peptides were presented.[143] It must first be mentioned again that when investigating isolated small peptides containing an aromatic ring, the dispersion interaction is the dominant attractive force in the peptide backbone–aromatic side chain intramolecular interaction and it plays an important role in determining the structure. Consequently, an accurate theoretical study of these systems requires the use of a methodology covering properly the London dispersion forces. Probably the most important step represents a localisation of energy minima at the PES. Since the empirical force field is not accurate enough (see later) the first screening should be performed using a more advanced technique, in this particular case the SCC-DF-TB-D molecular dynamics combined with quench was utilised. The recomended strategy for localisation of few energy minima from the first array of about 1000 structure is visualised in Figure 3.18.

The MD/Q calculations reveal a large number of energy minima (about 1000) and to further calculations only a small portion of structures (below some energy threshold) are considered. For these systems the QM gradient optimisations are performed. At the very end the high-level gradient optimisation followed by frequency and thermodynamic property calculations are performed. The final step represents the CCSD(T)/CBS calcutions of relative energies.

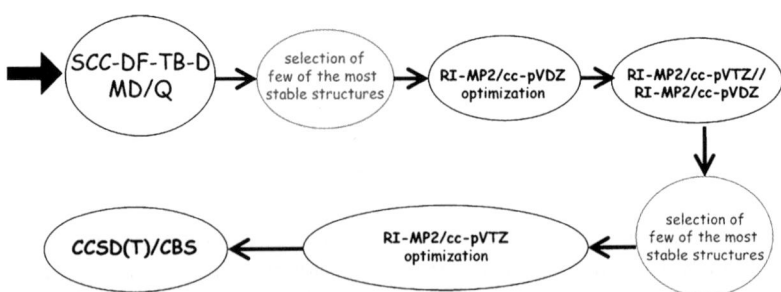

Figure 3.18 Strategy of calculation for localisation of lowest energy minima at the potential energy surface of oligopeptides.

The computational strategy shown above cannot evidently be applied for larger peptides and here lower-level methods should be utilised. It seems that the use of empirical potentials is limited in this field (see later) and, therefore, QM procedures sould be used. The highly accurate data collected for the above-mentioned peptides can be used for testing various WFT, DFT and empirical methods. The performance of the MP2, SCS-MP2, MP3, TPSS-D, PBE-D, M06-2X, BH&H, TPSS, B3LYP, tight-binding DFT-D methods and ff99 empirical force field compared to CCSD(T)/CBS limit benchmark data was carefully investigated and the final trends (in terms of mean unsigned error and standard deviations) for all 5 peptides studied are summarised in Table 3.10. The CCSD(T)/CBS energies were constructed similarly as in the case of molecular clusters and the only difference represents the fact that here the relative instead of interaction energies were systematically used.

All the DFT techniques with a "-D" symbol have been augmented by an empirical dispersion energy while the M06-2X functional was parameterised to cover the London dispersion energy. For the systems studied here we have concluded that the use of the ff99 force field is not recommended mainly due to problems concerning the assignment of reliable atomic charges (different structures are characterised by different charges). This is slightly surprising but it must be added that it is not improved even after performing the geometry optimisation. On the other hand, a tight-binding SCC-DFTB-D is efficient as a screening tool providing reliable geometries. Among the DFT functionals, the M06-2X and TPSS-D show the best performance that is explained by the fact that both procedures cover the dispersion energy. The B3LYP and TPSS functionals, not covering this energy, fail systematically. Both, electronic energies and geometries obtained by means of the wavefunction theory methods compare satisfactorily with the CCSD(T)/CBS benchmark data. The philosophy used in the present study was analogous to that applied to the S22 benchmark data set (see next section) and the present set of 26 structures of 5 peptides (P26 set) thus represents an analogy to the S22 set containing only intermolecular complexes. This set can be used for testing and/or parametrisation of new computational procedures.

3.6   JSCH-2005 and S22 Database Sets

There are several reasons for performing highly accurate QM calculations for model complexes. Probably most important among them is the necessity to have a suitable set for testing lower level computational methods. Among all computational procedures discussed in the methodical part only the CCSD(T)/CBS procedure represents the genuine *ab initio* technique. In this technique all quantities are calculated and no empirical parameter is adopted. The theory behind the procedure is capable of describing all different types of non-covalent interactions occurring in molecular clusters and the same is of course true for non-covalent interactions in isolated simple as well as complex molecular systems. This is a unique situation and the price we are paying is high. The

Table 3.10 Mean unsigned error (MUE), standard deviation (σ) and maximum MUE (Max) obtained from the comparison between each of the CCSD(T)/CBS benchmark relative energies and the relative energies calculated at different WFT and DFT levels of theory, and using the empirical *ff99* potential, for each individual peptide. All energies are given in kcal/mol. MP3/CBS was determined analogously as CCSD(T)/CBS ; LP means 6311++G(3df, 3pd).

	WFT			DFT-D			DFT				*ff99*[a]	
	MP2/CBS	SCS-MP2/CBS	MP3/CBS	TPSS-D/LP	PBE-D/LP	SCC-DF-TD-D	M06-2X/LP	BH&H/LP	B3LYP/LP	TPSS/LP	HF/6-31G*	B3LYP/cc-pVTZ
WG												
MUE	0.52	0.83	0.44	0.81	1.53	0.56	0.68	1.07	2.72	3.01	2.70	1.29
σ	0.28	0.53	0.28	0.73	0.71	0.38	0.50	0.71	1.35	1.71	1.98	0.87
Max	0.92	1.55	0.89	1.80	2.45	1.44	1.45	2.24	4.48	5.00	7.25	2.73
WGG												
MUE	0.47	1.52	0.51	0.80	1.13	1.10	0.62	1.39	3.15	3.06	3.41	2.28
σ	0.40	1.00	0.47	0.69	1.03	0.51	0.54	0.78	2.32	2.14	3.14	2.44
Max	1.15	3.14	1.51	2.78	3.69	2.06	1.71	2.87	6.67	7.08	11.72	8.79
FGG												
MUE	0.42	0.89	0.49	0.80	0.73	0.88	1.16	0.76	1.52	1.24	3.57	3.00
σ	0.32	0.67	0.34	0.42	0.66	0.63	0.65	0.50	1.27	1.27	1.93	1.59
Max	1.13	2.35	1.11	1.54	2.31	2.16	2.11	1.98	4.57	4.78	7.08	6.64
GGF												
MUE	0.38	0.54	0.16	1.20	0.92	0.69	0.58	1.13	1.24	1.56	4.99	3.86
σ	0.20	0.46	0.14	0.80	0.84	0.55	0.49	0.67	1.09	1.26	3.41	2.61
Max	0.78	1.21	0.40	2.75	2.70	1.83	1.67	2.13	4.42	4.59	10.93	8.17
GFA												
MUE	0.43	0.66	0.30	1.47	0.85	0.75	0.39	1.02	1.47	1.78	2.79	1.81
σ	0.25	0.50	0.27	0.94	0.79	0.46	0.21	0.84	0.91	1.21	2.20	1.32
Max	0.98	1.69	0.81	3.02	2.58	1.42	0.77	2.65	2.61	3.23	6.05	3.61

[a] atomic charges are determined either at HF/6-31G* or B3LYP/cc-pVTZ levels

computational cost of the CCSD(T)/CBS procedure for medium and extended complexes is very high, and despite enormous progress in computational hardware and software, the use of this technique for complexes with more than about 100 atoms is (and will be in the near future) still impractical. The same is true for dynamical calculations of smaller systems. Evidently, much faster computational procedures need to be introduced allowing accurate calculations for these type of complexes and complex molecular systems to be performed. As was shown in the previous parts of the book the standard lower-level computational procedures fail to describe various types of molecular complexes and the most frequent problem is connected with the description of dispersion energy. Let us recall again the problems of HF and DFT methods and the same is true for all semiempirical QM methods. The only chance is thus to para-metrise the energy term in these (and also other) methods toward highly accurate computational methods. The benchmark database set should cover all important bonding motifs and should be easily extended in the future for new structural motifs. MP2/CBS and CCSD(T)/CBS interaction energies and also geometries for more than 100 DNA base pairs, amino acid pairs and also of model complexes were presented in the JSCH-2005 benchmark set.[88] Extra-polation to the CBS limit was done by using the two-point extrapolation method and different basis-set extrapolation schemes (aug-cc-pVDZ → aug-cc-pVTZ, aug-cc-pVTZ → aug-cc-pVQZ, cc-pVTZ → cc-pVQZ) were adopted. The CCSD(T) correction term, determined as a difference between CCSD(T) and MP2 interaction energies, was evaluated with smaller basis sets (6-31G** and cc-pVDZ). We have seen that this procedure yields reasonable stabilisation energies well comparable to the most accurate ones obtained by direct extra-polation of the CCSD(T) energies. Two sets of complex geometries were used, optimised and experimental ones. Besides the large JSCH-2005 set a smaller S22 set was introduced. In the latter case larger basis sets were used for extrapolation to the CBS limit and also the CCSD(T) and counterpoise-cor-rected MP2 geometries were sometimes adopted. The S22 set can be recom-mended for the parametrisation or for the first screening, while the larger JSCH-2005 set can be used for verification of a computational procedure. Table 3.11 contains interaction energies for the S22 set containing 7 H-bonded complexes, 8 complexes with predominant dispersion contribution and finally 7 mixed complexes. It is important that the set spans a wide range of interaction strengths in order to represent the diversity of interactions in biomacromole-cules. In each of the above-mentioned subgroups the stability of complexes ranges between 3 and 20, 0.5 and 15, and 1.2 and 8 kcal/mol, respectively. In the set the dispersion-bonded complexes are as numerous as H-bonded ones, but they contribute to the sum of of the stabilisation energies by less than 40% which is in line with our previous calculations on DNA nucleic acid bases. It is to be mentioned that the original S22 set[88] contained some errors; the present version is upgrated and all these errors and uncertainities were removed.

Energies presented in Table 3.11 represent a very valuable set, which can be used for testing of different lower-level methods. A strong point represents the fact that different types of non-covalent complexes are presented. Table 3.12

Table 3.11 Benchmark stabilisation energies (ΔE) for 7 H-bonded, 8 dispersion controlled and 8 mixed complexes determined at the CCSD(T)/CBS level. MP2 stabilisation energies in smaller and larger basis sets and MP2 and CCSD(T) extrapolated (CBS) stabilisation energies (all in kcal/mol) are given. Deformation energies of monomers are not included. The basis set abbreviations TZ, aTZ, QZ, aQZ and 5Z (in brackets) stand for cc-pVTZ, aug-cc-pVTZ, cc-pVQZ, aug-cc-pVQZ and cc-pV5Z, respectively. In the modified cc-pVTZ set (tz-fd) one set of f- and one set of d-functions were removed (only the more diffuse d-function was kept) and the hydrogen basis set was modified analogically. Geometry was determined with full (counterpoise-corrected) optimisations with analytical (MP2) or numerical (CCSD(T)) gradients.

Hydrogen-bonded complexes (7)

No.	Complex	(symmetry)	ΔE(MP2)				CBS	CCSD(T)/CBS		Geometry
1.	(NH$_3$)$_2$	(C$_{2h}$)	−3.02	(QZ)	−3.10	(5Z)	−3.20	−3.17	(qz)	CCSD(T)/QZ
2.	(H$_2$O)$_2$	(C$_s$)	−4.75	(QZ)	−4.89	(5Z)	−5.03	−5.02	(qz)	CCSD(T)/QZ
3.	Formic acid dimer	(C$_{2h}$)	−17.88	(QZ)	−18.23	(5Z)	−18.60	−18.61	(tz)	CCSD(T)/TZ
5.	Uracil dimer	(C$_{2h}$)	−19.90	(TZ)	−20.28	(QZ)	−20.61	−20.65	(tz-fd)	MP2/TZ-CP
6.	2-pyridoxine–2-aminopyridine	(C$_1$)	−15.91	(TZ)	−16.77	(TZ)	−17.37	−16.71	(tz-fd)	MP2/TZ-CP
7.	Adenine–thymine WC	(C$_1$)	−14.92	(TZ)	−15.89	(TZ)	−16.54	−16.37	(dz)	MP2/TZ-CP

Complexes with predominant dispersion contribution (8)

No.	Complex	(symmetry)	ΔE(MP2)		CBS	CCSD(T)/CBS		Geometry
8.	(CH$_4$)$_2$	(D$_{3d}$)	−0.42	(QZ)	−0.51	−0.53	(qz)	CCSD(T)/TZ
9.	(C$_2$H$_4$)$_2$	(D$_{2d}$)	−1.43	(QZ)	−1.62	−1.51	(qz)	CCSD(T)/QZ
10.	Benzene–CH$_4$	(C$_3$)	−1.66	(QZ)	−1.86	−1.50	(tz-fd)	MP2/TZ–CP
11.	Benzene dimer	(C$_{2h}$)	−4.70	(aT)	−4.95	−2.73	(adz)	MP2/TZ–CP
12.	Pyrazine dimer	(C$_{2h}$)	−6.56	(aT)	−6.90	−4.42	(tz-fd)	MP2/TZ–CP
13.	Uracil dimer	(C$_2$)	−10.63	(TZ)	−11.39	−10.12	(tz-fd)	MP2/TZ–CP
14.	Indole–benzene	(C$_1$)	−6.44	(TZ)	−8.12	−5.22	(dz)	MP2/TZ–CP
15.	Adenine–thymine stacked	(C$_1$)	−12.30	(TZ)	−14.93	−12.23	(dz)	MP2/TZ–CP

Mixed complexes (7)

No.	Complex	(symmetry)	ΔE(MP2)		CBS	CCSD(T)/CBS		Geometry
16.	Ethene–ethine	(C$_{2v}$)	−1.57	(QZ)	−1.69	−1.53	(tz)	CCSD(T)/QZ
17.	Benzene–H$_2$O	(C$_S$)	−3.28	(QZ)	−3.61	−3.28	(tz-fd)	MP2/TZ–CP
18.	Benzene–NH$_3$	(C$_S$)	−2.44	(QZ)	−2.72	−2.35	(tz-fd)	MP2/TZ–CP
19.	Benzene–HCN	(C$_S$)	−4.92	(aT)	−5.16	−4.46	(tz-fd)	MP2/TZ–CP
20.	Benzene dimer	(C$_{2v}$)	−3.46	(aT)	−3.62	−2.74	(adz)	MP2/TZ–CP
21.	Indole–benzene T–shaped	(C$_1$)	−6.16	(TZ)	−7.03	−5.73	(dz)	MP2/TZ–CP
22.	Phenol dimer	(C$_1$)	−6.71	(TZ)	−7.76	−7.05	(tz-fd)	MP2/TZ–CP

Table 3.12 Performance of selected WFT (variation and perturbation) and DFT (without or with dispersion energy) and quantum Monte Carlo (QMC) methods. Standard deviations with respect to CCSD(T)/CBS values (in kcal/mol) for H-bonded and dispersion-dominated (stacked) structures of the S22 reference geometries set and the whole S22 set also containing mixed complexes.

Method	H-bonded structures	Stacked structures	The whole S22 set	Ref
MP2/cc-pVTZ	0.36	1.62	1.10	a
MP2/CBS	0.27	1.24	0.94	a
SCS-MP2/CBS	0.54	0.60	0.58	148
SAPT/aug-cc-pVDZ	0.90	0.83	0.91	a
SAPT/aug-cc-pVTZ	0.53	0.56	0.56	a
QMC	0.71	0.52	0.79	149
DFT/TPSS/LP	1.02	3.76	2.88	96
DFT-D/TPSS/LP	0.48	0.32	0.38	96
PLYP-D/TZVPP	0.23	0.23	0.34	150
OM3	2.22	2.85	2.27	151
OM3-D	1.71	0.73	1.64	151
PM6	3.00	2.48	2.52	a
PM6-D	2.49	0.59	2.16	a

[a]unpublished data from our laboratory.

gives the mean average errors of various DFT and WFT interaction energies for the entire S22 set of complexes and separately for H-bonded and dispersion-bound (stacked) complexes. The table clearly shows what was mentioned several times in the previous text and the main result can be summarised as follows:

  i) MP2 provides relatively large error even at the CBS level;
  ii) SCS-MP2 exhibits considerable improvement over MP2 (this concerns mainly stacked structures);
  iii) SAPT yields values as good as the SCS-MP2 but it must be kept in mind that SAPT does not contain any empirical parameter;
  iv) QMC works better than MP2 and here, similarly as with SAPT it must be stressed that no empirical parameters are included;
  v) DFT without dispersion energy provide large deviations that are dramatically reduced upon inclusion of dispersion energy.

To make the work with the benchmark energies and geometries easier, we decided to collect all our data in a database and make it accessible on the internet.[144] We have launched a website www.begdb.com (Benchmark Energy and Geometry DataBase) for easy access to the database. Currently, the following databases are included: S22 and JSCH-2005[88], S26 [145] (the S22 set was extended as S26 to emphasise the hydrogen-bonded complexes), complexes with halogen bonds[146] and small peptides.[140,143,147]

In summary:

- The BEGDB database allows easy access to high-quality molecular geometries and benchmark CCSD(T)/CBS calculations on them. Other methods are also added for comparison.
- The results are organised in logical datasets according to their nature and source.
- The interface allows simple browsing of the results as well as advanced search functions applicable across the datasets.
- The advanced search can be used to combine and analyse the results in new ways.

3.7 Experimental Methods for Exploring Stationary Points on the PES: Stimulated-Emission Pumping

A problem associated with the spectroscopy of molecular clusters or molecules of biological interest in the gas phase is associated with the complexity of the potential-energy surface. Complexity here means that the different conformers associated with stationary points on the potential-energy surface, *i.e.* the global minimum and local minima with comparable energy, lead to several conformers populated in the supersonic jet expansion. This problem is related to the folding of polymers and proteins, which depends on the flexibility of sites whose energetic barriers are dictated largely by "single bonds". Molecules with flexible side chains can have several low-energy conformational minima with no significant energy barriers between them. From the experimental viewpoint a difficulty arises when distinguishing between these different conformers. If the energy barriers are sufficiently high and vibrational spectra are well resolved, then a distinction can be made by infrared hole burning. By similar reasoning, this distinction may also be made by UV hole burning, for which a resonant excitation into an electronically excited state, for instance the S_1 state, is used to deplete the population of one conformer (just replace the IR photon in Figure 1.1 with a UV photon, thus transferring population from the S_0 into the S_1). However, for more complex molecules and clusters, these single-photon hole-burning experiments become more and more difficult due to the inherent signal-to-noise limitations of signal-depletion experiments. A recent experiment from the group of Zwier has shown that stimulated-emission pumping-hole-filling spectroscopy (SEP-HFS) can be used to determine different conformers in an elegant way by increasing their population.[152] This was done for tryptamine, a molecule for which several conformers exist, which are stabilised by non-covalent *intra*molecular interactions. The method is described in Figure 3.19. Very close to the nozzle of supersonic jet expansion, *i.e.* in a region with high collision rates, a single conformer (*e.g.*, conformer A) is selectively excited from its zero-point vibrational level to a certain vibrational level using SEP (λ_1 and

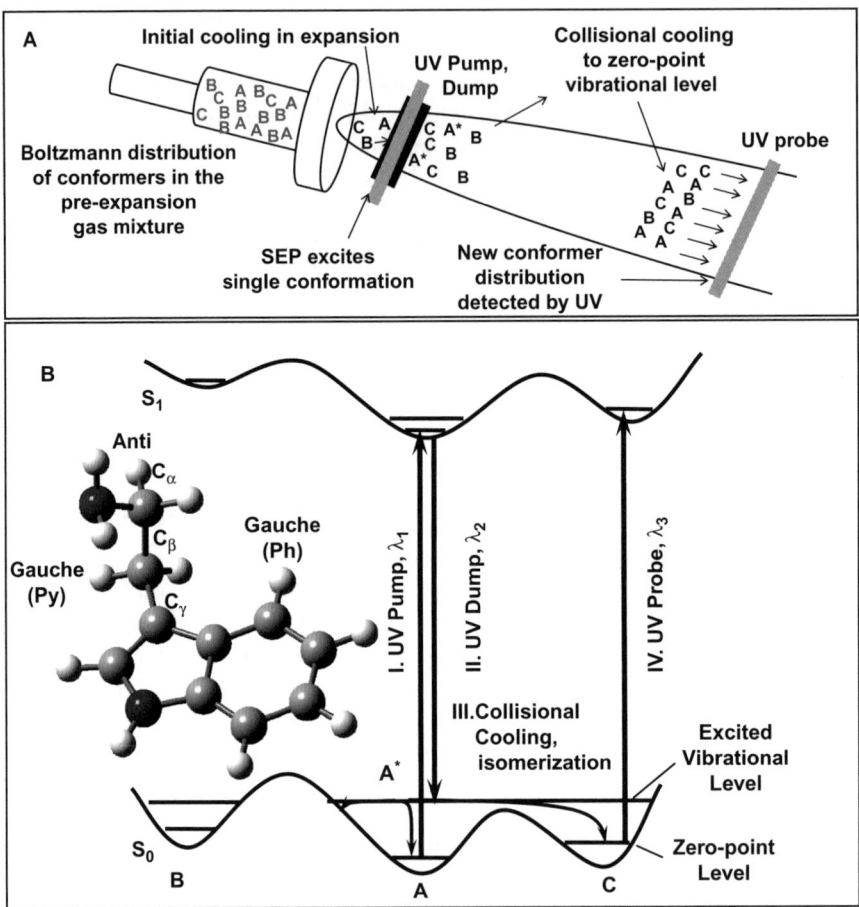

Figure 3.19 (**A**) Schematic diagram of the spatial and temporal arrangement and (**B**) energy-level diagram for the SEP-HFS (stimulated emission pumping-hole filling spectroscopy) experiment. A single conformer (*e.g.*, conformer A) is selectively excited to a certain vibrational level using SEP (λ_1 and λ_2) from its zero-point vibrational level. Once the barrier to isomerisation is exceeded to form a given product (*e.g.*, conformer C), the isomerisation followed by collisional cooling will result in an increase in the population in the zero-point level of conformer C downstream in the expansion. This change is detected via LIF using a third pulsed tunable UV laser (λ_3). As the λ_2 wavelength is tuned further towards higher energy, the rate-limiting barriers of other isomerisation pathways are overcome, producing gains in the population in other conformational zero-point levels (again, after collisional cooling). Reproduced from Ref. 152 with permission.

λ_2). Once the barrier to isomerisation is exceeded to form a given product (*e.g.*, conformer C), the isomerisation followed by collisional cooling results in an increase in the population of the zero-point level of that conformer (C) downstream in the expansion. This change is detected, in this case, by laser-induced fluorescence (LIF), using a third pulsed tuneable UV laser (λ_3). As λ_2 is tuned further towards higher energy, the rate-limiting barriers to other isomerisation pathways are overcome, producing gains in the population in other conformational zero-point levels (again, after collisional cooling). Hence, this method can also be used to access conformers that are not at all populated in the supersonics expansion. Clearly, the method can also be used with the third pulsed tunable UV laser (λ_3) ionising the molecule for mass-selected REMPI detection. The experimental trick of SEP-HFS is to carry out the pump/dump stimulated emission early in the supersonic expansion, making it possible to recool the isomerised products into their vibrational zero-point levels from which they can be interrogated downstream.

References

1. G. R. Denis and L. D. Richie, *J. Phys. Chem.*, 1991, **95**, 656.
2. G. Karlstrom, P. Linse, A. Wallquist and B. Jonsson, *J. Am. Chem. Soc.*, 1983, **105**, 3777.
3. C. A. Hunter and J. K. Sanders, *J. Am. Chem. Soc.*, 1990, **112**, 5525.
4. F. Cozzi, M. Cinquini, R. Annunziata, T. Dwyer and J. S. Siegel, *J. Am. Chem. Soc.*, 1992, **114**, 5729.
5. E. Kim, S. Paliwal and C. S. Wilcox, *J. Am. Chem. Soc*, 1998, **120**, 11192.
6. H. Krause, B. Ernstberger and H. J. Neusser, *Chem. Phys. Lett.*, 1991, **184**, 411.
7. H. Krause, B. Ernstberger and H. J. Neusser, *Ber. Bunsen-Ges.*, 1992, **96**, 1183.
8. W. Scherzer, O. Krätzschmar, H. L. Selzle and E. W. Schlag, *Z. Naturforsch., Phys. Sci.*, 1992, **47A**, 1248.
9. E. Arunan and H. S. Gutowsky, *J. Chem. Phys.*, 1993, **98**, 4294.
10. P. Hobza and R. Zahradník, *Int. J. Quant. Chem.*, 1992, **42**, 581.
11. P. Hobza, H. L. Selzle and E. W. Schlag, *Chem. Rev.*, 1994, **94**, 1767.
12. V. Špirko, O. Engvist, P. Soldán, H. L. Selzle, E. W. Schlag and P. Hobza, *J. Chem. Phys.*, 1999, **111**, 572.
13. S. Tsuzuki and H. P. Luthi, *J. Chem. Phys.*, 2001, **114**, 3949.
14. M. O. Sinnokrot, E. F. Valeev and C. D. Sherrill, *J. Am. Chem. Soc.*, 2002, **124**, 10887.
15. B. W. Hopkins and G. S. Tschumper, *J. Phys. Chem. A*, 2004, **108**, 2941.
16. M. O. Sinnokrot and C. D. Sherrill, *J. Phys. Chem. A*, 2004, **108**, 10200.
17. J. M. Steed, T. A. Dixon and W. Klemperer, *J. Chem. Phys.*, 1979, **70**, 4940.
18. K. C. Janda, J. C. Hemminger, J. S. Winn, S. E. Novick, S. J. Harris and W. Klemperer, *J. Chem. Phys.*, 1975, **63**, 1419.

19. O. Krätzschmar, H. L. Selzle and E. W. Schlag, *J. Phys. Chem.*, 1994, **98**, 3501.
20. A. Kiermeier, B. Ernstberger, H. J. Neusser and E. W. Schlag, *J. Phys. Chem.*, 1988, **92**, 3785.
21. K. H. Fung, H. L. Selzle and E. W. Schlag, *J. Phys. Chem.*, 1983, **87**, 5113.
22. K. O. Börnsen, H. L. Selzle and E. W. Schlag, *J. Chem. Phys.*, 1986, **85**, 1726.
23. B. F. Henson, G. V. Hartland, V. A. Venturo and P. M. Felker, *J. Chem. Phys.*, 1992, **97**, 2189.
24. V. A. Venturo and P. M. Felker, *J. Chem. Phys.*, 1993, **99**, 748.
25. P. M. Felker, P. M. Maxton and M. W. Schaeffer, *Chem. Rev.*, 1994, **94**, 1787.
26. J. R. Grover, E. A. Walters and E. T. Hui, *J. Phys. Chem.*, 1987, **91**, 3233.
27. E. A. Meyer, R. K. Castellano and F. Diederich, *Angew. Chem. Int. Ed.*, 2003, **42**, 1210.
28. F. J. M. Hoeben, P. Jonkheijm, E. W. Meijer and A. P. H. J. Schenning, *Chem. Rev.*, 2005, **105**, 1491.
29. B. H. Hong, J. Y. Lee, C.-W. Lee, J. C. Kim, S. C. Bae and K. S. Kim, *J. Am. Chem. Soc.*, 2001, **123**, 10748.
30. B. H. Hong, S. C. Bae, C.-W. Lee, S. Jeong and K. S. Kim, *Science*, 2001, **294**, 348.
31. H. Fenniri, P. Mathivanan, K. L. Vidale, D. M. Sherman, K. Hallenga, K. V. Wood and J. G. Stowell, *J. Am. Chem. Soc.*, 2001, **123**, 3854.
32. H. Fenniri, B.-L. Deng, A. E. Ribbe, K. Hallenga, J. Jacob and P. Thiyagarajan, *Proc. Natl. Acad. Sci. USA*, 2002, **99**, 6487.
33. R.-F. Dou, X.-C. Ma, L. Xi, H. L. Yip, K. Y. Wong, W. M. Lau, J.-F. Jia, Q.-K. Xue, W.-S. Yang, H. Ma and A. K.-Y. Jen, *Langmuir*, 2006, **22**, 3049.
34. G. Karlström, P. Linse, A. Wallqvist and B. Jonsson, *J. Am. Chem. Soc.*, 1983, **105**, 3777.
35. K. Müller-Dethlefs and P. Hobza, *Chem. Rev.*, 2000, **100**, 143.
36. P. Hobza, V. Špirko, H. L. Selzle and E. W. Schlag, *J. Phys. Chem.*, 1998, **102**, 2501.
37. O. Engkvist, P. Hobza, H. L. Selzle and E. W. Schlag, *J. Chem. Phys.*, 1999, **110**, 5758.
38. P. Hobza, H. L. Selzle and E. W. Schlag, *J. Phys. Chem.*, 1996, **100**, 18790.
39. P. Hobza, H. L. Selzle and E. W. Schlag, *J. Am. Chem. Soc.*, 1994, **116**, 3500.
40. P. Hobza, H. L. Selzle and E. W. Schlag, *J. Chem. Phys.*, 1990, **93**, 5893.
41. P. Hobza, H. L. Selzle and E. W. Schlag, *J. Phys. Chem.*, 1993, **97**, 3937.
42. E. C. Lee, B. H. Hong, J. Y. Lee, J. C. Kim, D. Kim, Y. Kim, P. Tarakeshwar and K. S. Kim, *J. Am. Chem. Soc.*, 2005, **127**, 4530.

43. B. H. Hong, J. Y. Lee, S. J. Cho, S. Yun and K. Kim, *Org.Chem.*, 1999, **64**, 5661.
44. K. S. Kim, P. Tarakeshwar and J. Y. Lee, *J. Am. Chem. Soc.*, 2001, **123**, 3323.
45. T. K. Manojkumar, H. S. Choi, B. H. Hong, P. Tarakeshwar and K. S. Kim, *J. Chem. Phys.*, 2004, **121**, 841.
46. T. K. Manojkumar, D. Kim and K. S. Kim, *J. Chem. Phys.*, 2005, **122**, 014305.
47. S. Tsuzuki, K. Honda, T. Uchimaru, M. Mikami and K. Tanabe, *J. Am. Chem. Soc.*, 2002, **124**, 104.
48. S. Tsuzuki, T. Uchimaru, K.-I. Sugawara and M. Mikami, *J. Chem. Phys.*, 2002, **117**, 11216.
49. S. Tsuzuki, T. Uchimaru, M. Mikami and K. Tanabe, *Chem. Phys. Lett.*, 1996, **252**, 206.
50. S. Tsuzuki, K. Honda, T. Uchimaru and M. Mikami, *J. Chem. Phys.*, 2005, **122**, 144323.
51. S. Tsuzuki, K. Honda, T. Uchimaru and M. Mikami, *J. Chem. Phys.*, 2006, **125**, 124304.
52. S. Tsuzuki, T. Uchimaru and M. Mikami, *J. Phys. Chem. A*, 2006, **110**, 2027.
53. S. Tsuzuki, T. Uchimaru, M. Mikami and K. Tanabe, *J. Phys. Chem. A*, 1998, **102**, 2091.
54. M. O. Sinnokrot and C. D. Sherrill, *J. Phys. Chem. A*, 2005, **109**, 10475.
55. T. P. Tauer and C. D. Sherrill, *J. Phys. Chem. A*, 2005, **109**, 10475.
56. M. O. Sinnokrot and C. D. Sherrill, *J. Phys. Chem. A*, 2003, **107**, 8377.
57. M. O. Sinnokrot and C. D. Sherrill, *J. Am. Chem. Soc.*, 2004, **126**, 7690.
58. A. L. Ringer, M. O. Sinnokrot, R. P. Lively and C. D. Sherrill, *Chem. Eur. J.*, 2006, **12**, 3821.
59. M. O. Sinnokrot, E. F. Valeev and C. D. Sherrill, *J. Am. Chem. Soc.*, 2002, **124**, 10887.
60. Y. Wang and X. Hu, *J. Am. Chem. Soc.*, 2002, **124**, 8445.
61. C. Chipot, R. Jaffe, B. Maigret, D. A. Pearlman and P. A. Kollman, *J. Am. Chem. Soc.*, 1996, **118**, 11217.
62. R. L. Jaffe and G. D. Smith, *J. Chem. Phys.*, 1996, **105**, 2780.
63. A. Puzder, M. Dion and D. C. Langreth, *J. Chem. Phys.*, 2006, **124**, 164105.
64. E. C. Lee, D. Kim, P. Jurečka, P. Tarakeshwar, P. Hobza and K. S. Kim, *J. Phys. Chem. A*, 2007, **111**, 3447.
65. R. Podeszwa, R. Bukowski and K. Szalewicz, *J. Phys. Chem. A*, 2006, **110**, 10345.
66. M. Pitoňák, P. Neogrády, J. Řezáč, P. Jurečka, M. Urban and P. Hobza, *J. Chem. Theory Comput.*, 2008, **4**, 1829.

67. C. A. Hunter, J. Singh and J. M. Thornton, *J. Mol. Biol.*, 1991, **218**, 837.
68. J. H. Williams, *Acc. Chem. Res.*, 1993, **26**, 593.
69. A. P. West Jr, S. Mecozzi and D. A. Dougherty, *J. Phys. Org. Chem.*, 1997, **10**, 347.
70. K. Pluháčková, P. Jurečka and P. Hobza, *Phys. Chem. Chem. Phys.*, 2007, **9**, 755.
71. S. A. Amstein and C. D. Sherrill, *Phys. Chem. Chem. Phys.*, 2008, **10**, 2646.
72. A. L. Ringer, M. O. Sinnokrot, R. P. Lively and C. D. Sherrill, *Chem. Eur. J.*, 2006, **12**, 3821.
73. E. C. Lee, B. H. Hong, J. Y. Lee, J. C. Kim, D. Kim, Y. Kim, P. Tarakeshwar and K. S. Kim, *J. Am. Chem. Soc.*, 2005, **127**, 4530.
74. E. C. Lee, B. H. Hong, J. Y. Lee, J. C. Kim, D. Kim, Y. Kim, P. Tarakeshwar and K. S. Kim, *J. Am. Chem. Soc.*, 2005, **127**, 4530.
75. D. J. Wales, *Energy Landscapes*, Cambridge University Press, Cambridge, 2003.
76. F. G. Amar and R. S. Berry, *J. Chem. Phys.*, 1986, **85**, 5943.
77. W. D. Cornell, P. Cieplak, C. I. Bayly, I. R. Gould, K. M. Merz Jr, D. M. Ferguson, D. C. Spellmeyer, T. Fox, J. W. Caldwell and P. A. Kollman, *J. Am. Chem. Soc.*, 1995, **117**, 5179.
78. P. Hobza and J. Šponer, *Chem. Rev.*, 1999, **99**, 3247; J. Šponer and P. Hobza, *Collect. Czech. Chem. Commun.*, 2003, **68**, 2231.
79. M. Kratochvíl, O. Engkvist, J. Šponer, P. Jungwirth and P. Hobza, *J. Phys. Chem.*, 1998, **102**, 6921.
80. F. Ryjáček, M. Kratochvíl and P. Hobza, *Chem. Phys. Lett.*, 1999, **313**, 393.
81. M. Kratochvíl, O. Engkvist, P. Jungwirth and P. Hobza, *Phys. Chem. Chem. Phys.*, 2000, **2**, 2419.
82. M. Kratochvíl, J. Šponer and P. Hobza, *J. Am. Chem. Soc.*, 2000, **122**, 3495.
83. F. Ryjáček, O. Engkvist, J. Vacek, M. Kratochvíl and P. Hobza, *J. Phys. Chem. A*, 2001, **105**, 1197.
84. M. Kabeláč and P. Hobza, *J. Phys. Chem. B*, 2001, **105**, 5804.
85. I. K. Yanson, A. B. Teplitsky and L. F. Sukhodub, *Biopolymers*, 1979, **18**, 1149.
86. P. Jurečka and P. Hobza, *J. Am. Chem. Soc.*, 2003, **125**, 15608.
87. P. Jurečka and P. Hobza, *Chem. Phys. Lett.*, 2002, **365**, 89; P. Hobza and J. Šponer, *J. Am. Chem. Soc.*, 2002, **124**, 11802.
88. P. Jurečka, J. Šponer, J. Černý and P. Hobza, *Phys. Chem. Chem. Phys.*, 2006, **8**, 1985.
89. J. Šponer, K. E. Riley and P. Hobza, *Phys. Chem. Chem. Phys.*, 2008, **10**, 2595.
90. P. Hobza, J. Šponer and M. Polášek, *J. Am. Chem. Soc.*, 1995, **117**, 792.
91. L. R. Rutledge and S. D. Wetmore, *J. Chem. Theory Comp.*, 2008, **4**, 1768.

92. A. Hesselmann, G. Jansen and M. Schütz, *J. Am. Chem. Soc.*, 2006, **128**, 11730.
93. M. Korth, A. Luchow and S. Grimme, *J. Phys. Chem. A*, 2008, **112**, 2104.
94. M. Pitoňák, K. E. Riley, P. Neogrady and P. Hobza, *ChemPhysChem*, 2008, **9**, 1636.
95. R. Sedlák, P. Jurečka and P. Hobza, *J. Chem. Phys.*, 2007, **127**, 075104.
96. P. Jurečka, J. Černý, P. Hobza and D. R. Salahub, *J. Comput. Chem.*, 2007, **28**, 555.
97. A. Fiethen, G. Jansen, A. Hesselmann and M. Schütz, *J. Am. Chem. Soc.*, 2008, **130**, 1802.
98. H. Cybulski and J. Sadlej, *J. Chem. Theory Comput.*, 2008, **4**, 892–897.
99. M. Kabeláč and P. Hobza, *Chem. Eur. J.*, 2001, **7**, 2067.
100. M. Kabeláč, L. Zendlová, D. Řeha and P. Hobza, *J. Phys. Chem.*, 2005, **109**, 12206.
101. M. Kabeláč and P. Hobza, *Phys. Chem. Chem. Phys.*, 2007, **9**, 903–917.
102. S. A. Trygubenko, T. V. Bogdan, M. Rueda, M. Orozco, F. J. Luque, J. Šponer, P. Slavíček and P. Hobza, *Phys. Chem. Chem. Phys.*, 2002, **4**, 4192.
103. M. Hanus, F. Ryjáček, M. Kabeláč, T. Kubar, T. V. Bogdan, S. A. Trygubenko and P. Hobza, *J. Am. Chem. Soc.*, 2003, **125**, 7678.
104. M. Hanus, M. Kabeláč, J. Rejnek, F. Ryjáček and P. Hobza, *J. Phys. Chem. B*, 2004, **108**, 2087.
105. J. Rejnek, M. Hanus, M. Kabeláč, F. Ryjáček and P. Hobza, *Phys. Chem. Chem. Phys.*, 2005, **7**, 2006.
106. L. Zendlová, P. Hobza and M. J. Kabeláč, *Phys. Chem. B*, 2007, **111**, 2591.
107. M. J. Doktycz, M. D. Morris, S. J. Dormady and K. L. Beattie, *J. Biol. Chem.*, 1995, **270**, 8439.
108. J. Řezáč and P. Hobza, *Chem. Eur. J.*, 2007, **13**, 2983.
109. J. Černý, M. Kabeláč and P. Hobza, *J. Am. Chem. Soc.*, 2008, **130**, 16055.
110. M. Elstner, T. Frauenheim and S. Suhai, *J. Molec. Struct.-Theochem.*, 2003, **632**, 29.
111. C. Branden and J. Tooze, *Introduction to Protein Structure*, Garland Publishing, New York, 1999.
112. C. D. Tatko and M. L Waters, *J. Am. Chem. Soc.*, 2004, **126**, 2028.
113. V. I. Abkevich, A. M. Gutin and E. I. Shaknovich, *Biochemistry*, 1994, **33**, 10026.
114. A. R. Fehrst, *Proc. Natl. Acad. Sci. USA*, 2000, **97**, 1525.
115. N. V. Dokholyan, S. V. Buldyrev, H. E. Stanley and E. I. Shakhnovich, *J. Mol. Biol.*, 2000, **296**, 1183.
116. E. I. Shakhnovich, V. Abkevich and O. Ptitsyn, *Nature*, 1996, **379**, 96.
117. L. S. Itzhaki, D. E. Otzen and A. Fehrst, *J. Mol. Biol.*, 1995, **254**, 260.
118. C. Tanford, *Science*, 1978, **200**, 1012.
119. G. D. Rose, A. R. Geselowitz, G. J. Lesser, R. H. Lee and M. H. Zehfus, *Science*, 1985, **229**, 834.

120. E. A. Meyer, R. K. Castellano and F. Diederich, *Angew. Chem. Int. Ed.*, 2003, **42**, 1210.

121. P. Hobza, R. Zahradník and K. Müller-Dethlefs, *Collect. Czech. Chem. Comm.*, 2006, **71**, 443.

122. J. Vondrášek, L. Bendová, V. Klusák and P. Hobza, *J. Am. Chem. Soc.*, 2005, **127**, 2615.

123. J. Vondrášek, T. Kubař, F. E. Jenney Jr, M. W. W. Adams, M. Kožíšek, J. Černý, V. Sklenář and P. Hobza, *Chem. Eur. J.*, 2007, **13**, 9022.

124. S. W. R. Malham, S. Johnstone, R. J. Bindgham, E. Barratt, S. E. V. Phillips, C. A. Laughton and S. W. Homans, *J. Am. Chem. Soc.*, 2005, **127**, 17061.

125. J. Černý, J. Vondrášek and P. Hobza, *J. Phys. Chem. B*, 2009, **113**, 5657.

126. M. Elstner, P. Hobza, T. Frauenheim, S. Suhai and E. Kaxiras, *J. Chem. Phys.*, 2001, **114**, 5149.

127. L. Bidermannová, K. E. Riley, K. Berka, P. Hobza and J. Vondrášek, *Phys. Chem. Chem. Phys.*, 2008, **10**, 6350.

128. J. Chocholoušová, J. Vacek and P. Hobza, *Phys. Chem. Chem. Phys.*, 2002, **4**, 2119.

129. J. Chocholoušová, J. Vacek and P. Hobza, *J. Phys. Chem. A*, 2003, **107**, 3086.

130. J. Chocholoušová, J. Vacek, F. Huisken, O. Werhahn and P. Hobza, *J. Phys. Chem. A*, 2002, **106**, 11540.

131. M. Gantenberg, M. Halupka and W. Sander, *Chem. Eur. J.*, 2000, **6**, 1985.

132. B. Brutschy, *personal communication*, 2000.

133. F. Madeja, M. Havenith, K. Nauta, R. E. Miller, J. Chocholoušová and P. Hobza, *J. Chem. Phys.*, 2004, **120**, 10554.

134. R. J. Petrella and M. Karplus, *J. Mol. Biol.*, 2001, **312**, 1161.

135. S. G. Pennathur and R. Anishetty, *BMC Struct. Biol.*, 2002, **2**, 9.

136. T. van Mourik, P. G. Karamertzanis and S. L. Price, *J. Phys. Chem. A*, 2006, **110**, 8.

137. L. F. Holroyd and T. van Mourik, *Chem. Phys. Lett.*, 2007, **442**, 42.

138. H. Valdes, V. Klusák, M. Pitoňák, O. Exner, I. Starý, P. Hobza and L. Rulíšek, *J. Comput. Chem.*, 2008, **29**, 861.

139. G. Duan, V. H. Smith Jr and D. F. Weaver, *Int. J. Quantum Chem.*, 2002, **90**, 669.

140. D. Řeha, H. Valdes, J. Vondrášek, P. Hobza, A. Abu-Riziq, B. Crews and M. S. de Vries, *Chem. Eur. J.*, 2005, **11**, 6803.

141. M. Iannuzi, A. Laio and M. Parrinello, *Phys. Rev. Lett.*, 2003, **90**, 238302.

142. H. Valdes, V. Spiwok, J. Řezáč, D. Řeha, A. G. Abo-Riziq, M. S. deVries and P. Hobza, *Chem. Eur. J.*, 2008, **14**, 4886.

143. H. Valdes, K. Pluháčková, M. Pitoňák, J. Řezáč and P. Hobza, *Phys. Chem. Chem. Phys.*, 2008, **10**, 2747.

144. J. Řezáč, P. Jurečka, K. E. Riley, J. Černý, H. Valdes, K. Pluháčková, K. Berka, T. Řezáč, M. Pitoňák, J. Vondrášek and P. Hobza, *Coll. Czech. Chem. Commun.*, 2008, **73**, 1261.

145. K. E. Riley and P. Hobza, *J. Phys. Chem. A*, 2007, **111**, 8257.
146. K. E. Riley and P. Hobza, *J. Chem. Theory Comput.*, 2008, **4**, 232.
147. H. Valdes, V. Spiwok, J. Řezáč, D. Řeha, A. G. Abo-Riziq, M. S. de Vries and P. Hobza, *Chem. Eur. J.*, 2008, **14**, 4886.
148. J. Antony and S. Grimme, *Phys. Chem. A*, 2007, **111**, 4862.
149. M. Korth, A. Luchow and S. Grimme, *J. Phys. Chem. A*, 2008, **112**, 2104.
150. T. Schwabe and S. Grimme, *Phys. Chem. Chem. Phys.*, 2007, **9**, 3397.
151. T. Tuttle and W. Thiel, *Phys. Chem. Chem. Phys.*, 2008, **10**, 2159.
152. B. C. Dian, J. R. Clarkson and T. S. Zwier, *Science*, 2004, **303**, 1169.

CHAPTER 4

Classification of Non-covalent Complexes

Non-covalent complexes can be classified with respect to the dominant contribution to stabilisation energy, by following a structural type, or simply on the basis of their size. Classification based on the first criterion is not unique since only very rarely is one particular energy term dominant. Probably the only complexes that fulfil this criterion are van der Waals complexes where the dominant attraction comes from the London dispersion energy; the only complexes that belong to this class are rare-gas dimers. Electrostatic complexes can be considered as the other class, but here also the other energy contributions (induction and dispersion) contribute considerably. Further, these complexes mostly contain hydrogens and are called H-bonded complexes (see later). The electrostatic term in H-bonded complexes is mostly dominant but an important feature of these complexes is the non-negligible role played by induction and dispersion contributions. In addition, we meet difficulties with another energy contribution, namely the charge-transfer term. In the perturbation-theory classification this energy contribution is missing but it is included in the induction term. The concept of charge transfer is quite intricate due to experimental difficulties, but we know from experience that electron density is transferred from one system to another and this transfer takes place *not only* if one of the subsystems is a good electron donor and the other a good electron acceptor. An important feature of a H-bonded complex (see Section 4.1) is just this electron-density transfer from the proton acceptor to the proton donor.

Classification of non-covalent complexes on the basis of structure is not unambiguous either and here we mostly recognise only the H-bonded complexes that are mostly planar, and stacked structures with vertical (*i.e.* vertical with respect to the main nodal plane of planar π-systems) π–π interactions. The dominant attraction in stacked structures comes from the London dispersion

RSC Theoretical and Computational Chemistry Series No. 2
Non-covalent Interactions: Theory and Experiment
By Pavel Hobza and Klaus Müller-Dethlefs
© Pavel Hobza and Klaus Müller-Dethlefs 2010
Published by the Royal Society of Chemistry, www.rsc.org

energy, but again the other energy terms are not negligible. The orientation of the subsystems in the stacked complex is due to the electrostatic term.

The specificity of H-bonded systems does not come from the nature of non-covalent interactions but it is associated rather with their abundance in nature and their very specific and easily detectable spectroscopic manifestation. Besides H-bonded complexes and similar improper H-bonded complexes, rather rare complexes with a dihydrogen bond will also be considered here. Finally, the recently introduced halogen bond will also be discussed.

4.1 Hydrogen Bonding and Improper Hydrogen Bonding

Hydrogen bonding is a subject that occupies quite a unique position in the realm of chemistry and it is linked to a story that started early in the 20th century. This is not an appropriate place to attempt a synthesis of all our knowledge; it seems, however, appropriate to try to pick up a few recent research tendencies. Moreover, it is virtually impossible to give a full account of hydrogen bonding. This is indicated, for instance, by Gálvez *et al.*,[1] who found that the keyword hydrogen bond, just in the period 1996–1999, is associated with about 16 000 works in the Chemical Abstracts Database.

The hydrogen bond (H-bond) is one of the strongest and the most common type of non-covalent bond. It is difficult to define H-bonds in terms of all the features ascribed to them in different branches of science; moreover, some older definitions (especially those taking the red spectral shift into account; see later) now appear ill-conceived. The most recent and general definition from the IUPAC Task Group[2] states: "The H-bond is an attractive interaction between the hydrogen from a group X–H and an atom or group of atoms Y, in the same or different molecule(s), where there is evidence of a partial bond formation. The evidence to be used should be experimental or theoretical, or better a combination of both. The greater the number of criteria satisfied, the more reliable is the characterisation as a hydrogen bond. In the criteria listed below, X–H represents a typical hydrogen bond donor and Y–Z a typical hydrogen bond acceptor. The acceptor may be an electron-rich region associated with the atom Y that is bonded to Z or the electrons from the Y–Z bond, either σ or π.

For a hydrogen bond X–H . . . Y–Z:

1) The forces involved in a hydrogen bond include, on balance, attractive quantum-mechanical orbital-overlap as well as classical electrostatic multipole interactions, also characterised by polarisation of the electron distribution and charge transfer between the electron donor and acceptor atoms.
2) The atoms X and H are covalently bonded to one another and the X–H bond is polarised with the H . . . Y bond strength increasing with the increase of electronegativity of X.

3) The length of the X–H bond increases on hydrogen-bond formation leading to a redshift in the infrared X–H stretching frequency with an increase in transition intensity. The greater the lengthening of the X–H bond in X–H . . . Y, the stronger the H . . . Y bond. New vibrational modes associated with the formation of the H . . . Y bond are generated. (In general the X–H bond length increases and there is an associated redshift of the X–H stretching frequency. There are, however, certain hydrogen bonds in which the X–H bond length decreases and a blueshift in X–H stretching frequency is observed. It is conceivable that a hydrogen bond could exist without redshift or blueshift. To a lesser extent, the Y–Z bond deviates from the length of the Y–Z bond in the isolated subunit. The Y–Z bond vibrational frequencies and spectral band intensities show corresponding changes on hydrogen-bond formation.)

4) The X–H . . . Y–Z hydrogen bond leads to characteristic NMR signatures, typically including pronounced proton deshielding and unusual spin–spin couplings and nuclear Overhauser enhancements indicative of "through H-bond" electronic coupling.

5) The stronger the hydrogen bond, the more linear the X–H . . . Y angle and the shorter the H . . . Y distance.

6) The Gibbs interaction energy per hydrogen bond should be greater than kT ($1\ kT = 600$ cal/mol $= 2.4$ kJ/mol at 300 K)."

In the following text we will discuss more thoroughly some of the characteristic features of H-bonding and also give a history of the discovery of blueshifting (also called improper) H-bonding.

The H-bond plays a key role in chemistry, physics, and biology and its consequences are enormous. Hydrogen bonds are responsible for the structure and properties of water, an essential compound for life. Further, H-bonds also play a key role in determining the structures and, consequently, shapes, properties and functions of biomolecules. For a survey, the reader is referred to three published monographs on H-bonding published at the end of the last century: *The Weak Hydrogen Bond*[3] by Desiraju and Steiner, *Hydrogen Bonding*[4] by Scheiner and *An Introduction to Hydrogen Bonding*[5] by Jeffrey. From the more recent literature the book *Hydrogen Bonding – New Insights* edited by S. Grabowski should be mentioned.[6] The term "hydrogen bond" was probably first used by Linus Pauling in his paper on the nature of the chemical bond.[7]

As mentioned above, the H-bond is a non-covalent bond between electron-deficient hydrogen and a region of high electron density. Most frequently, an H-bond is of the X–H . . . Y type, where X is the electronegative element and Y is the moiety with an excess of electrons. H-bonds with X, Y = F, O and N are the most frequent and most extensively studied, though X–H . . . π hydrogen bonds (for X=O and C) have also been observed.[8–14] CH H-bonds are weaker than OH or NH H-bonds. Nevertheless, CH H-bonds play an important role, especially in determining biomolecular structures, due to their large number.

Despite an enormous amount of literature dedicated to this subject an important question can still be raised: does an H-bond represent some special type of non-covalent interaction? The answer is unambiguous – no. Any type of H-bonded complex is stabilised by the same energy components as any other non-covalently bonded complex. The most important electrostatic contribution is accompanied by induction and dispersion contributions; these attractive terms are balanced by exchange-repulsion. There is nothing special about this energy decomposition and, from the point of view of intermolecular non-covalent bonding, H-bonds do not form a special class. However, the H-bond comes with a peculiar directionality: it is the sharing of the hydrogen atom between two electronegative atoms (the most frequent example) that causes the typical pseudolinearity of the X–H . . . Y arrangement. Furthermore, sharing of a very light hydrogen atom between two electronegative atoms results in a rather dramatic change of properties of the X–H covalent bond. This bond becomes mostly weaker upon formation of the H-bond and this weakening is the key factor in our understanding of the novel properties of the X–H stretching vibrational frequency.

We can summarise (see also the definition of H-bond) that the most characteristic features of the X–H . . . Y hydrogen bond are: (i) the X–H covalent bond stretches in correlation with the strength of the H-bond; (ii) a small amount of electron density (0.01–0.03 e) is transferred from the proton-acceptor (Y) to the proton-donor molecule (X–H); (iii) the band that corresponds to the X–H stretching shifts to lower frequency (redshift), increases in intensity and broadens. A shift to lower frequencies (redshift), is the most important, easily detectable manifestation of the formation of the H-bond. Indeed, it is the experimental basis for detection of H-bonding. A redshift represents a "fingerprint" of the H-bonding and a "no redshift – no H-bonding" relation was until very recently a statement of dogma. None of the three books[3–5] mentioned above on H-bonding and collecting more than 3000 references – which appeared at the very end of the last century – presents a single piece of evidence against this "rule".

However, reading the literature carefully, we found a few mostly experimental studies[15–24] showing that the X–H . . . Y arrangement (which is accompanied by attraction of the X–H proton donor and Y proton acceptor) can also exhibit different spectral manifestation. Instead of a redshift of the X–H stretching vibration, the *contraction* of this bond and an associated *blueshift* of the respective stretching vibration have been detected. Moreover, the intensity of the X–H stretching vibrational frequency decreased upon formation of the X–H . . . Y contact in strong contrast to "standard" H-bonding. The first systematic theoretical study on these nonstandard H-bonds was performed at the end of the 1990s[25] and because several features of this novel bonding differ from standard H-bonding, whilst others are similar, it was first called an anti-hydrogen bond and later on improper, blueshifting H-bonding.[26,27]

The history of the discovery of improper, blueshifting H-bonding is interesting and will be discussed in some detail. When investigating the benzene dimer we found the T-shaped structure to be the global minimum.[25] Because of the importance of dispersion energy the PES of the dimer should be studied at least at the (correlated) MP2 level. At the beginning of the 1990s the gradient optimisation at the MP2 level for a system of the size of the benzene dimer was demanding and this was even truer for performing vibration analysis. To our surprise we found that the C–H bond of the proton donor in the dimer was not elongated, as was expected, but contracted. Before, the T-shape dimer was believed to be stabilised by the C–H . . . π H-bond and, therefore, the C–H bond should be elongated. In the mid-1990s there was not a single piece of evidence about the opposite behaviour and it took us more than three years before our result – the proof of the existence of blueshifting H-bonding – was published. We were aware that without experimental proof the existence of this bonding would not have been accepted. The proof of improper H-bonding in the benzene dimer was and still is experimentally challenging, but in the same paper[25] we suggested an existence of this bond in the haloform . . . benzene complexes. It was not easy to convince experimentalists to investigate these novel complexes but Bernd Brutschy was an exception. After a rather short time he really found the blueshift of the C–H stretch frequency in the fluoro-form . . . benzene complex and, moreover, the absolute value of the shift nicely agreed with our prediction.[28] This experimental verification represents the origin of improper, blueshifting H-bonding. In the beginning we believed that this bond would be quite rare but soon it was found not only in the gas phase but also in liquid and solid phases and also other types of X–H . . . Y improper H-bonding were detected.[26,27]

Before discussing the nature of improper, blueshifting H-bonding we must elucidate the origin of the "standard" H-bonding. There are two models describing the origin of H-bonding: electrostatic and hyperconjugative charge-transfer (CT). The former model explains the formation of the H-bond on the grounds of energetics: elongation of the X–H bond increases the dipole moment of the proton donor and thus also the dipole–dipole attraction between proton donor and proton acceptor. Consequently, the total stabilisation energy becomes larger. This model explains the geometrical, energetical and also vibrational characteristics of an H-bonded complex. Is it thus necessary to include the concept of charge transfer? Let us mention that the phenomenon of charge transfer in hydrogen bonding is rather vague. We cannot detect it directly and its theoretical justification is not unambiguous. The first clear evidence supporting the CT concept comes from Coulson[29] who showed that without allowing for electron transfer between proton acceptor and proton donor, one cannot explain the dramatic intensity increase in the X–H stretching vibration upon formation of the H-bond. Later, the concept of CT was proved by using the natural bond orbital (NBO) analysis. Reed *et al.*[30] performed an NBO analysis for several typical H-bonded systems, demonstrating charge transfer from the lone pairs of the proton acceptors (Y) to the X–H σ* antibonding orbital of the proton donors. An increase in electron density in the antibonding orbitals results in a weakening of the X–H covalent bond and this is accompanied by a concomitant lowering of the X–H stretching frequency. The NBO analysis is thus a very useful technique for the study of the underlying principles of H-bonding. It becomes clear at this stage that a realistic picture of the H-bonding is based on a combination of both models; electrostatic and charge-transfer approaches complement each other, and only in borderline cases does one approach become dominant.

Are these models applicable also to the improper H-bond? Electrostatic models give, in many cases, qualitatively correct answers and can be used for a description of improper H-bonds. The necessary condition for a successful use of this model is the negative sign of the proton donor dipole moment derivatives with respect to the stretching coordinate;[31] this means that the dipole moment increases with contraction of the X–H bond. Such behaviour is not typical since in most cases the contraction of a bond is associated with a decrease in dipole moment. However, for some classes of system (*e.g.* halogenated hydrocarbons), this condition can be fulfilled. This means contraction of the X–H bond is associated with an increase of the dipole moment. We were the first[32] to demonstrate it for the fluoroform . . . ethylene oxide complex. However, improper H-bonding has also been detected in many complexes where the proton donor did not exhibit a negative gradient and in these cases the electrostatic model fails completely. So what is this telling us about the nature of the improper H-bond? First, the overall charge transfer between proton acceptor and proton donor is smaller than in the case of the H-bonding; in this instance, the charge transfer is mostly directed to the X–H σ^* antibonding orbitals. In improper H-bonding only a small fraction of charge transfer goes to this orbital, while the largest portion of CT finally lands in the remote part of the proton donor. (On the basis of this difference the so-called H-index[33] was defined that unambiguously discriminates between standard and improper H-bonding.) However, even a small increase in electron density in the X–H σ^* antibonding orbitals of the proton donor leads to a significant weakening and elongation of this bond and not to the observed contraction. Contraction of the X–H bond would require a decrease in electron density in this antibonding orbital, which seems to be impractical because the antibonding orbital should be the electron-donating and not the electron-accepting orbital.

This unusual effect (a decrease in electron density in the σ^* antibonding orbitals) was found[34] in isomers of the guanine dimer (K9K9-1, K9K7-1 and K7K7-1, where K9 indicates the canonical tautomer and K7 indicates the tautomer having the hydrogen at N7 instead at N9 (see Figure 4.1) possessing two N–H . . . O=C H-bonds. Dimers were optimised at the HF/6-31G** level and vibrational frequencies were calculated. Amino groups in these isomers were not directly involved in the H-bonding. Besides the redshifts of the N_1–H stretching vibrations (supporting the H-bonding character of these contacts), an unexpected blueshift of amino N–H stretching vibrations was found, which fully agrees with published experimental results.[35] The blueshift of the amino group N–H stretching vibrations in all the guanine dimer structures was rationalised by the planarisation of the guanine amino group (in the isolated guanine the amino group is strongly nonplanar). Absolute values of harmonic amino N–H stretching frequencies and their shifts upon planarisation were verified by performing two-dimensional anharmonic vibrational analyses. The planarisation of the guanine amino group cannot be interpreted on the basis of

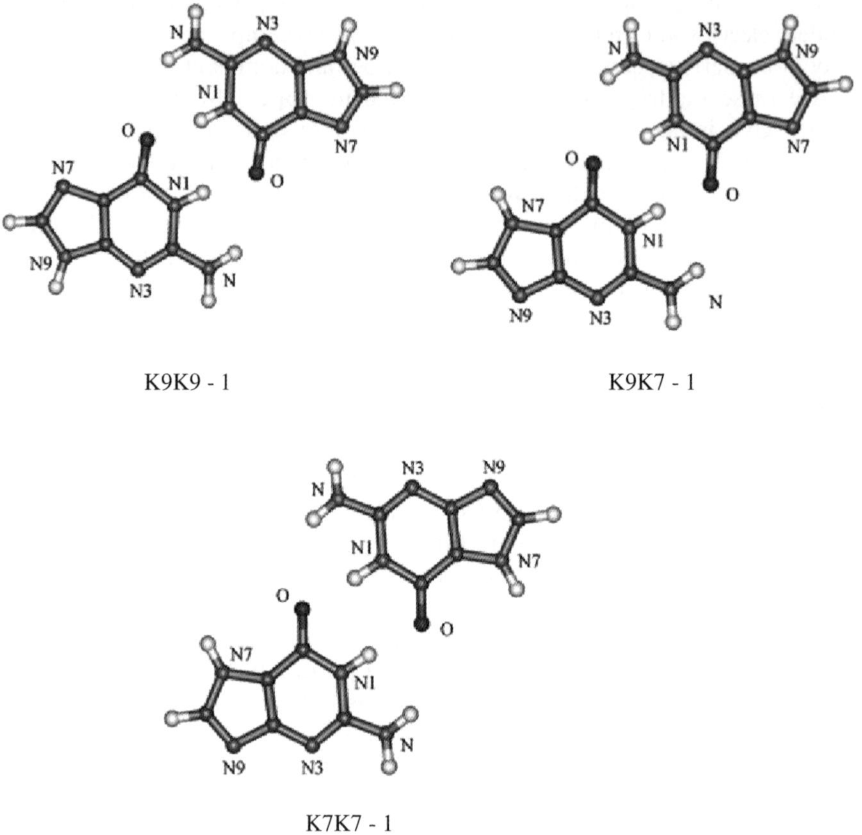

Figure 4.1 Structures of the guanine dimer. Redrawn from Ref. 34.

an electrostatic model and is due to the redistribution of electron density in the subsystems upon dimerisation; this redistribution occurs within the aromatic ring as well as the amino group nitrogen and leads to the formation of a new mesomeric structure. The electron density decrease in the lone electron pair of the amino nitrogen gives way to rehybridisation of the respective atomic orbitals (changing them from sp^3 to sp^2) resulting in planarisation of the amino group. The other consequence is a decrease of electron density in the amino N–H σ^* antibonding orbitals. The (amino) N–H . . . O=C contacts can thus be described as improper, blueshifting H-bonds. An increased amino N–H stretching frequency is the fingerprint of the planarisation of the guanine amino group and is the first spectroscopic manifestation of the fact that the amino group in nucleic acid bases is nonplanar. Blueshifts of the amino N–H stretching frequencies occur only if the amino group is bifurcated (see Figure 4.1). The guanine dimer is the first complex where one proton acceptor (the C=O group) is simultaneously linked to two proton donors (NH and NH_2) by a standard H-bond and an improper, blueshifting H-bond. Other types of

X–H bond contractions and blueshifts were detected in some X–H moieties even if H was not directly involved in the H-bonding. These H-bonds are called blueshifted (and not blueshifting).[36] The mechanism of the improper H-bonding was originally explained by charge transfer to the remote part of the proton donor, leading there to an elongation of the X–Y (mostly C–halogen) bond, which subsequently causes the contraction of the X–H bond. This suggested the two-step mechanism manifests itself by redshifts of the X–Y stretching frequencies and a blueshift of the X–H stretching frequency, and for the dimethyl ether . . . fluoroform complex this was indeed observed experimentally.[37]

An elegant interpretation of standard H-bonding and improper H-bonding, based on application of Bent's rule, was presented by Alabugin and co-workers.[38] These authors described the role of the two effects present in both types of H-bonded complexes: The hyperconjugative X–H bond weakening (by charge transfer) and the rehybridisation-promoted X–H bond strengthening. The former effect, described above, is well recognised, while the latter is not. The X–H strengthening is due to an increase in the s-character of the hybrid orbital of atom X in the X–H bond, which occurs upon a decrease in the X–H . . . Y distance. A direct consequence of Bent's rule is that a decrease in effective electronegativity of hydrogen (or an increase in hydrogen net charge) in an X–H bond leads to an increase in the s-character of the atom X hybrid orbital of this bond; the increase in s-character is associated with bond contraction. Since the total s-character at atom X is conserved, an increase in s-character in the X–H bond must be accompanied by a simultaneous increase in the p-character (or a decrease in s-character) of other bonds connected to the X atom: the increase in p-character is associated with bond elongation. Both effects, *i.e.* the hyperconjugative X–H weakening (by charge transfer) and the (rehybridisation-promoted) X–H bond strengthening are always present within any type of H-bonded complex and the final picture (elongation or contraction) depends on the balance of these effects. The theory mentioned is consistent with the structural reorganisation in the remote part of the molecule suggested by us, but does not require a two-step mechanism.

The direct consequence of the present theory is the fact that the blueshift decreases when going from sp^3- to sp^2-hybridised atoms and completely vanishes with sp-hybridised atoms. This is, however, not completely true and even the first complex for which improper H-bonding was theoretically predicted, the benzene dimer in the T-shaped structure,[25] belongs to this family. It is thus evident that other mechanisms still must exist. Let us recall that the C–H bond of the benzene proton donor contracts upon the formation of the T-shaped structure and the calculated blueshift is rather large. The original anharmonic calculations show a shift of about $50\,cm^{-1}$, while recent anharmonic calculations[39] predict a blueshift of $14\,cm^{-1}$ for the T-shaped C_{2v} structure (which corresponds to the transition-state structure) and a small redshift for the tilted T-shaped structure which is the global minimum. (For comparison with experiment see Section 5.5.) The shift is currently explained[28,40,41] by the balance between dispersion attraction and exchange-repulsion and in the absence of other data, this explanation is still valid. The movement of the X–H bond of the proton donor against a repulsion wall leads to contraction of the bond and a concomitant blueshift of its stretching frequency. This is the simplest and the most natural model for improper H-bonding. Similarities and differences between H-bonded and improper H-bonded complexes were recently studied using the perturbation SAPT calculations. In the redshifting complexes, the induction energy is mostly larger than the dispersion energy, while, in the case of blueshifting complexes, the situation is opposite. Dispersion as an attractive force increases the blueshift in the blueshifting complexes as it compresses the H-bond and, therefore, it increases the Pauli repulsion.[42]

The nature of X–H . . . Y H-bonding was investigated by Joseph and Jemmis[43] who explained redshifts and blueshifts as follows: all X–H bonds are exposed to opposing forces leading to their elongation and contraction. The former is due to attractive interaction between the positive H and the negative, electron-rich Y, while the latter originates from the electron affinity of X which causes a net gain of electron density at the X–H bond and leads to an X–H bond contraction. For electron-rich, highly polar X–H bonds the lengthening effect always dominates, whereas for less polar, electron-poor X–H bonds the shortening effect can be important, especially if Y is not a strong H-bond acceptor. An important consequence of this theory is the fact that it offers a possibility of H-bonds that do not show *any* IR frequency change upon formation of a complex. A similar explanation was suggested by Wang and Hobza[44] using Berlin's theory. Kryachko and Karpfen[45,46] explain blueshifting and blueshifted H-bonding on the basis of intramolecular mode coupling between C–H and C–halogen bonds and introduce the concept of intrinsic, intramolecular negative response.

Achievements in the realm of H-bonding were treated in a review[47] dealing mainly with hydrogen bonds in the gas phase and solution. Hydrogen bonding in the solid state deserves special attention.[48] In this review thorough attention was paid to the history of the discovery of the hydrogen bond and to pioneering monographs. Features related to DFT calculations in the solid state were discussed with respect to periodicity in a paper dealing with ammonia and urea.[49]

Recent papers of general importance deal with the decomposition of the interaction energy in H-bonded dimers,[50] with the role of nuclear quantum effects on the structure of H-bonded systems,[51] the interpretation of the nature of the hydrogen bond in terms of topological descriptors (location of the bond critical point and geometry of the lone pair),[52] and with charge transfer in H-bonded clusters.[53] The role that H-bonding plays in molecular recognition[54,55] and in spontaneous self-organisation,[56,57] not only within biological systems, is overwhelming. Nitrogen-, oxygen-, and fluorine-containing compounds represent systems for classical O–H . . . O, N–H . . . O, and N–H . . . N bonds; significantly weaker C–H . . . O, C–H . . . π, and N–H . . . π have been investigated only in recent years. Valuable references to these types of interactions are available in a publication.[58] Hydrogen bonding in water clusters is an evergreen.[59] A quarter of a century elapsed before the pioneering work on

H_2O and D_2O clusters in the gas phase[59a] was complete. Water clusters were studied in hydrophobic solvents (liquid-helium droplets, solid parahydrogen, and CCl_4) and also quantum chemically.[59b] Three-body interactions in water clusters were studied, which led to a new *ab initio* three-body potential.[59c,d] New water chains and layers have been observed, including $(H_2O)_{12}$ rings.[59e] $(H_2O)_{20}$ clusters have attracted attention for years; there are 30 026 symmetry-distinct ways of arranging twenty water molecules.[59f] The presence of ions in water has only a very small effect on the H-bond structure of liquid water.[59f,i,g] Increasing attention has been paid to the water structure in solid hosts.[59h,i] With regard to the roles that the hydrogen-bonded oligomers of HF have played, a classic work on $(HF)_2$, HFDF, and $(DF)_2$ should be quoted[60a] together with a work on J–J correlations in state-to-state photodissociation[60b] of $(HF)_2$.

Within the framework of classic H-bonds,[61] various complexes have been studied, *e.g.*, formamide . . . methanol,[61a] oligomers of formamide and thio-formamide,[61b] indole . . . pyrrole,[61c] together with the role of H-bonding in Schiff bases,[61d] furan . . . hydrogen halides (rotational spectroscopy),[61e] strong complexes (*e.g.*, H_3N . . . HF),[61f] vibrational coupling through individual H-bond chains[61g] in helices of pentapeptides, and a strong $Br–H–Br^-$ bond.[61h] It is possible to obtain evidence about the strength of the H-bonding from NMR isotope shifts.[61i] It is now widely recognised and accepted that correction for BSSE in calculations of H-bonded systems is highly desirable. The use of a function counterpoise procedure is reasonably efficient,[61j] though BSSE corrections are unnecessary when we are able to get extrapolated ΔE values for infinite basis sets.[61k] The existence of the C–H . . . O bond has been presumed intuitively since the middle of the last century. In 1962 a paper appeared on the presence of this bond in crystals.[62a] Since the turn of the century, however, the lack of literature in this area has changed significantly and the number of works has been increasing very rapidly.[62,63]

Most often, theoretical and experimental tools are used simultaneously. On the theoretical side, the MP type of calculations and economically attractive DFT procedures with, *e.g.*, B3LYP functional are most frequently used. Infrared and NMR spectroscopy are the most powerful tools for characterising hydrogen bonding.[62,63] In this illustrative set of systems, bonds such as C–H . . . O, C–H . . . N, C –H . . . F, C–H . . . π, O–H . . . π, are included. Also here, extrapolations to an infinite basis set were carried out,[63j] with regard to the formation of an acetylene dimer (ΔE) and its isomerisation ($\Delta E^{\neq}$) to an equivalent form.

What has been considered for years as "nonbonded steric repulsion" between hydrogen atoms (*e.g.*, in planar biphenyl), contributes in fact to the stabilisation of the system by about 10 kcal/mol.[64a] The authors stress that another type of H–H interaction also exists, the interaction labelled "di-hydrogen bonding" (see the next section), a bond with the hydride ion in the role of a base. Recently, the shortest H . . . H distance of 1.95 Å was found by neutron diffraction in a derivative of D-ribofuranose.[64b] For the sake of completeness, we need to add that a remarkably strong, very short H–H bond

(0.95 Å) exists in the radical cation of the methane dimer, which has the linear structure,[64c] $H_3C-H-H-CH_3^+$.

4.2   Dihydrogen Bonding

Complexes with a dihydrogen bond are much less numerous than the previous ones, but represent an interesting class of non-covalent complexes. The dihydrogen bond of the type M–H . . . H–Y was originally found[65] in metal complexes (M=metal), but was also later detected[66] in the H_3BNH_3 dimer. The unusually high boiling point of this system gave evidence for a strong intermolecular attraction that was finally shown to be of dihydrogen origin.

The explanation for this unconventional H-bond is surprising but straightforward:[67] two hydrogens may only interact if one is positive and the other negative. This can be realised if one hydrogen is bound to an electropositive element, whilst the other is bound to a very electronegative element. The most common elements that are more electropositive than hydrogen are boron, alkali metals and heavy transition metals; dihydrogen bonding was found for various complexes containing these elements. The hydrogen bound to this element becomes negative, whilst the one bound to a very electronegative element becomes positive. Thus, there is an electrostatic attraction between these hydrogens. The dihydrogen bonds are characterised by a short hydrogen–hydrogen distance in the range 1.7–2.2 Å and surprisingly large stabilisation energies of 6.1–7.6 kcal/mol per interaction.[68] The stabilisation of dihydrogen bonding mainly originates in the electrostatic interaction between oppositely charged hydrogen atoms. However, other energy terms are important as well with induction and dispersion terms showing similar contributions to overall stabilisation. The literature on this type of bonding is growing and many complexes with different types of dihydrogen bond are now known to exist.[69] (See also a recent book by Bakhmutov.[70])

Dihydrogen bonds were first detected in solid phase but soon also in the gas phase. Pioneering studies were performed at the beginning of this century in Mikami's laboratory at Tohoku University. Phenol . . . borane–dimethylamine, phenol . . . .borane–trimethylamine and phenol and aniline . . . borane–amines were investigated experimentally and theoretically and their structures with dihydrogen bonds as well as vibration characteristics were identified.[71] Forthcoming theoretical study[72] showed that under identical conditions the dihydrogen-bonded complex is about 16–20% weaker than the corresponding hydrogen-bonded complex.

The dihydrogen bond is mainly connected with boranes or carboranes and the latter systems become highly attractive since it was shown that substituted metallacarboranes act as potent and specific inhibitors of HIV-1 protease.[73] The question remains, however, what is the nature of attraction between these systems and biomolecules. It was shown[74,75] that it is the dihydrogen bond that is the dominant interaction between carboranes and biomolecules, and also that this interaction arises from an electrostatic attraction between negatively charged surface hydrogens of metallocarboranes and positively charged hydrogens of building blocks of biomolecules. The dihydrogen bonds formed were characterised by a short distance between hydrogen atoms (as close as 1.8 Å) and an average strength in the range of 4.2–5.8 kcal/mol.

4.3 Halogen Bonding

In recent years halogen bonding has been implicated as an important type of interaction in many different types of physical systems and is especially interesting within the fields of biochemistry[76–86] and materials science.[87–97] These interactions play important roles in a wide variety of biochemical phenomena such as protein–ligand complexation[76–77,79–81,84,94] and they are responsible for many novel properties of materials[89–92,94–96] These types of interactions, many believe, promise to be of great importance in the design of novel drugs and materials.

A halogen bond is defined as a C–X . . . Y–Z interaction (where X is typically chlorine, bromine, or iodine, Y is an electron donor such as oxygen, nitrogen, or sulfur, and Y–Z represents a side group such as a hydroxyl or carbonyl group), where the X . . . Y distance is less than the sum of the van der Waals radii of X and Y. Halogen bonds share numerous physical properties with the more commonly encountered hydrogen bonds and are often treated analogously to their ubiquitous counterparts.[76,94] There is a broad range of reported halogen-bond interaction energies with values varying from about 1.2 kcal/mol (Cl . . . Cl) to about 43.0 kcal/mol (I_3^- . . . I_2).[94]

Considering the fact that halogen atoms (X), as well as halogen-bond electron donors (Y), are negatively charged, the existence of non-covalent halogen bonds is surprising and counterintuitive. However, studies of the electrostatic potentials of halogen-bonding systems by Auffinger *et al.*,[76] Clark *et al.*,[98] and Politzer *et al.*[99] show that a large halogen atom bound to carbon tends to form an electropositive crown, which is distal to the carbon, an electroneutral ring, which surrounds the crown, and an electronegative belt, which goes around the circumference of the halogen atom in the plane that is perpendicular to the C–X bond (see Figure 4.2). In the works by Clark and Politzer, the electropositive crown is referred to as the σ-hole to denote the region of positive charge on the halogen surface. Halogen bonding can be, at least partially, attributed to the favourable interaction that exists between a halogen's electropositive σ-hole and an electronegative atom, such as oxygen.[76,94,98–101] A halogen's σ-hole becomes larger and gains a higher degree of electropositivity as the size of the halogen increases, with a corresponding tendency for the halogen bond to become stronger. Fluorine, the smallest (and most electronegative) halogen, does not form an electropositive crown, and thus does not participate in halogen bonding.[76,98,99] It has also been observed that the size and charge of the σ-hole tends to increase as electronegative substituents are added to a halogen-containing molecule.[76,98,99]

There have been several theoretical and experimental studies seeking to characterise the geometric and energetic properties of halogen bonds. For example,

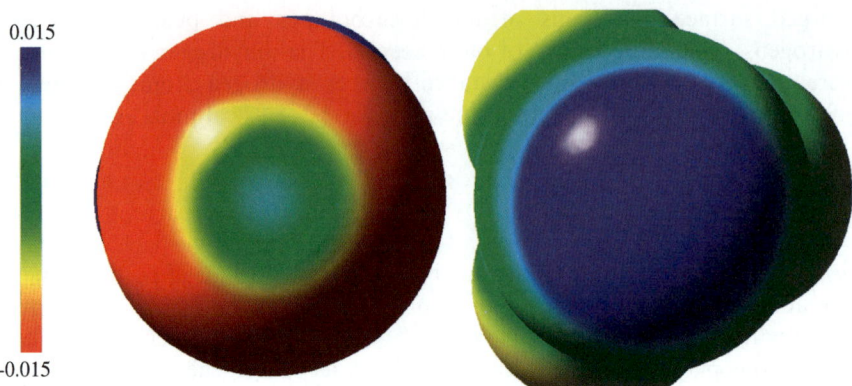

Figure 4.2 Molecular electrostatic potential for H_3CBr (left) and F_3CBr (right) at the 0.001 Bohr^{-3} electron isodensity surface; blue and red colour indicate highly positive and highly negative potentials. The σ-hole for the fluorine-substituted system is significantly larger than that of the unsubstituted molecule.

Valerio *et al.* performed *ab initio* calculations on the $CH_{n-3}FX \ldots NH_3$ (X = I, Br, Cl) halogen-bonded complexes and found that substitution of successive fluorine substituents results in X . . . N halogen bonds that are shorter and stronger.[102] The strongest halogen bond found in this study occurs for the $CF_3I \ldots NH_3$ complex with a binding energy of 5.8 kcal/mol. Riley and Merz characterised halogen bonds involving chlorine, bromine, and iodine, and carbonyl oxygens as a function of the halogen-bonding distance and the X . . . O–C halogen-bonding angle. In this work it was found that the optimum halogen bond angle is generally within the range 95° to 115°, corresponding to an interaction between the halogen σ-hole and the lone pair of electrons on oxygen.[101] Lommerse *et al.* carried out intermolecular perturbation theory calculations on several halogen-bonding systems containing chlorine as the halogen bond donor and both nitrogen and oxygen as the halogen-bond acceptors.[100] In this study, it was concluded that the attractive nature of halogen bonds is mostly attributable to electrostatic effects although dispersion, polarisation, and charge-transfer effects seem to also play a role in these interactions. It should be pointed out that these studies were carried out with the 6-31G basis set, which is not large enough to describe dispersion effects properly. It would be expected that the use of this small basis set would result in an underestimation of the dispersion energy by about an order of magnitude. On the experimental side, Corradi *et al.* determined the binding energy for a halogen-bonded complex of 1-iodoperfluorohexane and 2,2,6,6-tetramethylpiperidine to be 7.4 kcal/mol.[92]

Halogen bonds involving oxygen as the halogen bond acceptor are especially interesting in biochemistry because they are, by a large margin, the most common types of halogen bonds involved in protein–ligand interactions. Auffinger and coworkers carried out a database survey of short halogen . . . oxygen interactions; in this study it was found that 81 out of 113 X . . . O interactions involved carbonyl oxygens (data set contained 66 protein structures and 6 nucleic acid structures from the protein data bank).[76] These interactions generally involve a protein's backbone carbonyl group (78 out of 81 interactions). Interactions involving hydroxyl groups were also fairly common, with 18 X . . . O interactions involving hydroxyl oxygens. As has been shown in several studies, a nitrogen atom can act as an efficient halogen-bond acceptor and one might expect that nitrogen atoms found in proteins (both in the backbone and in side chains) might tend to be involved in halogen bonding with roughly the same frequency as oxygen atoms. Auffinger's work shows that there are only a handful of halogen bonds involving nitrogen: seemingly these atoms are somehow inaccessible to halogen atoms.

An important question concerns the origin of stabilisation in halogen bonding. Does the stabilisation come from electrostatic energy or are other energy terms important as well? To answer these questions, systematic study of C–X . . . O–Z halogen bonds was performed[103] where the O–Z group represents a carbonyl group. The model systems used are the halomethane . . . formaldehyde dimers. Because the binding energies of halogen bonds are comparable to those of hydrogen bonding, very accurate quantum-mechanical procedures should be adopted to describe them and the CCSD(T)/CBS and DFT-SAPT methods were applied. Table 4.1 shows the HF, MP2 and CCSD(T) interaction energies for complexes investigated while Table 4.2 presents the DFT-SAPT energy components.

Table 4.1 Interaction energies (kcal/mol) for H_3CX . . . OCH_2 complexes, X = Cl, Br. (a-pVXZ denotes an aug-cc-pVXZ basis set), X = D, T, Q.

	a-pVDZ	*a-pVTZ*	*a-pVQZ*	*CBS*
		H_3CCl . . . OCH_2		
HF	0.63	0.66	0.65	
MP2	−0.86	−1.11	−1.19	−1.25
CCSD(T)	−0.78	−1.05	−1.12	−1.18
		H_3CBr . . . OCH_2		
HF	0.29	0.36	0.37	
MP2	−1.37	−1.61	−1.69	−1.75
CCSD(T)	−1.24	−1.49	−1.58	−1.64
		H_3CBr . . . $OCH_2{}^a$		
HF	0.20	0.27	0.28	
MP2	−1.44	−1.68	−1.76	−1.82
CCSD(T)	−1.32	−1.57	−1.65	−1.71
		H_3CI . . . OCH_2		
HF	−0.33	−0.21	−0.21	
MP2	−2.08	−2.34	−2.43	−2.50
CCSD(T)	−1.87	−2.15	−2.25	−2.32

^{a}Br described by pseudopotentials

Table 4.2 SAPT decomposition of the interaction energies (kcal/mol) for the
H$_3$CCl . . . OCH$_2$, H$_3$CBr . . . OCH$_2$, and H$_3$CI . . . OCH$_2$ complexes. (Chlorine is described using the aug-cc-pVXZ basis sets,
while bromine and iodine are described using the aug-cc-pVXZ-PP
basis sets, CBS refers to the extrapolated complete basis-set limit, X
denotes D, T, Q).

		H$_3$CCl . . . OCH$_2$		
	aug-cc-pVDZ	*aug-cc-pVTZ*	*aug-cc-pVQZ*	*CBS*
E(elec.)	−1.01	−0.96	−0.96	−0.96
E(ind.)	−0.22	−0.23	−0.23	−0.23
E(disp.)	−1.55	−1.81	−1.89	−1.96
E(exch.)	2.03	2.02	2.02	2.02
ΔE_{int}^{SAPT}	−0.75	−0.98	−1.07	−1.13
		H$_3$CBr . . . OCH$_2$		
	aug-cc-pVDZ	*aug-cc-pVTZ*	*aug-cc-pVQZ*	*CBS*
E(elec.)	−1.56	−1.47	−1.46	−1.45
E(ind.)	−0.36	−0.37	−0.37	−0.37
E(disp.)	−1.69	−1.98	−2.08	−2.15
E(exch.)	2.12	2.12	2.11	2.11
ΔE_{int}^{SAPT}	−1.49	−1.70	−1.80	−1.86
		H$_3$CI . . . OCH$_2$		
	aug-cc-pVDZ	*aug-cc-pVTZ*	*aug-cc-pVQZ*	*CBS*
E(elec.)	−2.77	−2.61	−2.61	−2.60
E(ind.)	−0.77	−0.78	−0.78	−0.79
E(disp.)	−1.91	−2.31	−2.44	−2.54
E(exch.)	3.01	3.01	2.98	2.96
ΔE_{int}^{SAPT}	−2.45	−2.67	−2.85	−2.96

Coupled cluster (CCSD(T)/aug-cc-pVTZ) calculations indicate that the
binding energies for these types of interactions lie in the range between −1.05
kcal/mol (H$_3$CCl . . . OCH$_2$) and −3.72 kcal/mol (F$_3$CI . . . OCH$_2$). The HF
interaction energy is repulsive (with the exception of iodocomplexes) and
halogen bonding becomes attractive only when correlation energy is taken into
account. One of the most important findings is that, according to symmetry-
adapted perturbation theory (SAPT) analysis, halogen bonds are largely
dependent on both electrostatic and dispersion-type interactions. SAPT analyses of halogen bonds in systems containing chlorine and bromine indicate
that halogen-bonding interactions involving these halogen atoms are principally dispersive in nature, although electrostatic contributions to halogen
bonds are not negligible. The electrostatic contribution to the interaction
energy in halogen bonding increases as the size of the halogen-bonding halogen
increases. The most dominant physical component of interactions for systems
containing iodine is the electrostatic one, which accounts for slightly more than
half of the total binding energy. Upon substitution of fluorine atoms, which are
very electronegative, onto the halomethanes, halogen bonds become more
stable and more electrostatic (and less dispersive) in nature. Halogen-bonding

interactions also become stronger and more electrostatic upon substitution of (the very electronegative) fluorine into the halomethane molecule.

The occurrence of increased vibrational frequencies (blueshifts) and bond shortening *vs.* decreased frequencies (redshifts) and bond lengthening for the covalent bonds to the atoms having the σ-holes (the σ-hole donors) was investigated recently.[104] Both are possible, depending upon the properties of the donor and the acceptor. Our results are consistent with previous models in relation to blueshifting *vs.* redshifting in hydrogen-bond formation. These models invoke the derivatives of the permanent and the induced dipole moments of the donor molecule.

References

1. O. Gálvez, P. C. Gómez and L. F. Pacios, *J. Chem. Phys.*, 2001, **115**, 11166.
2. IUPAC Task Group 2004-026-2-100, August 24, 2007.
3. G. R. Desiraju and T. Steiner, *The Weak Hydrogen Bond*, Oxford University Press, Oxford, 1999.
4. S. Scheiner, *Hydrogen Bonding*, Oxford University Press, New York, 1997.
5. G. A. Jeffrey, *An Introduction to Hydrogen Bonding*, Oxford University Press, New York, 1997.
6. S.J. Grabowski, ed., *Hydrogen Bonding - New Insights. Series: Challenges and Advances in Computational Chemistry and Physics*, Vol. 3. Springer, Dordrecht, 2006.
7. L. Pauling, *J. Am. Chem. Soc.* 1931, **53**, 1367; L. Pauling, *The Nature of the Chemical Bond*, Cornell University Press, Ithaca, 1939.
8. S. Suzuki, P. G. Green, R. E. Bumgarner, S. Dasgumpta, W. A. Goddard III and G. A. Blake, *Science*, 1992, **257**, 942.
9. R. N. Pribble, A. W. Garret, K. Haber and T. S. Zwier, *J. Chem. Phys.*, 1995, **103**, 531.
10. S. Djafari, G. Lembach, H.-D. Barth and B. Brutschy, *Z. Phys. Chem.*, 1996, **195**, 253.
11. S. Djafari, H.-D. Barth, K. Buchhold and B. Brutschy, *J. Chem. Phys.*, 1997, **107**, 10573.
12. A. C. Legon, B. P. Roberts and A. L. Wallwork, *Chem. Phys. Lett.*, 1990, **173**, 107.
13. T. Steiner and G. R. Desiraju, *Chem. Commun.*, 1998, 891.
14. T. Steiner et al. *J. Chem. Soc. Perkin Trans. 2*, 1996, 2441; *J. Chem. Phys.* 1992, **96**, 1787; *J. Chem. Soc. Perkin Trans. 2*, 1995,1321; *J. Chem. Phys.*, 1992, **96**, 7321; *J. Am. Chem. Soc.*, 1997, **119**, 4232.
15. S. Pinchas, *Anal. Chem.*, 1957, **29**, 334.
16. N. P. Resaev and K. Szczepaniak, *Opt. Spectr. (USSR)*, 1964, **16**, 43.
17. G. T. Trudeau, J.-M. Dumas, P. Dupuis, M. Guerin and C. Sandorfy, *Top. Curr. Chem.*, 1980, **93**, 91.

18. H. Satonaka, K. Abe and M. Hirota, *Bull. Chem. Soc. Jpn.*, 1987, **60**, 953.

19. M. Budišínský, P. Fiedler and Z. Arnold, *Synthesis*, 1989, 858.

20. I. E. Boldeskul, I. F. Tsymbal, E. V. Ryltsev, Z. Latajka and A. J. Barnes, *J. Mol. Struct.*, 1997, **167**, 436.

21. A. V. Iogansen, *Spectrochim. Acta*, 1999, **55A**, 1585.

22. W. Caminati, S. Melandri, P. Moreschini and P. G. Favero, *Angew. Chem., Int. Ed.*, 1999, **38**, 2924.

23. N. Karger, A. M. Amorin da Costa and P. J. A. Riberio-Claro, *J. Phys. Chem. A*, 1999, **103**, 8672.

24. K. Miyumo, S. Imafuji, T. Ochi, T. Ohta and S. Maeda, *J. Phys. Chem. B*, 2000, **104**, 11001.

25. P. Hobza, V. Špirko, H. L. Selzle and E. W. Schlag, *J. Phys. Chem. A*, 1998, **102**, 2501.

26. P. Hobza and Z. Havlas, *Chem. Rev.*, 2000, **100**, 4253.

27. P. Hobza and Z. Havlas, *Theor. Chem. Acc.*, 2002, **108**, 325.

28. P. Hobza, V. Špirko, Z. Havlas, K. Buchhold, B. Reimann, H.-D. Barth and B. Brutschy, *Chem. Phys. Lett.*, 1999, **299**, 180.

29. C. A. Coulson, *Res. Appl. Ind.*, 1957, **10**, 149.

30. A. E. Reed, L. A. Curtiss and F. Weinhold, *Chem. Rev.*, 1988, **88**, 899.

31. K. Hermansson, *J. Phys. Chem.*, 2002, **106**, 4695.

32. P. Hobza and Z. Havlas, *Chem. Phys. Lett.*, 1999, **303**, 447.

33. P. Hobza, *Phys. Chem. Chem. Phys.*, 2001, **3**, 2555.

34. P. Hobza and V. Špirko, *Phys. Chem. Chem. Phys.*, 2003, **5**, 1290.

35. E. Nir, C. h. Janzen, P. Imhof, K. Kleinermanns and M. S. de Vries, *Phys. Chem. Chem. Phys.*, 2002, **4**, 740.

36. A. Karpfen and E.S. Kryachko, *J. Phys. Chem. A*, 2003, **107**, 9724; 2005, **109**, 8930.

37. B. J. van der Veken, W. A. Herrebout, R. Szostak, D. N. Shepkin, Z. Havlas and P. Hobza, *J. Am. Chem. Soc.*, 2001, **123**, 12290.

38. I. V. Alabugin, M. Manoharan, S. Peabody and F. Weinhold, *J. Am. Chem. Soc.*, 2003, **125**, 5973.

39. W. Wang, M. Pitoňák and P Hobza, *ChemPhysChem*, 2007, **8**, 2107.

40. B. Reimann, K. Vaupel, S. Buchhold, B. Brutschy, Z. Havlas and P. Hobza, *J. Phys. Chem. A*, 2001, **105**, 5560.

41. P. Hobza and Z. Havlas, *Chem. Phys. Lett.*, 1999, **303**, 447.

42. W. Zierkiewicz, P. Jurečka and P. Hobza, *ChemPhysChem*, 2005, **6**, 609.

43. J. Joseph and E. D. Jemmis, *J. Am. Chem. Soc.*, 2007, **129**, 4620.

44. W. Wang and P. Hobza, *Coll. Czech. Chem. Commun.*, 2008, **73**, 862.

45. E. S. Kryachko and A. Karpfen, *Chem. Phys.*, 2006, **329**, 313.

46. A. Karpfen and E. S. Kryachko, *J. Phys. Chem. A*, 2007, **111**, 8177.

47. S. Scheiner, *Reviews in Computational Chemistry*, ed. K.B. Lipkowitz, D.B. Boyd, VCH Publishers Inc., Weinheim, 1991.

48. T. Steiner, *Angew. Chem., Int. Ed.*, 2002, **41**, 48.

49. C. A. Morrison and M. M. Siddick, *Chem. Eur. J.*, 2003, **9**, 628.

50. J. Langlet, J. Caillet, J. Bergès and P. Reinhardt, *J. Chem. Phys.*, 2003, **118**, 6157.

51. S. Raugei and M. L. Klein, *J. Am. Chem. Soc.*, 2003, **125**, 8992.

52. A. Ranganathan, G. U. Kulkarni and C. N. R. Rao, *J. Phys. Chem. A*, 2003, **107**, 6073.

53. A. van der Vaart and K. M. Merz Jr, *J. Chem. Phys.*, 2002, **116**, 7380.

54. D. M. Rudkevich and J. Rebek Jr, *Eur. J. Org. Chem.*, 1999, 1991.

55. P. R. Ashton, V. Baldoni, V. Balzani, A. Credi, H. D. A. Hoffmann, M.-V. Martínez-Díaz, F. M. Raymo, J. F. Stoddart and M. Venturi, *Chem. Eur. J.*, 2001, **7**, 3482.

56. A. Jayaraman, V. Balasubramaniam and S. Valiyaveettil, *Cryst. Growth Design*, 2006, **6**, 360.

57. V. Berl, I. Huc, R. G. Khoury and J.-M. Lehn, *Chem. Eur. J.*, 2001, **7**, 2810.

58. V. T. Nguyen, P. D. Ahn, R. Bishop, M. L. Scudder and D. C. Craig, *Eur. J. Org. Chem.*, 2001, 4489.

59. (a) L. A. Curtiss, D. J. Frurip and M. Blander, *J. Chem. Phys.*, 1979, **71**, 2703; (b) T. Köddermann, F. Schulte, M. Huelsekopf and R. Ludwig, *Angew. Chem., Int. Ed.*, 2003, **42**, 4904; (c) E. M. Mas, R. Bukowski and K. Szalewicz, *J. Chem. Phys.*, 2003, **118**, 4386; (d) E. M. Mas, R. Bukowski and K. Szalewicz, *J. Chem. Phys.*, 2003, **118**, 4404; (e) B.-Q. Ma, H.-L. Sun and S. Gao, *Angew. Chem., Int. Ed.*, 2004, **43**, 1374; (f) J.-L. Kuo, C. V. Ciobanu, L. Ojamäe, I. Shavitt and S. J. Singer, *J. Chem. Phys.*, 2003, **118**, 3583; (g) A. W. Omta, M. F. Kropman, S. Woutersen and H. J. Bakker, *Science*, 2003, **301**, 347; (h) S. Pal, N. B. Sankaran and A. Samanta, *Angew. Chem., Int. Ed.*, 2003, **42**, 1741; (i) A. V. Larin, D. N. Trubnikov and D. P. Vercauteren, *Int. J. Quantum Chem.*, 2003, **92**, 71.

60. (a) B. J. Howard, T. R. Dyke and W. Klemperer, *J. Chem. Phys.*, 1984, **81**, 5417; (b) D. C. Dayton, K. W. Jucks and R. E. Miller, *J. Chem. Phys.*, 1989, **90**, 2631.

61. (a) A. Fu, D. Du and Z. Zhou, *Int. J. Quantum Chem.*, 2004, **97**, 865; (b) E. M. Cabaleiro-Lago and J. R. Otero, *J. Chem. Phys.*, 2002, **117**, 1621; (c) L. Pejov, *Int. J. Quantum Chem.*, 2003, **92**, 516; (d) P. M. Dominiak, E. Grech, G. Barr, S. Teat, P. Mallinson and K. Wozniak, *Chem. Eur. J.*, 2003, **9**, 963; (e) G. C. Cole, A. C. Legon and P. Ottaviani, *J. Chem. Phys.*, 2002, **117**, 2790; (f) S. W. Hunt, K. J. Higgins, M. B. Craddock, C. S. Brauer and K. R. Leopold, *J. Am. Chem. Soc.*, 2003, **125**, 13850; (g) R. Wieczorek and J. J. Dannenberg, *J. Am. Chem. Soc.*, 2003, **125**, 14065; (h) N. L. Pivonka, C. Kaposta, M. Brümmer, G. v. Helden, G. Meijer, L. Wöste, D. M. Neumark and K. R. Asmis, *J. Chem. Phys.*, 2003, **118**, 5275; (i) W. M. Westler, P. A. Frey, J. Lin, D. E. Wemmer, H. Morimoto, P. G. Williams and J. L. Markley, *J. Am. Chem. Soc.*, 2002, **124**, 4196; (j) S. F. Boys and F. Bernardi, *Mol. Phys.*, 1970, **19**, 372; (k) L. Šroubková and R. Zahradník, *Helv. Chim. Acta*, 2001, **84**, 1328.

62. (a) D. J. Sutor, *Nature*, 1962, **195**, 68; (b) B. Wang, J. F. Hinton and P. Pulay, *J. Phys. Chem. A*, 2003, **107**, 4683; (c) K. Oku, H. Watanabe, M.

Kubota, S. Fukuda, M. Kurimoto, Y. Tsujisaka, M. Komori, Y. Inoue and M. Sakurai, *J. Am. Chem. Soc.*, 2003, **125**, 12739; (d) Ch. Ramos, P. R. Winter, J. A. Stearns and T. S. Zwier, *J. Phys. Chem. A*, 2003, **107**, 10280; (e) L. George, E. Sanchez-García and W. Sander, *J. Phys. Chem. A*, 2003, **107**, 6850; (f) H. Matsuura, H. Yoshida, M. Hieda, S. Yamanaka, T. Harada, K. Shin-ya and K. Ohno, *J. Am. Chem. Soc.*, 2003, **125**, 13910; (g) S. N. Delanoye, W. A. Herrebout and B. J. van der Veken, *J. Am. Chem. Soc.*, 2002, **124**, 7490.

63. (a) M. A. Blatchford, P. Raveendran and S. L. Wallen, *J. Phys. Chem. A*, 2003, **107**, 13011; (b) R. D. Parra, S. Bulusu and X. C. Zeng, *J. Chem. Phys.*, 2003, **118**, 3499; (c) A. Noman, M. M. Rahman, R. Bishop, D. C. Craig and M. L. Scudder, *Eur. J. Org. Chem.*, 2003, 72; (d) J. Parsch and J. W. Engels, *J. Am. Chem. Soc.*, 2002, **124**, 5664; (e) X. Li, L. Liu and H. B. Schlegel, *J. Am. Chem. Soc.*, 2002, **124**, 9639; (f) A. Karpfen and E. S. Kryachko, *J. Phys. Chem. A.*, 2003, **107**, 9724; (g) A. Donati, S. Ristori, C. Bonechi, L. Panza, G. Martini and C. Rossi, *J. Am. Chem. Soc.*, 2002, **124**, 8778; (h) C. E. Cannizzaro and K. N. Houk, *J. Am. Chem. Soc.*, 2002, **124**, 7163; (i) S. A. C. McDowell, *J. Chem. Phys.*, 2003, **118**, 7283; (j) R. Zahradník and L. Šroubková, *Helv. Chim. Acta*, 2003, **86**, 979.

64. (a) Ch. F. Matta, J. Hernández-Trujillo, T.-H. Tang and R. F. W. Bader, *Chem. Eur. J.*, 2003, **9**, 1940; (b) P. Bombicz, M. Czugler, R. Tellgren and A. Kálmán, *Angew. Chem., Int. Ed.*, 2003, **42**, 1957; (c) Z. Havlas, E. Bauwe and R. Zahradník, *Chem. Phys. Lett.*, 1985, **121**, 330.

65. J. C. Lee, E. Peris, A. L. Rheingold and R. H. Crabtree, *J. Am. Chem. Soc.*, 1994, **116**, 11014.

66. (a) T. B. Richardson, S. De Gala and R. H. Crabtree, *J. Am. Chem. Soc.*, 1995, **117**, 12875; (b) W. T. Klooster, T. F. Koetzle, P. E. M. Siegbahn, T. B. Richardson and R. H. Crabtree, *J. Am. Chem. Soc.*, 1999, **121**, 6337.

67. Q. Liu and R. Hoffmann, *J. Am. Chem. Soc.*, 1995, **117**, 10108.

68. C. J. Cramer and W. L. Gladfelter, *Inorg. Chem.* 1997, **36**, 5358; P. L. A. Popelier, *J. Phys. Chem. A* 1998, **102**, 1873; T. B. Richardson, S. deGala, R. H. Crabtree and P. E. M. Siegbahn, *J. Am. Chem. Soc.* 1995, **117**, 12875.

69. W. Zierkiewicz and P. Hobza, *Phys. Chem. Chem. Phys.*, 2004, **6**, 5288.

70. V. Bakhmutov, *Dihydrogen Bonds: Principles, Experiments, and Applications*, Wiley-Interscience, Hoboken, NJ, 2008.

71. G. N. Patwari, T. Ebata and N. Mikami, *J. Chem. Phys.* 2000, **113**, 9885; 2002, **116**, 6056; G. N. Patwari, T. Ebata and N. Mikami, *Chem. Phys.* 2002, **283**, 193.

72. P. C. Singh and G. N. Patwari, *J. Phys. Chem. A*, 2007, **111**, 3178.

73. P. Cigler, *et al. Proc. Natl. Acad. Sci. USA*, 2005, **102**, 15394.

74. J. Fanfrlík, D. Hnyk, M. Lepšík and P. Hobza, *Phys. Chem. Chem. Phys.*, 2007, **9**, 2085.

75. J. Fanfrlík, M. Lepšík, D. Horinek, Z. Havlas and P. Hobza, *ChemPhysChem*, 2006, **7**, 1100.

76. P. Auffinger, F. A. Hays, E. Westhof and P. S. Ho, *Proc. Nat. Acad. Sci. USA*, 2004, **101**, 16789.
77. R. Battistutta, M. Mazzorana, S. Sarno, Z. Kazimierczuk, G. Zanotti and L. A. Pinna, *Chem. Biol.*, 2005, **12**, 1211.
78. A. Brouwer, D. C. Morse, M. C. Lans, A. G. Schuur, A. J. Murk, E. Klasson-Wehler, A. Bergman and T. J. Visser, *Toxicol. Ind. Health*, 1998, **14**, 59.
79. M. Ghosh, I. A. T. M. Meerts, A. Cook, A. Bergman, A. Brouwer and L. N. Johnson, *Acta Crystallogr. Section D-Biol. Crystallogr.*, 2000, **56**, 1085.
80. D. M. Himmel, K. Das, A. D. Clark, S. H. Hughes, A. Benjahad, S. Oumouch, J. Guillemont, S. Coupa, A. Poncelet, I. Csoka, C. Meyer, K. Andries, C. H. Nguyen, D. S. Grierson and E. Arnold, *J. Med. Chem.*, 2005, **48**, 7582.
81. Y. Jiang, A. A. Alcaraz, J. M. Chen, H. Kobayashi, Y. J. Lu and J. P. Snyder, *J. Med. Chem.*, 2006, **49**, 1891.
82. M. C. Lans, E. Klassonwehler, M. Willemsen, E. Meussen, S. Safe and A. Brouwer, *Chemico-Biol. Interact.*, 1993, **88**, 7.
83. M. C. Lans, C. Spiertz, A. Brouwer and J. H. Koeman, *Eur. J. Pharmacol.-Environ. Toxicol. Pharmacol. Sect.*, 1994, **270**, 129.
84. M. L. Lopez-Rodriguez, M. Murcia, B. Benhamu, A. Viso, M. Campillo and L. Pardo, *J. Med. Chem.*, 2002, **45**, 4806.
85. G. Trogdon, J. S. Murray, M. C. Concha and P. Politzer, *J. Molec. Mod.*, 2007, **13**, 313.
86. A. R. Voth, F. A. Hays and P. S. Ho, *Proc. Nat. Acad. Sci. USA*, 2007, **104**, 6188.
87. R. Bertani, F. Chaux, M. Gleria, P. Metrangolo, R. Milani, T. Pilati, G. Resnati, M. Sansotera and A. Venzo, *Inorg. Chim. Acta*, 2007, **360**, 1191.
88. K. Boubekeur, J. L. Syssa-Magale, P. Palvadeau and B. Schollhorn, *Tetrahedron Lett.*, 2006, **47**, 1249.
89. E. Cariati, A. Forni, S. Biella, P. Metrangolo, F. Meyer, G. Resnati, S. Righetto, E. Tordin and R. Ugo, *Chem. Commun.*, 2007, 2590.
90. T. Caronna, R. Liantonio, T. A. Logothetis, P. Metrangolo, T. Pilati and G. Resnati, *J. Am. Chem. Soc.*, 2004, **126**, 4500.
91. D. Chopra, V. Thiruvenkatam, S. G. Manjunath and T. N. G. Row, *Cryst. Growth Des.*, 2007, **7**, 868.
92. E. Corradi, S. V. Meille, M. T. Messina, P. Metrangolo and G. Resnati, *Angew. Chem., Int. Ed.*, 2000, **39**, 1782.
93. G. Marras, P. Metrangolo, F. Meyer, T. Pilati, G. Resnati and A. Vij, *New J. Chem.*, 2006, **30**, 1397.
94. P. Metrangolo, H. Neukirch, T. Pilati and G. Resnati, *Acc. Chem. Res.*, 2005, **38**, 386.
95. P. Metrangolo, G. Resnati, T. Pilati, R. Liantonio and F. Meyer, *J. Polym. Sci. Part A-Polym. Chem.*, 2007, **45**, 1.
96. S. Sourisseau, N. Louvain, W. H. Bi, N. Mercier, D. Rondeau, F. Boucher, J. Buzare and C. Legein, *Chem. Mater.*, 2007, **19**, 600.

97. P. Politzer, J. S. Murray and M. C. Concha, *J. Molec. Mod.*, 2007, **13**, 643.
98. T. Clark, M. Hennemann, J. S. Murray and P. Politzer, *J. Molec. Mod.*, 2007, **13**, 291.
99. P. Politzer, P. Lane, M. C. Concha, Y. G. Ma and J. S. Murray, *J. Molec. Mod.*, 2007, **13**, 305.
100. J. P. M. Lommerse, A. J. Stone, R. Taylor and F. H. Allen, *J. Am. Chem. Soc.*, 1996, **118**, 3108.
101. K. E. Riley and K. M. Merz, *J. Phys. Chem. A*, 2007, **111**, 1688.
102. G. Valerio, G. Raos, S. V. Meille, P. Metrangolo and G. Resnati, *J. Phys. Chem. A*, 2000, **104**, 1617.
103. K. E. Riley and P. Hobza, *J. Chem. Theory Comput.*, 2008, **4**, 232.
104. J. S. Murray, M. C. Concha, P. Lane, P. Hobza and P. Politzer, *J. Molec. Mod.*, 2008, **14**, 699.

CHAPTER 5

Interpretation of Experimental Results and Types of Molecular Clusters

Interpretation of experimental data of non-covalent systems is not straightforward and requires a detailed knowledge of the studied non-covalent systems; moreover, several additional factors should be taken into consideration. The first factor concerns the temperature of the experiment. This is associated with the fact that in the world of non-covalent species entropy always plays an important, sometimes even decisive role. When translational, vibrational and rotational temperatures are low we can safely use information obtained from the PES. When, however, the temperature is nonzero, then it is inevitable that the system will pass from the PES to the free-energy surface (FES). Determination of temperature in beam experiments is sometimes difficult. It is clear that rotational and vibrational temperature after expansion should be close to 0 K. The freezing is, however, so fast that conformers existing at high temperature do not have time to relax to low-temperature conformers. It is thus necessary to consider the conformers at higher (closer to room) temperatures and not at very low temperatures close to 0 K. The second factor is associated with the fact that the whole surface should be known, which means that all energy minima should be considered and must be properly weighted. The experimental information on, *e.g.* stabilisation enthalpy should be thus considered not only for the most stable or most highly populated structure but also other structures should be taken into account. This is important because the character of the PES and FES can be different. Specifically, the global minimum at the PES frequently becomes a local minimum at the FES and, further, the order of energy minima depends on the temperature. Finally, the role of the environment should be properly understood and adequate theoretical calculations should be performed. It is no longer considered appropriate to perform calculations

RSC Theoretical and Computational Chemistry Series No. 2
Non-covalent Interactions: Theory and Experiment
By Pavel Hobza and Klaus Müller-Dethlefs
© Pavel Hobza and Klaus Müller-Dethlefs 2010
Published by the Royal Society of Chemistry, www.rsc.org

in the gas phase and to use these results for interpretation of liquid-phase experiments, a fact that is now more or less accepted in the scientific community. We will show, however, some less-frequent examples concerning the interpretation of experimental results from He-droplets experiments.

5.1 Molecule . . . Rare-Gas Atom Clusters

Molecular non-covalent complexes have attracted much interest, because of their low binding energies, large equilibrium distances and very low frequency intermolecular vibrations. In particular, vibrationally resolved spectra of aromatic noble-gas complexes provide important benchmarks for the understanding and modelling of intermolecular forces. In particular, upon ionisation many of these intermolecular vibrations become Franck–Condon allowed and can be observed, whereas in electronic excitation spectra the geometry change from the neutral ground state to the electronically excited state is often small and hence not many intermolecular vibrations are seen. The first ZEKE study of a van der Waals complex was carried out by Chewter *et al.*[1] on the benzene . . . argon cluster, from which the determination of the ionisation energy of the cluster was possible. Van der Waals vibrations in a ZEKE spectrum have been observed for the first time for the NO . . . Ar and the aniline . . . Ar_2 complexes.[2–4] Complexes containing aniline or phenol are interesting because there are two principle ligand binding sites; hydrogen bonding via the NH_2 or OH group (H-bound), and van der Waals bonding via interaction with the aromatic π system (π-bound). Therefore the investigation of their spectra is very interesting as to which kind of bond is formed or if there are different stable configurations.

For the n-butylbenzene monomer four stable isomers have been detected. The influence of these different isomers on the covalent interactions is very subtle and an important difference exists for the *gauche*- and the *anti*-conformers.

If more than one argon atom is participating in the cluster formation then additional configurations are possible. For π-bound aromatic systems these possible isomers will be characterised by (n/m) where n and m represent the number of rare gas atoms attached to each side of the aromatic ring. The question of possible isomers will be discussed for the examples of aniline . . . Ar_2 and phenol . . . Ar_2. For both complexes recent investigations have shown the additional possibility that a π–H-isomeration occurs. The investigation of dynamics such as IVR and predissociation will be discussed later for the example of aniline . . . Ar.

5.1.1 NO . . . Ar

Weakly bound complexes of small inorganic molecules with rare-gas atoms are of fundamental interest since these complexes represent prototypes for characterising simple multipole-induced dipole interactions.[5–8] Clusters containing open-shell molecules, *e.g.* Ar . . . NO, can be used to study the perturbation of the open-shell molecule by the rare-gas atom by observing the coupling and quenching of angular momenta. A number of REMPI and ZEKE studies have been performed on Ar...NO, and for earlier work the reader is referred to a 1994 review.[9-11]

Bush *et al.*[5] performed a detailed REMPI and ZEKE study that has concentrated on the A $^2\Sigma^+$ state of the neutral. A simulation of the A $^2\Sigma^+ \leftarrow X \,^2\Pi$ REMPI spectrum (using the rare-gas-open-shell molecule model of Hutson[12]) suggested that the ground vibrationless level of the A $^2\Sigma^+$ state has a linear geometry, while some of the higher vibrational levels display a skewed T-shaped structure. This assignment was confirmed by obtaining ZEKE spectra via both sets of vibrations. Spectra recorded *via* the ground vibrationless level show progressions that correlate with transitions to highly excited vdW stretching and bending levels, while spectra recorded *via* intermediate levels with a T-shaped geometry show progressions that correlate principally with the vdW stretching mode. In both cases, no vibrational structure was detected close to the ionisation threshold but peaks at higher excitation energy were observed, corresponding to high-lying vibrational levels of the Ar...NO$^+$ ion. These observations were interpreted in terms of large changes in the intermolecular bond length and bond angle upon ionisation and discussed in terms of the vdW interactions that are present in the neutral and cationic systems. Analysis of the vibrational levels of the A state indicate that the vdW bond is considerably longer than that calculated for Ar...NO$^+$, and the A state is therefore non-Rydberg in character. This behaviour was attributed to the Ar atom being too large to fit within the average orbit of the NO 3s Rydberg electron. The spectra also reveal that Ar...NO adopts a linear geometry in the A state, indicating that dipole–induced-dipole and induced-dipole–induced-dipole forces dominate over the quadrupole–induced-dipole and exchange-repulsion interactions.

5.1.2 Benzene...Ar

Electronic excitation spectra. Van der Waals complexes consisting of benzene and a noble-gas atom were the subjects of many experimental[1,13-17] and theoretical[18-20] investigations. For these complexes it could be shown by rotational resolved UV spectroscopy that the noble-gas atom is bound above the aromatic ring system.[16,21]

ZEKE spectroscopy. The ZEKE spectrum observed by Chewter *et al.*[1] revealed only one peak, assigned to the ionisation threshold and being shifted with respect to the benzene origin by 172 cm^{-1}. From that spectrum and the rather small change in ionisation energy compared to the benzene monomer, the conclusion was drawn that ionisation of the complex does not lead to a drastic increase in the binding energy of the argon atom. Using MATI dissociation spectroscopy Krause and Neusser were able to obtain an upper limit of 629 cm^{-1} for the binding energy in the ionic ground state.[22] Due to comparison with the results obtained for benzene...Kr and theoretical calculations performed by Hobza *et al.*[23] the dissociation threshold in the neutral ground state was determined to be smaller than 340 cm^{-1}.

Aiger *et al.* showed that the application of a fast-switching electric pulse enhances the lifetime of Rydberg states in the benzene...Ar complex below $n = 55$ to many tens of microseconds.[24]

5.1.3 C_6H_5X...Ar Above-Ring π-Bound Clusters

Above-ring π-binding was also observed for mono substituted molecules of the type C_6H_5X (X=F, Cl, CH_3, OH, NH_2) [25–29] and for *para* disubstituted molecules of the type $C_6H_4Y_2$ (Y=Cl, F, CH_3).[30–33] For all these systems a redshift occurs for the transition energy into the first electronically excited state S_1 as well as for the ionisation energy. Compared to the monomer this shift reflects a stronger stabilisation of the complex in the first excited state as well as in the cation ground state. The frequencies of the in-plane vibrations were found to be hardly affected by the presence of the noble-gas atom. However, several complexes showed a small blueshift for the frequencies of the out of plane vibrations.[30,34–36] Additionally, the influence of the polarisability of the noble-gas atom on the binding energy and the structure was investigated by changing from argon to krypton and xenon. Even for the large and hence strongly polarisable xenon atom only binding above the aromatic π-system was found[34] and not at the substituted halogen atom. However, an increasing redshift of the excitation energy and a frequency decrease of the intermolecular stretching mode s_z were observed by increasing the polarisability, because the higher polarisability is responsible for a stronger binding. The strength of the vdW binding energy (BE) also follows this order:

BE(benzene...Ar) < BE(fluorobenzene...Ar) < BE(chlorobenzene...Ar).[37–39]

Hence, the magnitude of the polarisability of the halogen atom is the cause of the increasing binding energy. An overview of the shift in the transition and ionisation energies as well as the strength of the binding energy of some so far experimentally examined molecules is given in Table 5.1.

The change from benzene to monosubstituted benzene derivates causes an intensity increase of the intermolecular bending vibrations, in particular of the symmetric bending vibration b_x[26] during excitation in the S_1 state as well during ionisation. While these vibrations were hardly observed for benzene noble-gas complexes,[13] they were seen with high intensity for instance for fluorobenzene.[25] Additionally a shift of the Franck–Condon maximum towards three or four quanta was observed for fluorobenzene...Ar. This is due to a shift of the argon atom in the direction of the halogen atom during excitation and ionisation. It could be shown that the polarisability of the noble-gas atom, especially in the first excited state, strongly affects the shift of the Franck–Condon maximum, while it is less important for the ionisation process.

5.1.4 $C_6H_4Cl_2$...Ar: (*p,m,o*-dichlorobenzene)...Ar Clusters

R2PI and MATI spectroscopic investigations were performed for the non-covalent clusters of all three dichlorobenzene (DCB) isomers with argon, to

Table 5.1 Overview of experimental results for some non-covalent clusters of type molecule ... Ar. All values are given in cm^{-1}. ΔEE = change in excitation energy; ΔIE = change in ionisation energy; BE(S$_0$) = binding energy in (S$_0$); BE(D$_0$) = binding energy in (D$_0$).

Argon ... molecule	Energy shifts		Binding energies		Ref.
	ΔEE	ΔIE	BE(S$_0$)	BE(D$_0$)	
benzene	21	172	314 ± 7	486 ± 5	40
toluene	26	161			27,41
phenol	33	152	364 ± 13	535 ± 5	42,76
ethylbenzene	20	128			43
fluorobenzene	24	223	346	568	25
chlorobenzene	26	189	337–525	<714	44
p-difluorobenzene	30	237	339 ± 4	576 ± 4	45,31,33
p-fluorotoluene	35	181	329 ± 20	510 ± 20	46
p-dichlorobenzene	30				30
dibenzo-p-dioxine	49	212	505	717	47

investigate the influence of the position of the chlorine atoms especially on the binding energies.[48] Figure 5.1 shows optimised structures of the neutral ground of the clusters. Optimisations were performed using MP2 theory with Dunning's aug-cc-pVDZ basis set. Calculated binding energies were corrected for BSSE by using the counterpoise method of Boys and Bernardi and the positions of the argon atom are listed in Table 5.2.

The excitation energies (EE) of the first excited state of the *o*-, *m*- and *p*-dichlorobenzene ... argon clusters were determined as $36212 \pm 3\,\text{cm}^{-1}$, $36170 \pm 3\,\text{cm}^{-1}$ and $35732 \pm 3\,\text{cm}^{-1}$, respectively, redshifted by 26 cm^{-1}, 22 cm^{-1} and 30 cm^{-1} compared to the corresponding monomers. The lowering of the excitation energies can be explained by a stronger stabilisation of the clusters investigated in the first excited state compared to the ground state.

From the MATI spectra it could be found that the ionisation energies follow the order IE(*para*) < IE(*ortho*) < IE(*meta*), which seems to be a general trend for disubstituted benzene monomers.[49]

5.1.5 N-Butylbenzene ... Ar (BB ... Ar)

REMPI and MATI spectroscopic investigations, supported by extensive *ab initio* computations, of the BB ... Ar complex where performed by Tong *et al.*[50] Figure 5.2 (lower trace) displays the REMPI spectrum of BB ... Ar, while the corresponding spectrum of the BB monomers is shown in Figure 5.2 (upper trace).[51] The cluster REMPI spectrum is similar to the one for the monomer, which can be divided into *gauche*- and *anti*-ranges. The origins of the four detected conformers together with the spectral shifts to the corresponding monomers are listed in Table 5.3. Figure 5.3 shows the calculated structures for the four conformers in the neutral electronic ground state. Comparing the shifts for the different conformers (Table 5.3) one can see that the two different types

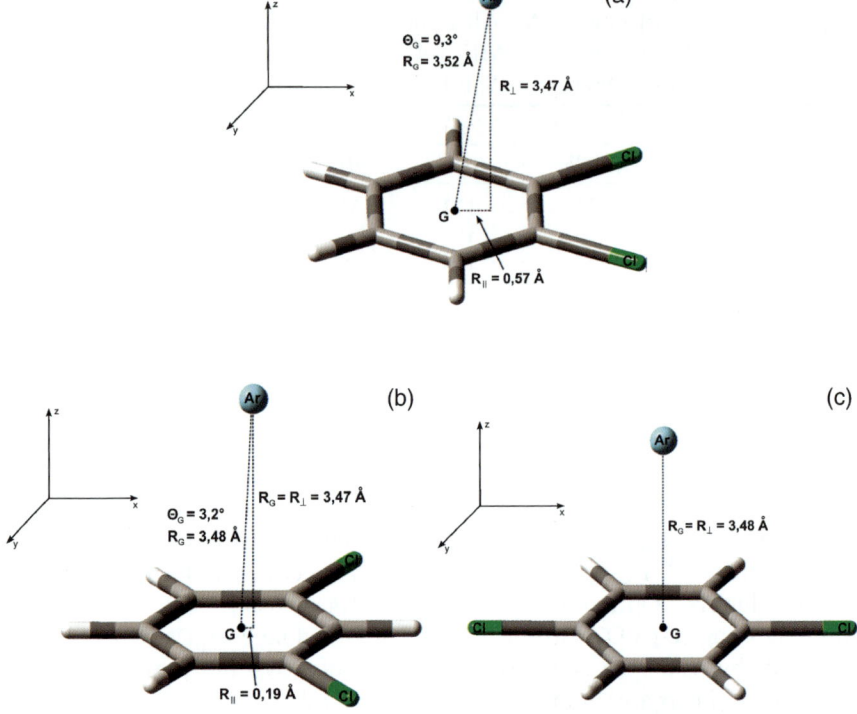

Figure 5.1   Calculated Geometries in the neutral ground state (a) *o*-DCB...Ar, (b) *m*-DCB...Ar and (c) *p*-DCB...Ar. Redrawn from Ref. 48.

Table 5.2   Calculated geometries and binding energies in the neutral ground state.

Argon vdW molecule	$BE(S_0)$ $[cm^{-1}]$	$R_\perp$ $[\mathring{A}]$	$R_\parallel$ $[\mathring{A}]$
benzene	316	3.6	0
chlorobenzene	355	3.51	0.43
o-dichlorobenzene	451	3.47	0.57
m-dichlorobenzene	433	3.47	0.19
p-dichlorobenzene	412	3.48	0

of **BB** conformers, *gauche*- and *anti*-, give different shift directions for the $S_1 \leftarrow S_0$ excitation energies, which was observed here for the first time for 1:1 aromatic...rare-gas clusters. The most probable explanation is that the enhanced attractive interaction between the alkyl chain and aromatic ring in the *gauche*-conformer in the S_1 state has been weakened by the presence of the argon atom. In addition, it also indicates that the argon is attached to the side of the aromatic ring where the alkyl chain bends towards.

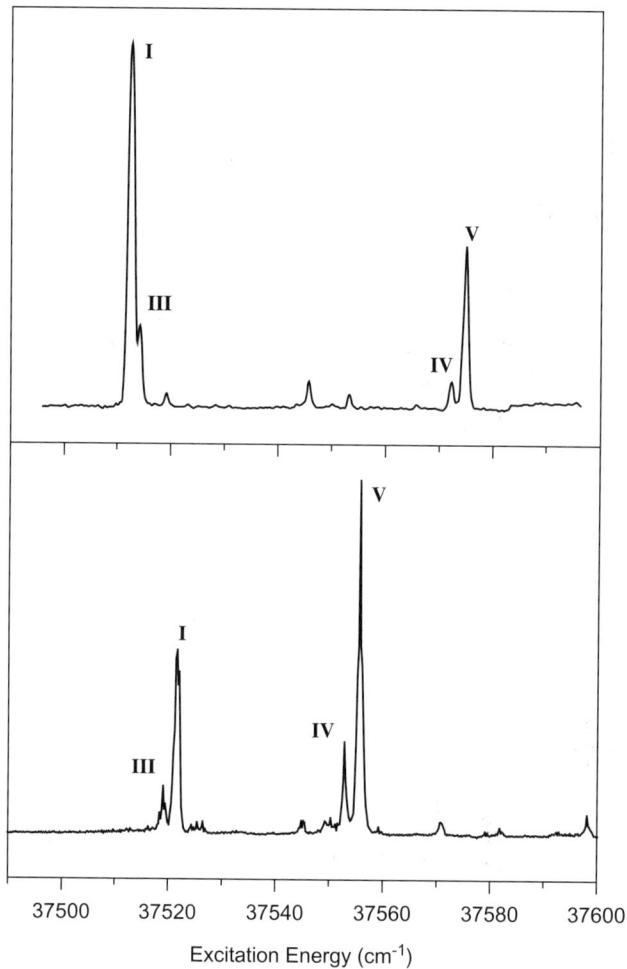

Figure 5.2 Two-colour $(1 + 1')$ $S_1 \leftarrow S_0$ REMPI spectrum of (upper trace) BB and (lower trace) BB...Ar, recorded with the ionisation laser set to $32\,700\,\mathrm{cm}^{-1}$. Assignments of conformers I, III, IV and V of BB...Ar are included. Redrawn from Ref. 50.

Table 5.3 Adiabatic ionisation energies (in cm^{-1}) and respective complexation shifts (in cm^{-1}) of the alkylbenzenes and their argon clusters obtained using ZEKE/MATI spectroscopy.

Molecule			IE(monomer)	IE(argon cluster)	IE shift
benzene[52,1]		$-$H	74 555	74 383	-172
toluene[53]		$-$CH$_3$	71 203	71 037	-166
ethylbenzene[43]		$-$CH$_2$CH$_3$	70 762	70 634	-128
n-propylbenzene[54]		$-$(CH$_2$)$_2$CH$_3$	70 272	70 160	-112
n-butyl-benzene[50,51]	(I)	$-$(CH$_2$)$_3$CH$_3$	70 148	70 052	-96
	(V)	$-$(CH$_2$)$_3$CH$_3$	69 955	69 845	-110

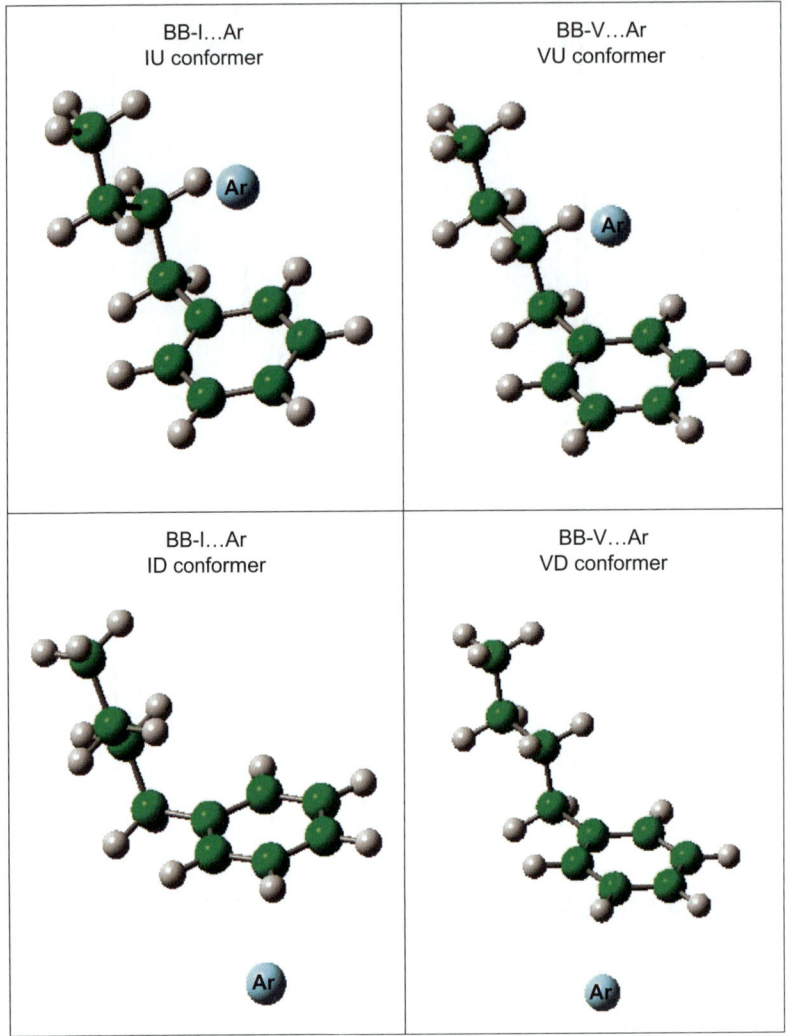

Figure 5.3 Geometric structures of the S_0 state of BB...Ar clusters calculated at the MP2/cc-pVTZ(BB)/aug-cc-pVTZ(Ar) level. Redrawn from Ref. 50.

MATI spectra were recorded for the conformers BB-I...Ar and BB-V...Ar, from which amongst others the ionisation energies could be determined and the assignment could be verified. Table 5.3 lists the ionisation energies of a series of alkylbenzene...Ar clusters. All ionisation energies are redshifted compared to the corresponding monomers, which is due to changes in the intermolecular interactions upon ionisation. The ion-induced dipole interaction, which is present in the cationic state, enhances the vdW bonding strength compared to the neutral ground state, where dipole-induced interactions are present. Increasing the alkyl side chain causes a decrease of the redshift, because the charge of the aromatic ring is more efficiently transferred to the alkyl side chain. This is because the conjugation effect is increased with increase of the side chain. That the argon atom is more weakly bound for the *gauche*-conformer compared to the *anti*-conformer is indicated by the smaller redshift for the *gauche*-conformer. A possible explanation is that the ionisation energy of the *gauche*-conformer is increased by repulsive interactions between some hydrogen atoms in the alkyl side chain and the ion-induced positive side of the argon atom. Another explanation is that the alkyl side chain of conformer I can have stronger conjugation effects than the fully extended side chain of conformer V. The stronger charge transfer to the alkyl side chain leads to a weakening of the ion-induced dipole interactions between Ar and the aromatic ring.

5.1.6 Clustering Dynamics of Bis(η^6-benzene)chromium ... Ar$_n$ Clusters ($n = 1$–15)

Unlike the hydrogen-bonded cluster of which the geometrical layout is determined by the local charge distribution of the solute core, the structures of non-covalent clusters are often considered to be less sensitive to the microscopic structure of the solute because the dispersive force is usually not oriented in a particular direction with respect to the specific nuclear layout.[55,56] Accordingly, the effect of the microscopic solute structure on the whole geometrical layout of the associated van der Waals cluster has been little studied to date. The metal ... benzene sandwich complex, because of its highly symmetric structure, plays an important role in the determination of the cluster structure. The understanding of non-covalent interactions in the close proximity of the organometallic compound might be quite useful for the understanding of the catalytic process at the molecular level.[57] Additionally, because the organometallic compound in porous composites is currently being considered as a potential container for the hydrogen storage,[58,59] van der Waals interaction in the vicinity of the metal ... organic sandwich complex could be quite informative in this regard.

The Ar clustering dynamics of the metal ... benzene sandwich complex, bis(η^6-benzene)chromium or Cr(Bz)$_2$ has been investigated by Choi *et al.*[60] For the Cr(Bz)$_2$ ... Ar$_n$ ($n = 1$–15) cluster the adiabatic ionisation potential (IP) has been determined by recording photoionisation efficiency (PIE) spectra. Additionally MATI spectra have been recorded for the cluster with $n = 0$–2. For $n = 1$–6 the ionisation energy is always redshifted by about 151 cm^{-1} upon addition of a single argon atom, indicating that the binding energy of the cluster becomes higher as the cluster is ionised. The linear dependence of the shift for $n = 1$–6 strongly suggests that the binding energy of the solute–solvent cluster is equally increased as the additional Ar is attached to the solute core until the number of the solvent molecules becomes six. For $n = 7$–12 the shift suddenly reduces to 82 cm^{-1} but retain its linearity. This indicates that the Ar cluster binding site becomes different and its associated binding energy probably weaker.

5.2 Fluorobenzene . . . Ar: Simulation of Rotational ZEKE/MATI Spectra

Rotational band contour analysis of $S_1 \leftarrow S_0$ transitions has been widely applied to determine the geometric structure of both vdW complexes and hydrogen bonding of aromatic molecules.[61–65]

The difficulty in interpreting ZEKE/MATI spectra arises from the fact that angular momentum can be transferred from the molecular core to the Rydberg electron during a ZEKE transition. Different approaches have been introduced such as the *ab initio* method of McKoy[66] and the compound state model of Müller-Dethlefs and coworkers[67,68]. Because these methods are complicated to apply a simpler approach was introduced by Ford *et al.*, named the spectator model.[69]

In this model, the intermediate state (with an N electron wavefunction Ψ') in the ZEKE transition is treated as a Rydberg-like state, having a separate spectator electron coupled to the ionic core with $N–1$ electron wavefunction, Ψ^+. This arises as the first term in an expansion of the intermediate state wavefunction in terms of all the possible states of the ion, Ψ_n^+, and a single electron wavefunction, that arises from the projection of a given state of the ion on the intermediate state wavefunction. This expansion gives rise to a set of orbitals that are nonorthogonal, the spectator orbital is therefore quite different to the orbitals obtained in a Hartree–Fock calculation. As the ionising photon interacts only with the spectator orbital (the ion core, Ψ^+, is already fully electronically relaxed) the ionisation process is effectively reduced from a complex multielectron process to a single-electron process. Rotational angular momentum in the initial state can be decoupled into the angular momentum of the spectator electron and rotational angular momentum of the ionic core.

For a detailed derivation and how the simulation of ZEKE spectra is performed in detail see Ref. 69. One of the most important results so far is that the ZEKE spectra of fluorobenzene . . . Ar could be interpreted with a set of rotational constants for the S_0, S_1 and cation D_0 states plus a set of intensity parameters. The method has thus been demonstrated to be powerful enough to elucidate the rotational structure of larger molecular cation clusters.

5.3 Aniline . . . Ar_n and Phenol . . . Ar_n

In the following section we will demonstrate the problems in interpreting experimental data for rather simple and extensively studied non-covalent complex, phenol . . . Ar. A very important general problem concerns the question "will a non-covalently bound system assume a planar or above-ring structure"? This question is of particular importance concerning the structure and dynamics of nucleic acid–base pairs (see Section 5.7). We want to illustrate this with a very simple example, the phenol . . . argon complex. In contrast to benzene . . . argon, extensively studied by REMPI and other excited state (*e.g.*, LIF) spectroscopy,[1,70–73] large benzene clusters and benzene . . . Ar_n[74,75] clusters, complexes containing phenol[76,77] are interesting because two principal ligand binding sites are available: *via* the OH group and *via* interaction with the aromatic π system. The phenol...argon complex can be expected to show two main structural motifs: a plane hydrogen-bonded geometry with the argon bonded to the OH group and an above- ring van der Waals-bonded geometry with the argon bonded to the aromatic π-system.

5.3.1　Aniline...Ar and Phenol...Ar

5.3.1.1　*Discussion of the π-bound Structure*

For aniline...Ar a π-bound structure has been found in the S_1 state that is redshifted by $36\,cm^{-1}$ compared to the aniline monomer.[78] From the spectrum the vdW modes s_z (symmetric stretch), b_x (symmetric bend) and b_y (asymmetric bend) could be determined to be $47\,cm^{-1}$, $21\,cm^{-1}$ and $19\,cm^{-1}$. The motion of the argon along the C_2 rotation axis (in C_{2v} planar aniline) corresponds to the symmetric bend. The asymmetric bending motion is across the aromatic plane perpendicular to the C_2 axis. Fundamentals and overtones of these two symmetric modes are spectroscopically allowed and have been observed in the S_1 spectrum, although not very intense, thus indicating only a slight geometry change in the $S_1 \leftarrow S_0$ excitation. Figure 5.4 shows the ZEKE spectra of aniline...Ar and aniline...Ar_2 as recorded by Knee and coworkers.[79] The ZEKE spectra are normalised with respect to the origin of the aniline ion. For aniline...Ar the ionisation energy was determined to be $62\ 168\pm4\,cm^{-1}$, indicating that the cation is more tightly bound than the S_0 and S_1 states by 113 and $60\,cm^{-1}$, respectively. The observed spectral shifts compared to the monomer are listed in Table 5.4. From a comparison of the ZEKE spectrum and the $S_1 \leftarrow S_0$ transition in the neutral[78] it is apparent that the ZEKE spectrum shows more pronounced vibrational structure, specifically a single progression of $15\,cm^{-1}$. By comparison to the S_1 spectrum Knee and coworkers assigned the $15\,cm^{-1}$ mode as the symmetric bending vibration in the cation. The relatively long progression observed, invoking Franck–Condon arguments, indicates a significant change in the symmetric bend coordinates upon ionisation. This was explained by a positive charge on the nitrogen shifting the equilibrium position of the argon towards the nitrogen. The observed intermolecular vibrations are consistent with an above-ring structure.

　　The dissociation energy of aniline...Ar was determined by Piest *et al.*[80] to be between $386\,cm^{-1}$ and $442\,cm^{-1}$ based on IR-UV double resonance spectroscopy. A very recent work by Gu and Knee[81] determined the dissociation energy to be $495\pm15\,cm^{-1}$ based on MATI spectroscopy, which is significantly higher than the value determined by Piest *et al.* The MATI spectra recorded by Gu and Knee display vibrational features in the An^+...Ar mass channel as well as in the An^+ mass channel over the range from $\sim360\,cm^{-1}$ to $480\,cm^{-1}$. The signal of the complex disappears between $480\,cm^{-1}$ to $510\,cm^{-1}$. The simultaneous observation of a signal in the cluster as well as the fragment mass channel can be explained to occur due to coupling to lower Rydberg states that occur when large voltage extraction pulses are used (see Section 2.3.1.4).

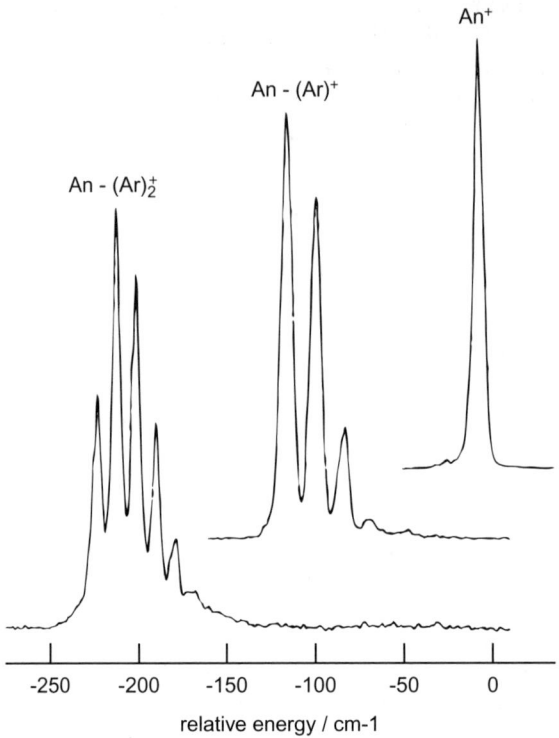

Figure 5.4   ZEKE spectra of aniline...Ar and aniline...Ar$_2$. Redrawn from Ref. 79.

Table 5.4   Observed spectral shifts (in cm^{-1}) for molecule...Ar complexes compared to the monomer in different states: S$_1$ (neutral) excited state and D$_0$ cation ground state.

N	Phenol [115,111]		Aniline [79,81,101]		Fluorobenzene [69,82]	
	S$_1$	D$_0$	S$_1$	D$_0$	S$_1$	D$_0$
1	−33	−176	−53	−113	−23	−224
2 (1/1)	−67	−335	−108	−220	−46	
(2/0)			−21	−364		
3 (1/2)			+22		+4.6	
(3/0)			−74			
(?/?)	+25	−553				

Because the threshold of the disappearance of the complex signal is not affected by the field, they stated that this value is more reliable to determine the dissociation threshold.

For phenol...Ar IR dip[83] and stimulated Raman[84] spectroscopy have shown the existence of a π-bound van der Waals structure in the S$_0$ state.

Haines *et al.* found a π-bound structure in the first excited state, which is redshifted by $33\,\mathrm{cm}^{-1}$ compared to the monomer.[85] We investigated the phenol...argon complex in the neutral ground state and the cation ground state by *ab initio* methods and also studied it experimentally using REMPI and ZEKE spectroscopy.[85] The energetics obtained are shown in Figure 5.5.

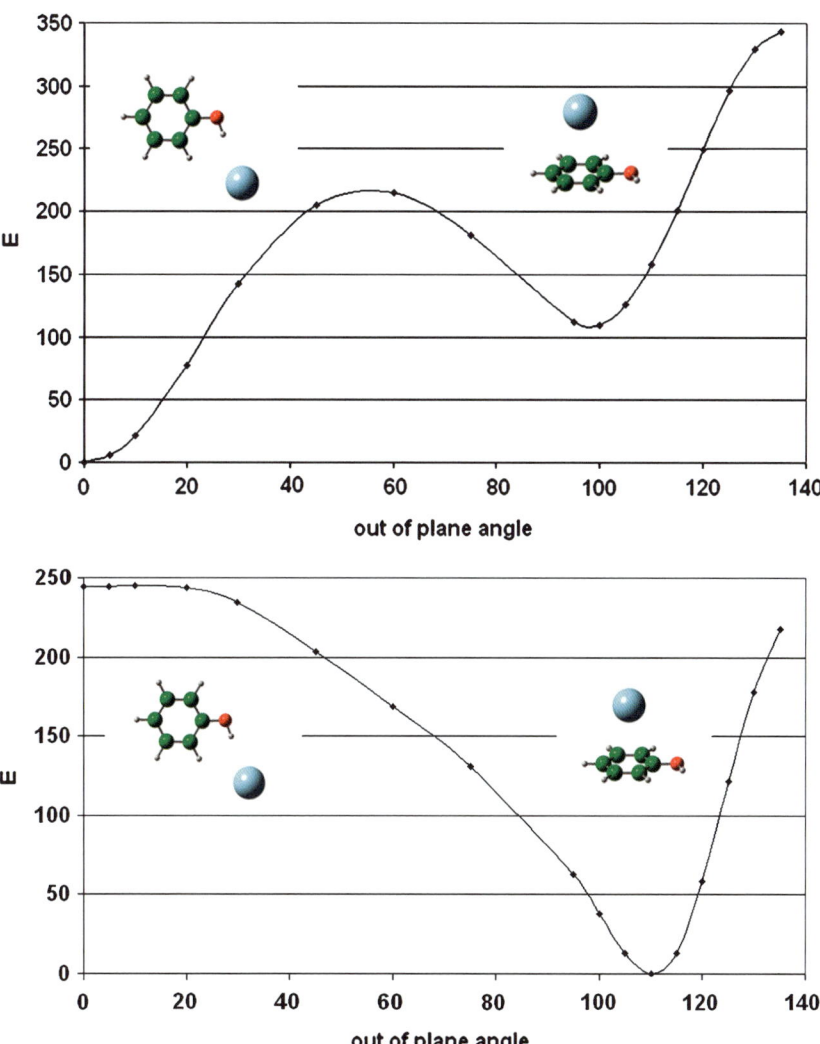

Figure 5.5 Calculated (minimum-energy path scan) potential-energy curves for the phenol...argon complex: (bottom) the S_0 and (top) D_0 state, starting from the H-bound minimum structure. For the S_0 neutral ground state, the above-ring structure has a substantially higher stabilisation energy than the hydrogen-bonded conformer, but the energy ordering is reversed in the cation and the H-bounded structure is a transition state in the neutral. Redrawn from Ref. 96

Table 5.5   Intermolecular vibrational frequencies for phenol...Ar calculated at MP2 (aug-cc-pVDZ) level and comparison with experiment. From Ref. 96.

Intermolecular vibration (cm^{-1})		b_x	b_y	S_z
S_0	vdW-bonded isomer	25	40	48
neutral ground state	hydrogen-bonded TS	–5	21	35
D_0	vdW-bonded isomer	5	37	50
cationic ground state	hydrogen-bonded isomer	18	28	61
	experimental value	14	25	66

Experimental vibrational frequencies for the S_1 state and the cationic ground state D_0 and calculated frequencies for the S_0 and the D_0 are summarised in Table 5.5. For the neutral cluster, the above-ring π-bound structure is the global minimum and the Ar...H-O-Ph hydrogen-bonded conformer does show up as a very shallow local minimum or, more likely, as a transition state, according to the vibrational frequency calculation at the MP2/cc-pVDZ level of theory (Table 5.5). However, for the ionic state, the calculation confirms that the hydrogen-bonded conformer is indeed the global minimum and the above-ring π-bound structure is a local minimum. A highly accurate study at the CCSD(T)/CBS-level proved that the π-bound phenol...Ar is indeed the global minimum in the neutral ground state, whilst the H-bound structure is only a transition state.[86] It should be noted that phenol...argon is, at the simplest level, comparable with the concept of going from hydrogen-bonded to stacked structures in the nucleic acid–base pairs.

The experimental study of the phenol...argon system showed very conclusively that for the $S_1 \leftarrow S_0$ transition, as shown in Figure 5.6, the above-ring structure is observed in the R2PI and MATI experiment.[85,87] This result was conclusively obtained from the simulation of the rotational band structure of the $S_1 \leftarrow S_0$ transition.[85] For this procedure, which has been pioneered by Simons and coworkers for molecules of biological interest in the gas phase,[88,89] different conformers, such as inplane or above-ring, do show spectral bands with a rather different rotational contour. The fitting procedure is carried out by simultaneously optimising the direction of the transition moment and the two sets of rotational constants for the S_0 and the S_1. An improved result for this fitting procedure, compared with the one in Ref. 85, is shown in Figure 5.6. Even with only partial rotational resolution, the fitting procedure results in a very satisfactory accuracy for the rotational constants. For the cation, the ZEKE spectra obtained via several different intermediate vibrational states of phenol...argon showed very clearly that the cation structure observed in the experiment must also be the van der Waals above-ring structure.[85,90]

Dissociation MATI spectroscopy yielded a value of 535 ± 3 cm^{-1} for the dissociation energy in the ion ground state.[91] While for aniline...Ar the spectral shifts (see Table 5.4) for the $S_1 \leftarrow S_0$ and the $D_0 \leftarrow S_1$ transition are nearly the same, the situation is very different for phenol...Ar as detected by

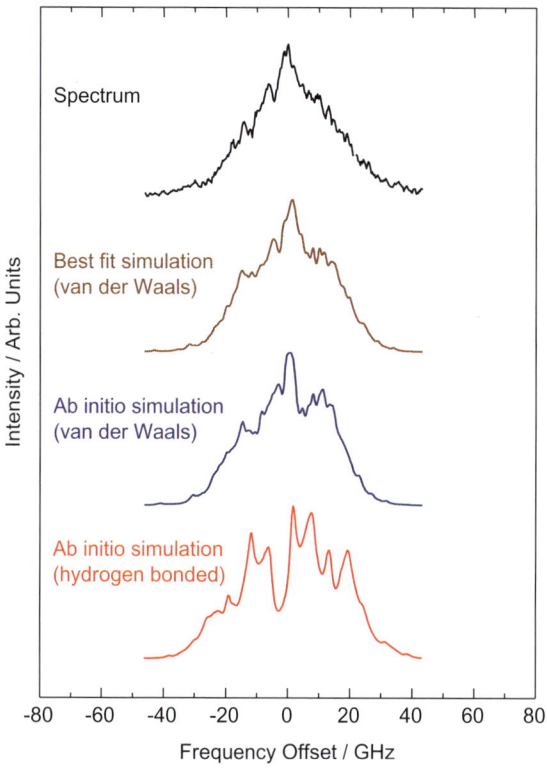

Figure 5.6 Rotational contours for the phenol ... argon $S_1 \leftarrow S_0$ transition: (a) for the van der Waals above-ring structure, (b) for the hydrogen-bonded local minimum structure. Redrawn from Ref. 85.

Haines *et al.*[85] Here, the spectral shift due to ionisation is more than 4 times the shift due to excitation.

5.3.1.2 Discussion of the H-bound Structure

Spectroscopic investigations of the aniline ... Ar and phenol ... Ar cations show evidence for a π-bound cluster. However, the most stable form for both cations produced by electron-impact ionisation is a H-bound structure, as shown by Dopfer and coworkers,[92–95] which is also supported by *ab initio* calculations.[86,96] Figure 5.5 shows the potential-energy curves (minimum-energy path at the RI-MP2 level for the neutral and RI-ROMP2 level for the cation) for phenol ... argon in the S_0 (Figure 5.5, bottom) and D_0 (Figure 5.5, top) state to explain this observed result (S_1 potential-energy curve not displayed). The most stable form in the neutral ground and excited state is the π-bound structure, while the most stable form in the cationic state is the H-bound structure. Using photoionisation only the π-bound structure can be formed due to the Franck–Condon principle. However, electron-impact ionisation, which does not have such a restriction, produces the most stable H-bound structure of the cation.[92,93] One should note that the H-bound structure in the neutral is actually a transition state.

5.3.2 Aniline ... Ar_2 and Phenol ... Ar_2

5.3.2.1 $\pi\pi$-bound Structure

For aniline ... Ar_2 the ionisation energy was determined to be $62\,061 \pm 4\,cm^{-1}$, indicating that the cation is more tightly bound than the S_0 and S_1 states by $220\,cm^{-1}$ and $113\,cm^{-1}$, respectively.[79] The change in bond energy is almost exactly twice that of the 1:1 complex (see Table 5.4), producing a considerably extended Franck–Condon envelope. The most intense peak within the vibrational van der Waals progression is not found at the ion origin but at the second member of the $11\,cm^{-1}$ vibrational progression. Following the symmetry analysis of Bieske et al.[97] for the S_1 spectrum of aniline ... Ar_2 and their classification, the progression observed in the ZEKE spectrum was assigned to the symmetric bending vibration along the C_{2v} axis, denoted b_x. This vibration has a frequency of $15\,cm^{-1}$ in the S_1 state, and upon ionisation, a reduction in frequency is again observed.

The fact that the spectral shift for the $S_1 \leftarrow S_0$ transition for aniline ... Ar_2 is exactly double that for aniline ... Ar and the spectral shift for $D_0 \leftarrow S_1$ is also nearly double ($113\,cm^{-1}$ to $60\,cm^{-1}$) indicates that the argon atoms are both π-bound in a (1/1) configuration and that the binding sites are equivalent.

Further investigations of the $S_1 \leftarrow S_0$ spectrum of aniline ... Ar_2 by Bréchignac et al.[98–100] showed the existence of a (2/0) isomer, which is blue-shifted by $87\,cm^{-1}$ compared to the origin of the (1/1) isomer. A deeper investigation of the structure and dynamics of aniline ... Ar_n ($n = 1$–6) clusters can be found in Ref. 101.

While for aniline ... Ar_2 two isomers were observed in the $S_1 \leftarrow S_0$ transition, hole-burning spectra of phenol ... Ar_2 showed evidence for the presence of only one conformer.[102] The question of which structure, (2/0) or (1/1) is present took quite a long time to resolve. REMPI spectra revealed that the origin is red-shifted by $67\,cm^{-1}$ compared to the phenol monomer, which is exactly twice the shift for phenol ... Ar, indicating a (1/1) structure.[85] From the REMPI spectrum of phenol ... Ar_2 the frequencies of the intermolecular vibrations b_x and s_z were determined to be $14\,cm^{-1}$ and $36\,cm^{-1}$ in the S_1 state, respectively.[85,103] These frequencies also support a (1/1) structure. A number of ZEKE spectra have been recorded and the ionisation shift and the vibrational structure observed confirms the existence of the π-bound conformer in the cationic state.[104] Comparing the IE of the phenol ... Ar_2 cluster with the phenol monomer IE produces a redshift of $335\,cm^{-1}$, which is almost double the redshift for the phenol ... Ar cluster ($176\,cm^{-1}$). This indicates again the similarity of the intermolecular bonding type in phenol ... Ar and phenol $\cdot\cdot\cdot Ar_2$. However, strong evidence for a (2/0) structure was also found. Band-contour analysis of the origin of the S_1 state supported the presence of this structure.[104] The fact that a very fast $\pi \rightarrow H$ isomeration occurs in the ionic ground state of phenol...Ar_2, but not in phenol...Ar (see below), could also be interpreted in terms of a (2/0) structure.[105,106] Later, the MATI dissociation spectra revealed a dissociation energy of less than $233 \, cm^{-1}$, much smaller than the dissociation energy for phenol...Ar, thus suggesting a (2/0) structure.[104] Very recently, a definite answer was given to this problem by very high resolution LIF spectroscopy of phenol...Ar_2[107] for which the rotational structure was fully resolved and assigned. The simulated spectra clearly prove that phenol...Ar_2 as seen in all experiments adheres to a (1/1) structure.

This example demonstrates that the interpretation of experimental results for non-covalent interactions is not always straightforward, even for such a simple cluster as phenol...Ar_2.

5.3.2.2 *HH and πH-bound Structures*

While all investigation of optically prepared aniline...Ar_2 ions suggest a (1/1) π-bound structure, the IR photodissociation spectrum of aniline...Ar_2 ion generated by electron impact in the range of the N–H stretching mode displays that both argon atoms are H-bound.[93] The explanation for this result is the same as for the aniline...Ar cation. Noteworthy are the results of an IR-UV double resonance investigation performed by Nakanaga *et al.*[108,109] Here, the recorded spectrum of the ion ground state in the region of the N–H stretching mode displays two sharp peaks that can be attributed to a structure where both argon atoms are π-bound. However, next to these sharp peaks there are two broad peaks, which could not unambiguously be assigned. Because these peaks are very near to the peaks which represent a HH-bound it was concluded that they may result from a structure where one argon atom is π-bound and one H-bound, or from a hot HH-bound structure.[95]

5.3.2.3 *π → H isomeration*

For the dissociation energy of the (1/1) conformer of aniline$^+$...Ar_2 into aniline$^+$...$Ar + Ar$, Gu and Knee[81] found a value of $380 \pm 5 \, cm^{-1}$. Surprisingly this value for the loss of one argon atom is substantial smaller (by about $115 \, cm^{-1}$) for the aniline$^+$...Ar_2 complex than for the aniline$^+$...Ar complex. In contrast, the dissociation energy for the loss of two argon atoms is $1020 \, cm^{-1}$, which is about twice the dissociation energy of aniline$^+$...Ar. As mentioned above the spectral shifts of aniline...Ar_2 compared to aniline ...Ar indicate that the binding sites are equivalent. If the binding sites are equivalent in the neutral ground state, then an optical transition into the ionic ground state cannot lead to transitions into significantly different conformers due to the Franck–Condon principle. Figure 5.7 shows a schematic energy diagram to understand this behaviour. The initial conformer of the aniline... Ar_2 complex upon photoionisation has both argon atoms π-bound on opposite sides and the first dissociation channel at $380 \, cm^{-1}$ represents the loss of a π-bound argon atom involving the rearrangement of the other argon atom from

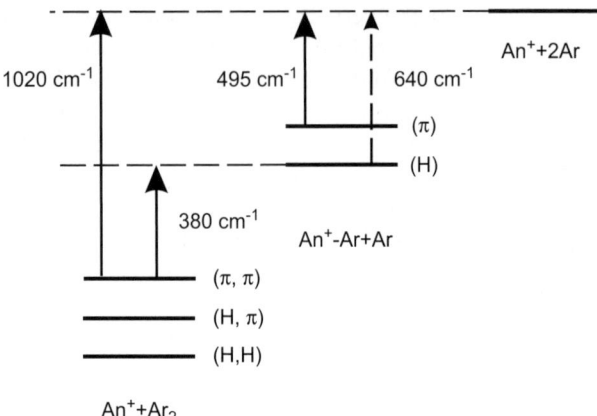

Figure 5.7 Energy schematic diagram for aniline...Ar$_2$. Redrawn from Ref. 81.

π- to H-bound. This rearrangement releases energy into the cluster, which then contributes to the dissociation energy of the second π-bound Ar. In contrast to this first dissociation channel at 380 cm^{-1} of the (1/1) π-bonded aniline$^+$...Ar$_2$ cluster into H-bound aniline$^+$...Ar + Ar, the second dissociation channel at 525 cm^{-1} (calculated from the measured dissociation energies) into π-bound aniline$^+$...Ar + Ar does not involve such a rearrangement. Assuming that both argon atoms stay more or less on the opposite side of the ring during dissociation the argon atoms will exchange very little energy, which is consistent with the observation of a dissociation energy similar to the phenol$^+$ ...Ar dimer. Similarly, it can be understood why the dissociation process leading to the loss of both argon atoms (1020 cm^{-1}) is close to double the dissociation energy of the π-bonded aniline$^+$...Ar dimer.

The MATI spectrum via the origin of the (2/0) conformer of aniline...Ar$_2$ shows one very broad peak. They assigned a value of 27 898 cm^{-1} for the ionisation threshold, which is in good agreement with the value of 27 908 cm^{-1} obtained by Douin *et al.* using photoionisation efficiency spectroscopy.[100] The ionisation energy of the (2/0) isomer is hence 144 cm^{-1} redshifted compared to that of the (1/1) species. Gu and Knee interpreted the strong redshift by invoking that one argon atom is H-bound in the cationic state. Because this configuration is optically accessible this would suggest that it is also H-bound in the neutral state. The second-lowest calculated neutral (2/0) geometry has an argon atom in a near H-bonding configuration, so that an optical transition into a cation structure with one H-bound argon atom may be possible. A significant geometry shift is also indicated by the very broad MATI peak.

In contrast to phenol...Ar,[110] for phenol...Ar$_2$ a most surprising result has been obtained: *upon ionisation one of the two argon atoms moves to the hydrogen-bonded site.*[104,111] Upon cation formation, the barrier between the π-bound and the hydrogen-bonded geometry seems to be very low, resulting in a fast geometry change. The ZEKE/MATI spectra of phenol...Ar$_2$ reveal this

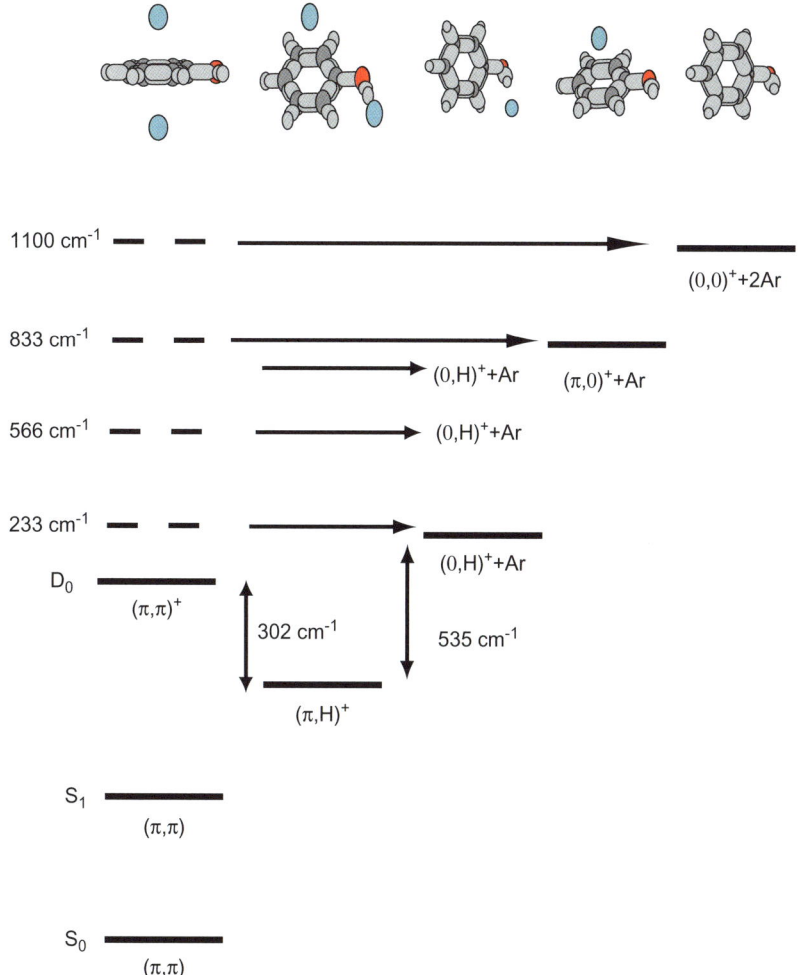

Figure 5.8 Energy diagram for the states and dissociation pathways of phenol...Ar$_2$.

geometry change in great detail.[104,111] Together with the results from pico-second IR-ion dip spectroscopy[105,106] the energy diagram for the dissociation dynamics as shown in Figure 5.8 has been drawn.

The investigation of phenol...Ar$_2$ by picosecond IR-ion dip spectroscopy has shown that photoinduced $\pi \rightarrow$ H site switching occurs in the cation state with a time constant of $\sim$7 ns and roughly independent from the available internal vibrational energy.[105,106] The barrier for this isomerization was deter-mined to be low ($<100\,\text{cm}^{-1}$) and the position as well as the width of the H-bound ν_{OH} band changes with the delay time, indicating an intracluster vibrational energy distribution upon site switching. This indicates that the less-stable π-bound conformer produced by photoionisation is isomerised promptly to the H-bound conformer, which is the global minimum of the cation state. Later, these investigations showed that by ionising with an excess energy of up to 566 cm^{-1} only the H-bound fragment can be observed, while for an excess energy of 833 cm^{-1} or more the H-bound as well as the π-bound fragment can be observed.

As mentioned above, the comparison of the spectral shifts of phenol...Ar and phenol...Ar$_2$ suggests a similar bonding type between the two vdW bonded argons. However, the dissociation energy of phenol...Ar$_2$ for the loss of one argon atom in the cation state was determined to be below 232 cm^{-1} by Tong *et al.*[104,111] using MATI spectroscopy (see Figure 5.9). As for aniline...Ar$_2$ this value is significantly lower than the dissociation energy of phenol$^+$...Ar (535 cm^{-1}). For phenol$^+$...Ar$_2$ this can be explained by an additional dissociation channel, involving π → H isomerisation, with a threshold of less than 232 cm^{-1}. The barrier between the structure where both argon atoms are π-bound (π, π), and the one where one argon atom is π-bound while the other one H-bound (π, H), is lower than the vibrational level of mode 11. This allows the (π, π) phenol...Ar$_2^+$ with 232 cm^{-1} excess internal energy to isomerise to phenol...Ar$_2^+$ in (π, H) configuration, which releases energy into the cluster. With this excess internal energy, the dissociation limit is reached for the π-bound argon atom of the phenol...Ar$_2^+$ in the (π, H) configuration. If we assume that the additional H-bound argon atom does not affect the binding

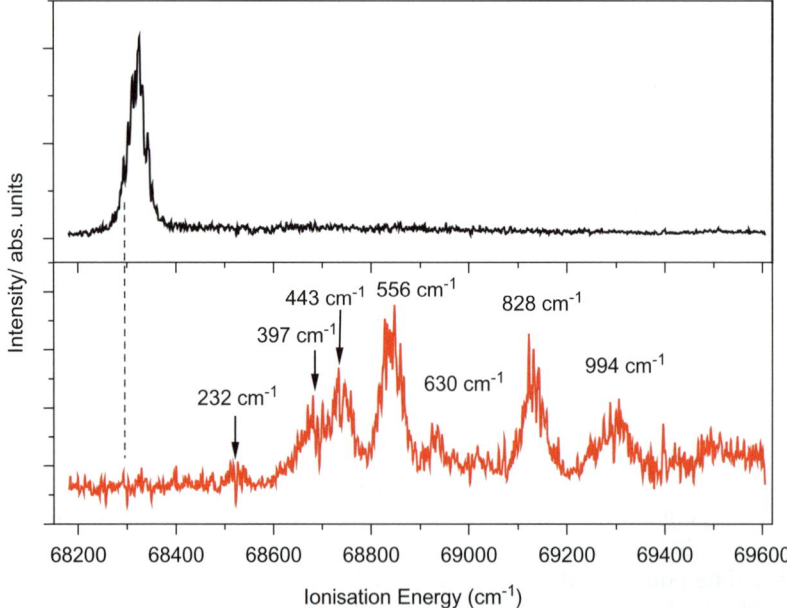

Figure 5.9   MATI dissociation spectrum of phenol...Ar$_2$. Reproduced from Ref. 104 with permission. Top: phenol$^+$...Ar$_2$ parent ion, bottom: phenol$^+$...Ar fragment.

energy of the π-bound argon atom (binding energy for phenol...Ar^+ is $535\,cm^{-1}$) very strongly, than we can estimate the difference between the (π, π) and (π, H) configuration in the ionic ground state to be around $300\,cm^{-1}$. From the MATI and PIE spectra it can be concluded that the dissociation energy for the loss of two argon atoms is around $1100\,cm^{-1}$, which is nearly exactly twice the dissociation energy for the phenol...Ar cluster. This observation is consistent with the observations for aniline...Ar_2.

5.3.3 Phenol...Ar_n Clusters with $n > 2$

Phenol...Ar_n clusters have been investigated by R2PI ($n = 3$–6) and MATI ($n = 3,4$) spectroscopy. The recorded R2PI spectra are displayed in Figure 5.10. While the spectra up to phenol...Ar_4 are relatively clear, the spectra for $n = 5,6$ clusters show a strong background, which is believed to result from dissociation of higher-mass clusters. All transitions are redshifted compared to the monomer, except for the phenol...Ar_3 cluster, where a blueshift was detected. Such a behavior has also been observed for aniline...Ar_3 and fluorobenzene...Ar_3 (see Table 5.4). However, in the case of aniline a peak with a redshift was also observed. The redshifted peak was attributed to the origin of the (2/1) structure, while the blueshifted peak was assigned as the origin of the (3/0) structure based on calculations.[112–114] It was shown that the (2/1) structure can only be observed, with relevant intensity, if the argon pressure is very low (best 10% argon in helium at a backing pressure of 3 bar).[98] For phenol...Ar_3 different experimental conditions have not yet been fully explored and the question which structure is seen in the R2PI spectrum of phenol...Ar_3 is still under investigation; this should be resolved by simulating the rotational contour of very recently measured high-resolution R2PI spectra.[115]

The MATI spectra of the phenol...Ar_3 cluster *via* two different intermediate states are displayed in Figure 5.11.[115] The spectrum obtained is extremely broad, indicating a strong geometry change upon ionisation. As mentioned above such a behavior was also observed in the MATI spectrum of the (2/0) isomer of aniline.

The aniline...Ar_2 results can be explained by an π-H isomeration upon ionisation, as the calculations have shown that one of the argon atoms is already nearly H-bound in the S_0 ground state. It is concluded that the same argument may be used here for phenol...Ar_3, however, calculations for phenol...Ar_3 are missing at this time.

5.4 Trimer Clusters with Hydrogen and π-Bonding

5.4.1 Phenol...Water...Ar

MATI experiments on the mixed hydrogen/van der Waals bonded complex, phenol...water...argon should prove interesting for studying the interplay

REMPI of Phenol...Arn clusters

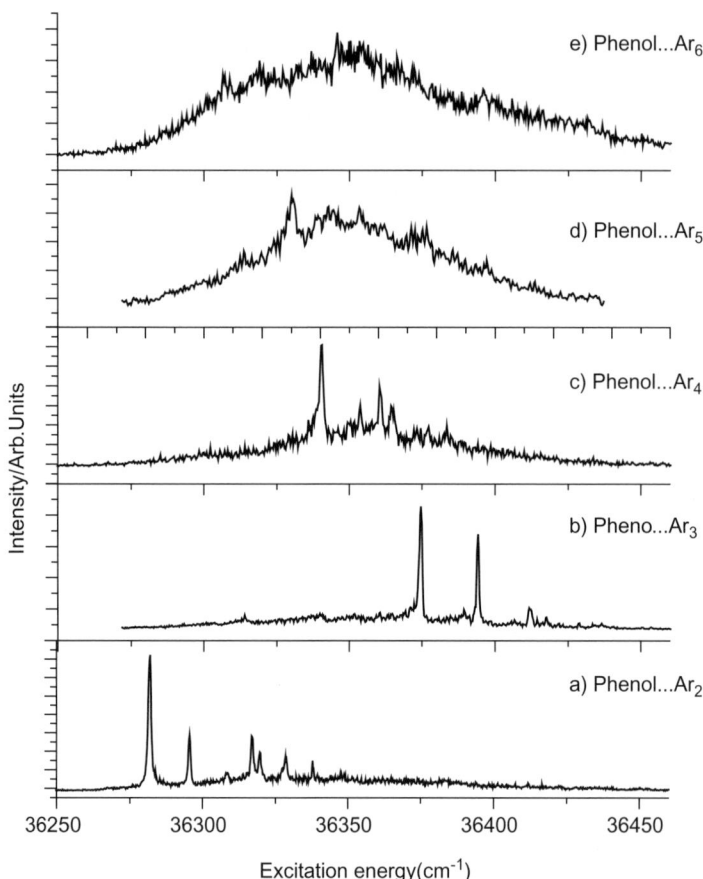

Figure 5.10   REMPI spectra of phenol...Ar_n ($n = 2$–6). Redrawn from Ref. 115.

between different types of intermolecular bonds during fragmentation. Rapid IVR of vibrational energy, which is initially localised at the intermolecular hydrogen bond, into the weaker above-ring non-covalent bond has already been observed in this system,[116] leading to dissociation of the weaker above-ring bond. The photoionisation efficiency (PIE) curve for phenol$^+$...water...argon displays steps that correlate to each strong peak in the ZEKE spectrum due to *delayed* ions that arise from pulsed-field ionisation of long-lived, high-lying Rydberg states. Notably, the steps in the PIE spectrum are rather broadened since the low-frequency intermolecular van der Waals vibrations of the argon atom (19, 24 and 33 cm^{-1}, and their combinations with the low-frequency vibrations of the hydrogen bond) produce a high density of

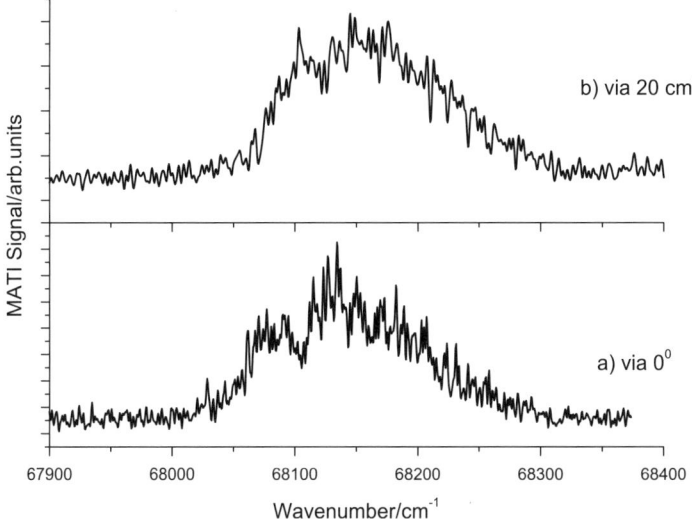

Figure 5.11 MATI spectra of phenol...Ar$_3$. Redrawn from Ref. 115.

vibronic states close to each dissociation threshold. As the photoionisation energy increases, the phenol$^+$...water...argon ion signal begins to drop, while phenol$^+$...water production increases, reflecting fragmentation of the parent cluster by ejection of argon. The energy difference between the fragmentation onset in the phenol$^+$...water channel and the ionisation onset provides an upper limit for the dissociation energy of phenol$^+$...water...argon into phenol$^+$...water. While this value represents an upper limit, it should be reasonably close to the true dissociation energy as the density of optically accessible vibrational states close to the dissociation threshold are so high.

5.4.2 Benzene...Water...Ar

The benzene$^+$...water cation, Bz$^+$...W, is one of the prototype systems to explore the charge-dipole interaction. Infrared spectroscopy and theoretical calculations of Bz$^+$...W have revealed that the water molecule locates at the side of the benzene ring, and the charge-dipole interaction as well as C–H...O hydrogen bond dominates the cluster structure.[117–123] In contrast, the cluster structure of the neutral Bz...W is totally different from that of Bz$^+$...W; the neutral Bz...W cluster is known to have an on-top structure bound by π-hydrogen bonds.[124] Such a drastic change of cluster structures upon photoionisation has been the subject of theoretical and experimental studies.[117–127] Theoretical calculations at B3LYP/6-31G(d,p), B3LYP/6-311G(d,p), and MP2/6-31G(d) predicted binding energies of 3290, 4410, and 4795 cm^{-1} (9.40, 12.6, and 13.7 kcal/mol), respectively.[117,119,122,123] Solcá and Dopfer experimentally evaluated the binding energy to be 4900 ± 1100 cm^{-1}

(14 ± 3 kcal/mol) on the basis of the OH stretching frequency shifts of the water moiety.[119,120] Though this experimental value well agrees with the latter two theoretical predictions, their evaluation assumed a correlation between the binding energy and the free-OH frequency shifts seen in typical hydrogen-bonded systems. Such a correlation may not be accurate in charge–dipole interaction systems, so an alternative experimental approach to the binding energy of $Bz^+\ldots W$ on the basis of infrared photodissociation (IRPD) experiments using an "Ar messenger" has been reported by Miyazaki *et al.*[128] The attachment of an Ar atom to the cluster cation not only suppresses the internal energy of the cluster cation but also opens multiple dissociation channels for the vibrational predissociation process. Their IR spectrum of the $Bz^+\ldots W\ldots Ar$ cluster cation was recorded by monitoring the $Bz^+\ldots W$ and the Bz^+ fragment and compared to the IR spectrum of $Bz^+\ldots W$ monitored on the Bz^+ mass. All the band positions of the spectra in $Bz^+\ldots W\ldots Ar$ are the same as those of $Bz^+\ldots W$, though the bandwidths are remarkably narrowed as a result of a restriction of the cluster internal energy due to the Ar attachment. The same IR spectral feature in these spectra demonstrated that the Ar atom locates on the benzene ring, while the benzene cation and water moieties hold the side structure as the $Bz^+\ldots H_2O$ cluster. The branching ratio of the two dissociation fragments shows a remarkable vibrational energy dependence. The Bz^+ fragment channel is open only in the higher-frequency region, while the $Bz^+\ldots H_2O$ channel is closed with an increase of the vibrational energy. The binding energy of $Bz^+\ldots H_2O$ was evaluated as $D_0^+ = 3290 \pm 120\,cm^{-1}$ by analysing the vibrational-energy dependence of the dissociation of $Bz^+\ldots W\ldots Ar$ with respect to the $Ar + Bz^+\ldots H_2O$ channel. In combination with a thermodynamic energy cycle, the adiabatic ionisation potential of $Bz\ldots H_2O$ was estimated to be 72 $160\pm150\,cm^{-1}$. This value is lower by $\sim2400\,cm^{-1}$ than that of bare benzene.

5.5 Benzene Dimer

Hole-burning[129] and microwave[130] studies supported the T-shaped global minimum structures and from rotational constants the distance of 4.96 Å (centre of mass – centre of mass) was deduced. Further, the IR studies of von Helden *et al.*[131] provided information on such structures of the dimer where both monomers were nonequivalent (the parallel-displaced (PD)-structure has equivalent monomers, while the T-structure has nonequivalent monomers). The authors also found a small redshift ($\sim3\,cm^{-1}$) of C–H stretch in benzene monomer upon dimerisation. Do these data agree with theoretical values? It must be immediately mentioned that comparison of experimental and theoretical values is not straightforward and care should be paid when comparing comparable characteristics. The main problem concerns the character of the PES of the benzene dimer (PD- and T- structures are practically isoenergetic and transition barriers separating these minima are very low, see above). Evidently, the concept of equilibrium structure of the dimer should be replaced by a concept of dynamical structure and since experiments were performed at nonzero temperatures the entropy should also be included. From the above-mentioned description of the PES of the benzene dimer it is evident that the standard harmonic approach (like rigid rotor – harmonic oscillator – ideal gas approximation) has several shortcomings. One possibility to solve the problem is the application of molecular-dynamics simulations. It is, however, evident that widely used empirical potentials cannot describe the fine features of the PES and more accurate quantum-mechanical procedures are required. The choice of a suitable method is limited since the dominant stabilisation energy term in all benzene dimer structures is the London dispersion energy. The method should be on the one hand fast enough since it will be utilised in the on-the-fly MD simulations and, on the other hand, accurate enough to describe all the fine features of the dimer. On-the-fly MD simulations of the benzene dimer were recently performed[132,133] and the latter study[133] will be discussed below.

The MD simulations were carried out with the DFT-D/BLYP/TZVP method; parameters in the dispersion correction term were fitted to exactly mimic benchmark CCSD(T)/complete basis-set limit potential-energy curves for both parallel-displaced and T-shaped structures of the dimer (see above). Let us mention here that the benzene dimer was included in the original parametrisation of the DFT-D procedure (set S22) but the calculated characteristics differ from accurate CCSD(T) values. The damping-function parameters were therefore reparametrised to fit the benchmark data. The resulting DFT-D/BLYP/TZVP procedure is thus highly accurate but it is a single-purpose computational procedure – it can be used only the for benzene dimer. Use of BLYP functional and TZVP basis set allows evaluating long enough trajectories. Probability distribution (histogram) of the angle between benzene plane rings (α, see Figure 5.12), a coordinate distinguishing best between these two structures, is plotted in Figure 5.12. Zero and 90 degrees correspond to PD- and TS-structures, respectively. In this coordinate, the peak at 90 degrees covers all possible T-shaped structures, which may differ in the tilt angle (see later).

From Figure 5.12 it is evident that the dynamic description becomes important at temperatures above 10 K, where interconversion between T- and PD-structures becomes accessible. At low temperatures (20–50 K), there exist a mixture of these two configurations, but the T-shaped structure becomes dominant when the temperature increases, because it is favoured by the entropic contribution to the free energy. This finding explains experimental studies, which detected only the T-structure (provided the vibrational temperature of the cluster is above 0 K; this is, however, in jet-cooling experiments highly probable). Deeper insight into populations of T-structures further showed that the fully symmetric C_{2v} T-shaped transition state remains unpopulated even at high temperatures and only the tilted T-structure was visible.

This finding is important since it explains the spectral characteristics of the dimer. It is in full agreement with infrared spectroscopy and calculations that assigned the spectrum to the tilted T-structure. This also explains the already mentioned fact that no blueshift and only a redshift of the C–H stretch

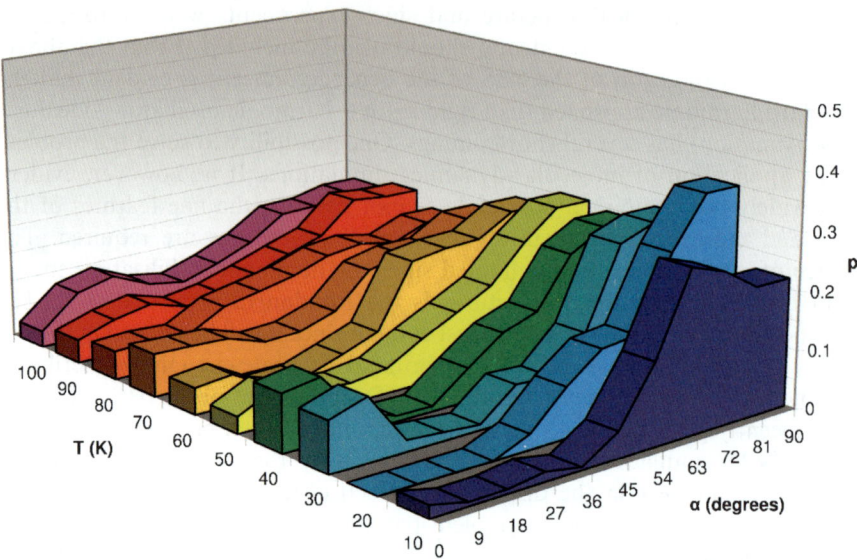

Figure 5.12 Probability distribution (histogram) of angle α between benzene rings in the dimer, plotted for various temperatures. 0 and 90 degrees correspond to PD- and T-shaped structures, respectively. Reproduced from Ref. 133 with permission.

vibration was detected experimentally. Let us recall the that benzene dimer (T-shaped C_{2v} structure) was the first system where we predicted[134] the improper blueshifting hydrogen bond. While various types of the blueshifting-bond were later detected experimentally[135] the C–H...π blueshifting-bond in the benzene dimer was not confirmed. On the contrary, von Helden *et al.*[131] detected a small redshift ($\sim 3\,\text{cm}^{-1}$) of the C–H stretch in the benzene monomer upon dimerisation. This contradiction can be explained by the above-mentioned dominant population of tilted T-shaped structure and negligible population of the T-shaped one. Recently we namely showed[136] the existence of blueshift in C_{2v} T-shaped structure and redshift in tilted C_s T-shaped structure.

Another parameter is the distance of the centre of masses of the benzene monomers. It perfectly distinguishes between parallel displaced (3.90 Å in minimum) and T-shape (4.90 Å in minimum) structures. The value for the T-shaped structure, calculated by our DFT-D method, which was fitted to accurate CCSD(T) energies, agree well with the experimentally measured[130] distance of 4.9 ± 0.01 Å determined from the rotational spectrum. From molecular dynamics, we can extract the average distance for each structure at different temperatures. Let us recall that the average distance determined in this way goes behind the harmonic approach and covers the vibrational averaging of distance reflecting the anharmonic nature of the PES. The results, listed in Table 5.6, show an increase of the distance with temperature. This is evidence of

Table 5.6 Average distance of benzene centres of mass in parallel-displaced (PD) and T-shaped (TS) dimer from MD simulations at different temperatures.

	distance (A)	
T(K)	*PD*	*TS*
10	3.95	4.87
20	3.98	4.88
30	4.00	4.87
40	4.03	4.89
50	4.07	4.93
60	—[a]	4.88
70	4.12	4.98
80	4.10	4.99
90	4.13	5.01
100	4.16	4.99

[a]no parallel-displaced structure was observed in the simulation.

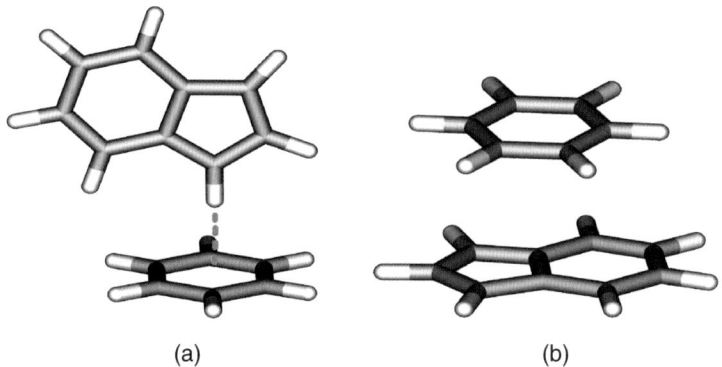

(a)　　　　　　　　　　(b)

Figure 5.13 Stacked (b) and N–H...π hydrogen-bonded structure (a) of the benzene...indole dimer.

the anharmonic nature of this intermolecular mode, and the rate of this increase (higher in the PD structure) could serve as a measure of anharmonicity.

5.6 Benzene...Indole Complex

This complex is one of very few complexes for which the stabilisation enthalpy is known from the MATI experiments.[137] As mentioned before, the technique does yield an experimental stabilisation enthalpy but does not provide information on the structure of a complex. We considered therefore two possible arrangements of the dimer, an N–H...π H-bonded structure (Figure 5.13(a)) and a stacked structure (Figure 5.13(b)). In the first step, both structural motifs were optimised at the RI-MP2/TZVPP level; the optimised geometries are presented in Figure 5.13. The RI-MP2 stabilisation energy of the stacked structure was slightly larger than that of the H-bonded structure (6.57 and 6.20 kcal/mol) and this difference was further enlarged when passing to stabilisation enthalpies at 0 K (the ZPE was added). Performing the CCSD(T) calculations with the 6-31G**(0.25,0.15) basis set, the preference of both structures is reversed and the N–H...π H-bonded structure becomes more stable (4.24 and 2.81 kcal/mol). These results are in line with data obtained for all stacked complexes; the CCSD(T) correction term is considerably larger for stacking structures than that for H-bonded structures (3.8 vs. 2.0 kcal/mol). This gives again clear evidence of the necessity of exploring the PES of multistructural molecular clusters at the highly correlated QM level (e.g. at CCSD(T)). In other words, when comparing stabilisation energies of various structural motifs, the higher-order correlation-energy contributions should be properly covered. Finally, the CBS limit of the MP2 stabilisation energy was determined for the global minimum and putting together all the contributions we obtained an estimate of the true stabilisation enthalpy of the N–H...O H-bonded structure (5.3 kcal/mol). This value nicely agrees with the experimental estimate of 5.2 kcal/mol and enhances confidence in the use of this procedure for other extended non-covalent complexes. Let us finish, however, by saying that further calculations for this complex are required. The question remains whether it is correct to limit the calculations only to the more stable H-bonded structure or to consider the less-stable stacked structure also. In other words, is the energy difference between these two structures (about 1.4 kcal/mol) large enough to justify the neglect of the less-stable structure?

5.7 Nucleic Acid–Base Pairs *in Vacuo*

To study the intrinsic interactions of DNA bases experimentally, one needs to carry out accurate gas-phase experiments. Gas-phase experiments provide data giving insight into the physicochemical origin of H-bonding and stacking. However, experiments on DNA base pairs yielding stabilisation energy (enthalpy) *in vacuo* are very difficult to perform. At the moment, we still have to rely on the field ionisation mass spectroscopy data provided by Yanson *et al.*[138] Let us further stress that there are still no other reliable gas-phase experiments reporting on energetics of base pairing, even though any such experimental data would be of enormous value. The state-of-the-art gas-phase experiments[139–144] from the laboratories of de Vries and Kleinermanns give evidence only about the vibrational spectrum of a selected nucleic acid–base pair but do not yield direct information on its structure or stabilisation energy.

Experimental results on 9-methyladenine...1-methylthymine and 9-methylguanine...1-methylcytosine pairs were obtained from the temperature dependence of equilibrium constants measured at rather high temperatures (average temperatures were 323 and 381 K for mA...mT and mG...mC, respectively) and thus correspond to stabilisation enthalpies at these temperatures.[138] After performing the first correlated *ab initio* calculations on the mA...mT and mG...mC complexes in the 1990s,[145,146] the resulting stabilisation energies of single Hoogsteen and Watson–Crick structures, respectively, were compared with experimental data. The very good agreement obtained between theoretical and experimental data was claimed as evidence of the reliability of the theoretical procedure used. The question arises, however, whether this treatment was justified. Is it correct to compare the experimental value only with the (expected) most-stable structure? The answer is no. The data[138] show no evidence that only one structure was populated at the experimental temperatures. It must be expected that not only the most stable structure of the base pair but also many other structures should be indispensably populated. The situation with the present complexes is more complicated than in the case of the benzene dimer and benzene...indole because in these cases only two structures coexisted, while here a much larger number of energy minima can be expected. In the following section we will show a correct procedure for comparison of theoretical data with the above-mentioned experiment in the case where a large number of structures can coexist.[147]

In order to compare the theoretical with the experimental data (both determined in the gas phase), it is necessary to pass from interaction energies to interaction enthalpies and, furthermore, to cover temperature-dependent enthalpy corrections. The main problem (see later) concerns the fact that due to high experimental temperatures several isomers (and not just the most stable one) of the base pairs should be present. This information is, however, not available from experiment and should be obtained theoretically.

Planar H-bonded and stacked structures of the 9-methylguanine...1-methylcytosine and 9-methyladenine...1-methylthymine were optimised at the RI-MP2 level using the TZVPP (5s3p2d1f/3s2p1d) basis set. The planar H-bonded structure of the mG...mC corresponds to the Watson–Crick (WC) arrangement, while the mA...mT possesses the Hoogsteen (H) structure. The final stabilisation energies (in kcal/mol) for base pairs studied were very large (mA...mT H 16.3, mA...mT stacked 13.1, mG...mC WC 28.5 and mG ...mC stacked 18.0 kcal/mol), much larger than the values previously published. These stabilisation energies should be first corrected for the zero-point energies and then the temperature-dependent enthalpy terms should be added. Since, however, more structures exist at the PES, each structure should be weighted according to its population under experimental conditions. On the basis of our previous calculations[148] using MD/Q simulations with the Cornell *et al.* empirical potential, stacked structures of mG...mC and mA...mT pairs are populated about 21% and 81% ($T = 300$ and 400 K, respectively, close to experimental conditions). It is thus evident that the experimental stabilisation enthalpy should be compared with the weighted average of stabilisation enthalpies of all non-negligibly populated structures rather than with the energy of the most stable (in terms of ΔH_T^0) but rarely present structure. The global minimum of mA...mT at the PES corresponds to the Hoogsteen structure; this structure is, however, at 400 K populated only at the 4% level.

The situation with mG ... mC is different. The global minimum corresponds to a Watson–Crick arrangement (which exists in DNA) and at 300 K it is populated at a level of 28%. This means that in both cases the stacked structures are populated dominantly. The explanation is simple – entropy favours loosely bound stacked structures over more strongly bound H-bonded structures. Populations (weighting factors) were taken from MD/Q results.[148] H-bonding and stacking energies were scaled[147] by the ratio of the calculated ΔH_T^0 and the respective low-level value, while the T-shaped structures were scaled by a factor taken as the average of the factors found for H-bonding and stacking. The resulting interaction enthalpies, $\Delta H^0_{323K} = -11.3$ kcal/mol (mA ...mT) and $\Delta H^0_{381K} = -18.0$ kcal/mol (mG ... mC), are in good agreement with experiment ($\Delta H^0_{323K} = -13.0$ kcal/mol (mA ...mT) and $\Delta H^0_{381K} = -21.0$ kcal/mol (mG ...mC)).

Highly accurate CCSD(T) calculations on the stabilisation energies of the uracil dimer discussed above have shown that the S22 results (including the above-mentioned calculations on mA ...mT and mG ...mC) represent an upper limit of the real values. This means that the real stabilisation energies of complexes discussed in the preceding section will be smaller, which makes the agreement with experiment a little worse. We believe that this is not due to the quality of calculated stabilisation energies but rather due to evaluation of the relative populations of single structures existing at experimental temperature. Let us recall that these populations were based on MD/Q simulations performed with the empirical Cornell *et al.* potential. A more reliable procedure should be based on on-the-fly *ab initio* MD simulations providing accurate relative populations that will be used in the second step for scaling of highly accurate CCSD(T) interaction energies.

The MD/Q simulations discussed in the previous section represent the most general method providing thermodynamic characteristics of molecular clusters. The method is, however, not too effective since extremely long trajectories are required for sampling the whole PES. There exist several faster converging methods that were applied to association of DNA bases. Cieplak and Kollman,[149] in 1988, already used the free-energy perturbation/molecular dynamics simulation method for the study of association of DNA bases *in vacuo* and water environments and showed that base pairs in a water environment possess stacked orientation. A few years later Friedman and Honig[150] showed that Lennard-Jones energy (*i.e.* mainly dispersion energy) and non-polar solvation contributions favour stacking of DNA bases over H-bonding, while electrostatic contributions oppose association. The potential of the mean force combined with umbrella sampling was applied[151] for the calculation of the free energies of stacking of DNA dimers. Finally, the same process was investigated by MD simulations based on replica-exchange umbrella sampling.[152] The latter technique is probably the most advanced since it allows sampling of a much wider conformational space than the previous methods. The free-energy profile was calculated for all 16 possible combinations of DNA dimers in a water environment (possessing the stacked base pair) and results obtained agreed reasonably well with experimental values. Stacked DNA base pairs were also studied using the QM calculations (*i.e.* interaction energy was determined) and the GG dimer was found to be the most favourable while UU was the least favourable.[146] This order basically agrees with the free-energy predictions mentioned above with exception of the AA dimer. It is definitely not easy to interpret this difference but one of the main reasons is the use of an empirical potential in all free-energy simulations mentioned. It would be thus highly desirable to combine these fast-converging methods for sampling the configuration space with accurate *ab initio* calculations as done in the case of the benzene dimer where on-the-fly MD simulations combined with DFT-D calculations were utilised.[133]

5.8 Ultrafast Hydrogen-Atom Transfer in Clusters of Aromatic Molecules Including Base Pairs

Presently, only very few groups worldwide have been able to show that nucleic acid–base pairs can be produced in a supersonic jet and that hole burning can be used to help distinguish the concomitant conformers. So far no threshold ionisation spectra have been attempted except one experiment by Kim and coworkers, who have reported the photoionisation efficiency spectra for adenine via the S_1 state.[153] The photochemical stability of DNA and nucleic acid–base pairs has been the subject of considerable discussion for some time; it poses very interesting questions as to exactly why nature has chosen to make some species undergo very rapid nonradiative processes upon electronic excitation to the S_1 state.[154–162] Lifetimes and decay channels reported in the literature are quite controversial,[163,164] due in part to the currently available methods of study and the concomitant amenability of available data to various equally valid but contradictory interpretations; all that is currently clear is that we are dealing with some very interesting and important dynamical processes.

This quandary can be related to the very surprising discovery by Jouvet and coworkers of hydrogen atom transfer (in contrast to proton transfer) in phenol...ammonia clusters[165–167] and other systems.[161,168–171] This hydrogen-detachment channel seems to be a general feature of a variety of aromatic molecules, including nucleic acid base prototypes such as indoles. It can be contemplated that this discovery will certainly have an important impact on the understanding of the excited-state dynamics of biologically important systems. Sobolewski and co-workers[168] (see Figure 5.14) have shown that hydrogen detachment and hydrogen transfer is driven by repulsive $n\sigma^*$ states and that the cross section for hydrogen-atom detachment and transfer depends on the position of the conical intersections produced by the interaction of the $n\sigma^*$ surface with the $\pi\pi^*$, $\pi\sigma^*$ and S_0 potential-energy surfaces. Time-resolved studies by Fujii and coworkers on phenol...ammonia clusters have monitored the hydrogen atom transfer on the picosecond time-scale.[172–175] Such dynamics will be further accessible experimentally by femtosecond and picosecond photoelectron spectroscopy and picosecond ZEKE spectroscopy.

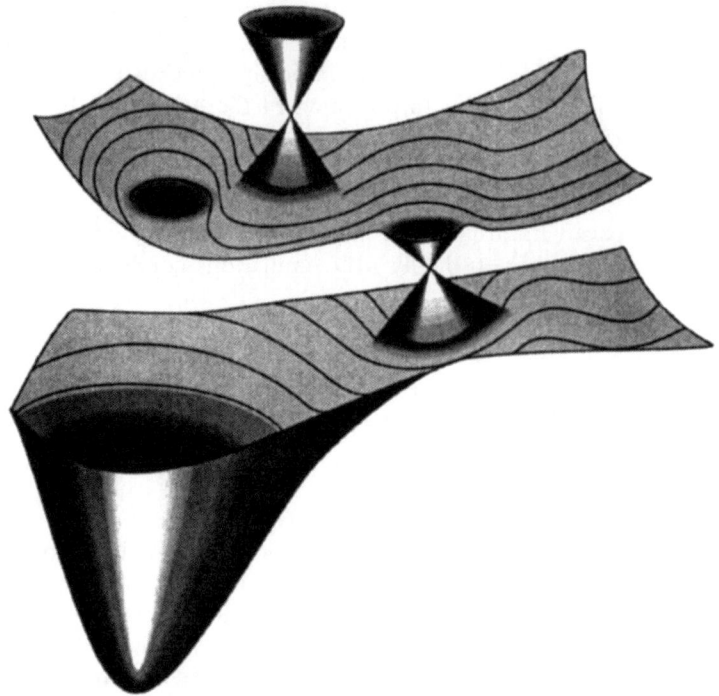

Figure 5.14   Conical intersections produced by the interaction of the nσ* surface with the ππ*, πσ* and S$_0$ potential-energy surfaces. Reproduced from Ref. 155 with permission.

Another very important dynamic process is charge transfer (CT), which may be studied by time-resolved femtosecond photoelectron spectroscopy (fs-TRPES). This method, which has been pioneered by Stolow,[176] allows one to follow the charge-transfer dynamics from an electronically excited doorway state α to CT state β (see Figure 5.15). The probe laser is then used to project this dynamics onto the surface of the corresponding cation, as depicted in Figure 5.15 for conformers α$^+$ and β$^+$. Especially for hydrogen atom and proton-transfer dynamics, this projection will result in the observation of photoelectrons of different kinetic energy, associated with either conformer α$^+$ or β$^+$. The evolution of the wave packet on the excited state surface will result in a time evolution of the transition probabilities to ionic conformers α$^+$ and β$^+$.

5.9   Photochemical Selectivity in Nucleic Acid Bases

Prebiotic chemistry presumably took place before formation of an oxygen-rich atmosphere and thus under conditions of intense short-wavelength UV irradiation. Therefore, the UV photochemical stability of the molecular building blocks of life may have been an important selective factor in determining the

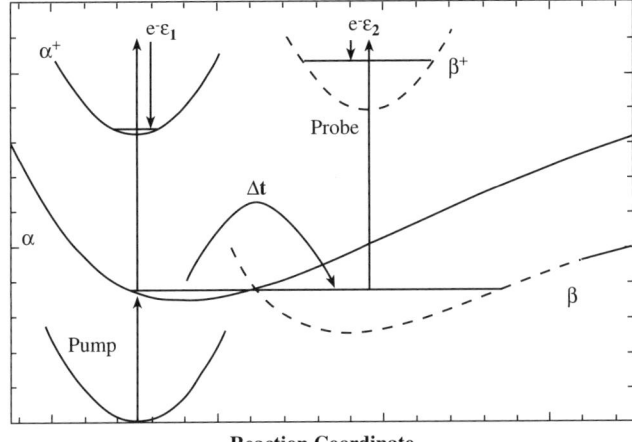

Figure 5.15   Charge-transfer dynamics from an electronically excited doorway state α to CT state β. The probe laser is then used to project these dynamics onto the surface of the corresponding cation, as depicted for conformers α^+ and β^+. Redrawn from Ref. 176.

eventual chemical makeup of critical biomolecules. The photochemical stability of the genetic material and the components of DNA is quite surprising considering that, in comparison to a large variety of aromatic molecules, strong decay channels are present. A very topical question in this context is what is responsible for the photochemical stability: is it a property of the nucleic acid bases themselves, is it their attachment to sugar or phosphoric acid, or is it aided by the hydrogen bonds formed in the nucleic acid–base pairs? The nucleic acid bases composing DNA exhibit rather short excited state lifetimes of the order of one picosecond or less. It has been argued that this rapid decay to the electronic ground state serves to protect these bases against photochemical damage because they do not cross into the reactive triplet state.

Twenty-seven different guanine...cytosine base pairs were studied[177] by high-resolution UV spectroscopy and correlated *ab initio* calculations. Of these, 24 pairs exhibit sharp UV spectra and the remaining three pairs broad UV spectra. This broad absorption may be explained by a rapid internal conversion that makes this specific base-pair arrangement uniquely photochemically stable. Theoretical calculations have proven that all these three structures are Watson–Crick ones and are characterised by rapid internal conversion that makes these pairs uniquely stable. This finding suggests that, when constructing possible scenarios for prebiotic chemistry, the photostability of different DNA base pair structures should be considered. The special photostability of the Watson–Crick arrangements could have made an important contribution to the evolution of the recognition mechanism for the transfer of genetic code.

5.10 Proton/Hydrogen Transfer and Hydrogen-Bonded Water Wires

The Grötthuss mechanism explains the anomalous proton mobility in liquid water or solution by successive proton hopping along a hydrogen-bonded water wire. Leutwyler and coworkers have shown that such a proton transfer occurs in cis-7-hydroxyquinoline...$(NH_3)_3$, [7HQ...$(NH_3)_3$], which is a microscopic model of the Grötthuss mechanism.[178,179]

The calculated structures of the educt (enol form) and product (keto form) for this proton transfer in 7HQ...$(NH_3)_3$ are shown in Figure 5.16(A). The proton transfer was proven by investigating this system with R2PI, hole-burning and laser induced fluorescence (LIF) spectroscopy. Figure 5.16(B) shows the potential-energy curves to illustrate the performed investigations. While the recorded R2PI spectra showed that by exciting with an excess energy above 200 cm^{-1} into the S_1 state no ion can be observed, the hole burning spectra displayed a variety of peaks above this value. Also, UV fluorescence with a lifetime of $\tau_{fl} = 1$ ns observed up the 200 cm^{-1} vanishes at this onset. This demonstrates that a fast process must exist with an onset of 200 cm^{-1} excess energy in the S_1 state. Recorded LIF spectra, by monitoring in the visible range (544.4 nm), reproduced the same vibronic band pattern observed in the hole-burning spectra when excited with an excess energy of 200 cm^{-1} or higher. For lower excitation energy no visible fluorescence was observed. These results are quite spectacular and clearly demonstrate the proton transfer along a NH_3 wire in 7HQ...$(NH_3)_3$.

The driving force of this microscopic Grötthuss mechanism is an excited-state proton[178,180,181]/hydrogen[182–196] transfer (ESPT/ESHT) in which the generality for aromatic acids has been suggested from both theoretical and experimental investigations. Thus, the same proton/H-atom transfer along the hydrogen-bonded wire could be possible in a hydrogen-bonded cluster of molecules that have proton/hydrogen donor and acceptor groups simultaneously. The molecule 7-azaindole (abbreviated 7AzI) has a N–H donor and a N-atom acceptor site. The excited-state proton-transfer reaction has been well studied in the 7AzI dimer, and the visible fluorescence is emitted from its tautomer form after UV excitation of the normal form. Therefore, the proton/hydrogen transfer in the 7-AzI hydrogen-bonded clusters is thought to be triggered by photoexcitation to the S_1, and a proton/hydrogen-transfer reaction can be easily detected by monitoring the visible fluorescence. The formation of a hydrogen-bonded wire between the two sites was found in 7AzI...$(H_2O)_n$ ($n = 1$–3) clusters by IR dip spectra and quantum chemical calculations.[197,198] One would then expect proton/hydrogen transfer along the water mole-cular wire in the clusters. However, because no visible fluorescence was detected, it was concluded that such a photoinduced transfer does not occur in 7AzI-$(H_2O)_n$ clusters.[199] In contrast, visible emission was found in 7AzI...$(CH_3OH)_2$ clusters, which has been assigned to a triple proton transfer in the S_1 state. This shows the possibility of successive proton/hydrogen transfer in hydrogen-bonded 7AzI clusters, although other 7AzI...$(CH_3OH)_n$

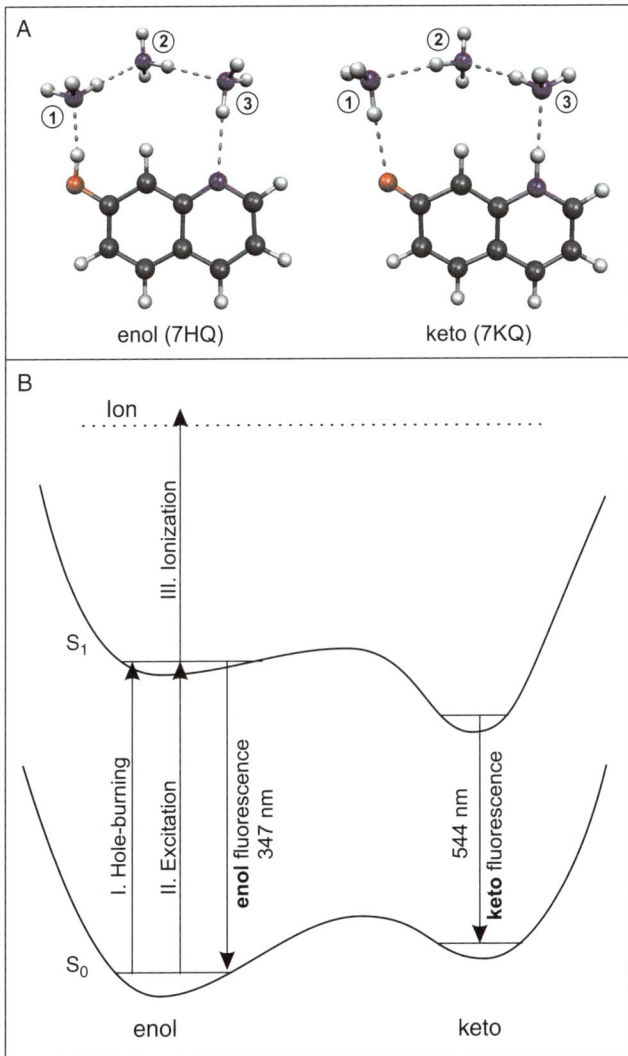

Figure 5.16   (A) Calculated structures of the enol and keto forms of 7-hydroxy-quinoline ... (NH$_3$)$_3$. (B) Schematic potential-energy curves for the neutral ground state and the first excited electronically excited state of 7HQ to illustrate the principle of detecting the hydrogen transfer (for details see text). Reproduced from Ref. 179 with permission.

clusters ($n = 1$ and 3) are not reactive.[200] In analogy to 7HQ...ammonia clusters, proton/hydrogen transfer along the hydrogen-bonded wire is also expected to occur efficiently in 7AzI hydrogen-bonded clusters, if water or methanol molecules are replaced by ammonia molecules, which are stronger bases. 7AzI...(NH3)$_n$ ($n = 1$, 2) clusters have been studied by Kim and

Bernstein by mass-selected two-colour R2PI spectroscopy, and clear S_1–S_0 transitions were found.[201]

5.10.1 Experimental and Theoretical Study of the Activity of Proton/Hydrogen Transfer in the 7-Azaindole ... Ammonia Clusters

R2PI spectra for the 7AzI...$(NH_3)_n$ clusters for n=1–3 were measured and the electronic origins of the first excited state was easily determined as 33 405 cm^{-1} and 32 929 cm^{-1} for $n = 1$ and $n = 2$, respectively.[202] The spectrum of the $n = 3$ cluster is much more complicated than the other two spectra and the origin was tentatively assigned to 32 015 cm^{-1}. By recording a hole-burning spectrum it became clear that only one isomer is present. The complexity of the observed R2PI spectrum was interpreted due to mixing of electronic states or a large geometrical change upon excitation.

For the investigation of proton/hydrogen-transfer LIF and action spectroscopy was applied. The LIF spectrum clearly shows the origin of the S_1 state for the $n = 1$ and $n = 2$ cluster, while the origin of the $n = 3$ cluster could not be observed.

The action spectrum showed no visible emission for neither of the clusters, indicating that the proton/hydrogen transfer reaction from the S_1 state does not occur or has negligible yield.

The geometry of 7AzI...$(NH_3)_n$ for $n = 1$ and 2 was estimated to be a van der Waals conformation, in which ammonia molecules locate on the π ring.[201] The van der Waals conformation may explain the absence of proton/hydrogen transfer, because there is no molecule attached to the acidic N–H bond of the 7AzI molecule. On the other hand, this structure seems to contradict the recent studies on 7AzI-solvated clusters. For example, the 7AzI...$(H_2O)_n$ clusters form cyclic hydrogen-bond networks, which bridge the N–H donor site to the N acceptor site,[197] and ammonia molecules are also expected to form hydrogen bonds with 7AzI. For the investigation of the structure of $n = 1,2$ clusters IR-dip spectra where recorded. The observed intense IR transitions are clearly redshifted from the free N–H stretching vibrations, indicating the presence of a hydrogen-bond network.

The hydrogen-bonded structure was confirmed by density-functional theory calculations (B3LYP/6-31G) and the optimised structure and calculated IR spectrum of 7AzI...(NH_3) compares well to experiment. No stable van der Waals conformation in which an ammonia molecule is attached above the ring was found. From the correspondence, the bands at 3225, 3312, and 3403 cm^{-1} are assigned to the NH stretching, symmetric NH stretching, and antisymmetric stretching in 7AzI, respectively. Thus, the geometry of 7AzI...(NH_3) is assigned to a cyclic hydrogen-bonded structure.

For the $n = 2$ cluster the IR-dip spectrum as well as the calculation indicate a hydrogen-bond network. From the comparison of the recorded spectrum with the simulated spectra (bridge structure and chain structure) it is concluded that

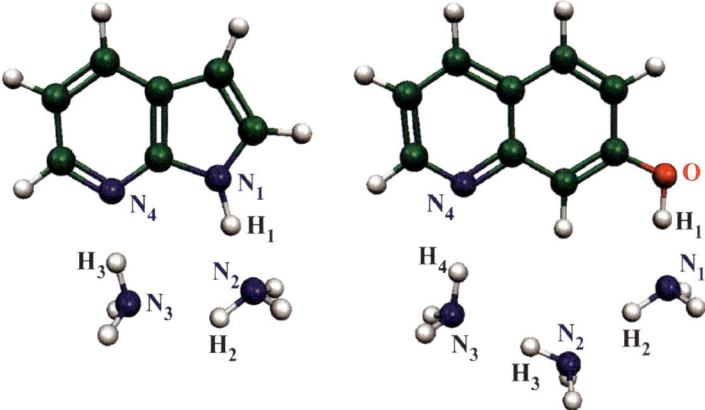

Figure 5.17 Wire structures for 7AzI-$(NH_3)_2$ (left) and 7HQ-$(NH_3)_3$ (right). Reproduced from Ref. 202 with permission.

the $n=2$ cluster yields a bridge structure, seen in Figure 5.17. For the $n=3$ cluster the same kind of structure is suggested from the performed calculations.

Vertical transitions energies to the $\pi\pi^*$ and $\pi\sigma^*$ were calculated for 7Az ... $(NH_3)_2$ and 7HQ ... $(NH_3)_3$. In 7HQ ... $(NH_3)_3$, the $\pi\pi$ state in the tautomeric form is located 0.8 eV lower in energy than the $\pi\pi^*$ excited state in the normal form. This is consistent with the fact that 7HQ ... $(NH_3)_3$ isomerises to the tautomer upon excitation to the $\pi\pi^*$ state. For 7Az ... $(NH_3)_2$ the calculations yield comparable results and the calculated geometry show that the bond length in the cluster is short enough to form a similar structure to 7HQ ... $(NH_3)_3$. Therefore, a proton/hydrogen transfer should also be expected to occur. To understand why proton/hydrogen transfer is observed for 7HQ ... $(NH_3)_3$ but not for 7Az ... $(NH_3)_2$, energies for optimised structures along the reaction coordinate were calculated and vertical transition energies were compared. The difference in reactivity is suggested to be due to an additional barrier in the 7Az ... $(NH_3)_2$ reaction path.

5.11 Helium Nanodroplets: Formic Acid Dimer and Glycine Dimer

As discussed earlier, the most stable structure of the formic acid dimer is without any doubt a cyclic structure with two strong O–H ... O hydrogen bonds.[203] This structure exists dominantly in the gas phase and it was expected that it must also predominantly exist in the He-droplet environment since this phase was believed to be close to a gas phase. The IR spectra of the dimer detected in He-droplets[204] were, however, not consistent with the gas-phase global minimum but supported the existence of a local minimum, also cyclic in structure but, instead, with one O–H ... O and one C–H ... O contact.

We have shown that the dimer formed in the He-droplet environment is controlled by electrostatic forces and the most stable structure corresponds to that with the most favourable orientation of dipoles.[204] These theoretical calculations revealed several reaction channels and the energetically most favourable one suggests a structure with one O–H . . . O and one C–H . . . O contact. Evidently, formation of a complex in a He-droplet is governed by factors other than those prevalent in molecular beams.

The situation for the glycine dimer, the IR spectra of which were also detected in He-droplets, is probably very similar. In Section 3.4 we discussed the disagreement between theoretical calculations and experimental findings.[205] We have shown that despite increasing the level of calculations, we were not able to interpret the experimental findings that showed the existence of a free OH group. We now believe the problem is not caused by inadequate theoretical treatment but it is due to the dimer, in He-droplet experiments, being formed in a completely different structure compared with that deduced from isolated-molecule, gas-phase calculations. By analogy with the previous complex, the structure of the glycine dimer is believed to be determined by long-range electrostatic interactions. These electrostatic forces are of long range and hence the two subsystems are already oriented preferentially according to their dipole (and quadrupole moments) when separated by very large distances. This preferential orientation leads to the formation of a dimer with a different energy minimum structure than that found in the gas phase; a similar mechanism of complex formation in the He-droplet environment has been previously described.[206]

5.11.1 Formation of Cyclic Water Complexes: $(H_2O)_n$, $n = 3$–6 (Cyclic Water)

Nauta and Miller showed that the growth of water clusters in helium nanodroplets results in the formation of cyclic water clusters up to $n = 6$.[206] The formation of water rings involves the insertion of water monomers into preformed cyclic water clusters. In a combined study of experiment and theory Burnham *et al.* investigated the process of insertion in a more detailed way.[207]

Figure 5.18 shows the existence of cyclic water complexes in helium droplets. All bands observed in the helium droplet spectrum could also be observed in the supersonic jet spectrum, except one. The bands of interest correspond to the OH ring modes of the cyclic complexes, which are particularly intense according to the results obtained from calculations. Comparing the measured frequencies with the calculated ones supports the performed assignment of the bands.

Further support for the performed assignment comes from the measurement of the spectrum by varying the water pressure in the pick-up cell. This picture clearly shows the growth of the peaks, assigned as large water clusters, with increasing water pressure. For a deeper investigation the pressure dependence of each peak the laser was tuned to the centre of this peak and the signal intensity dependence on the pressure was recorded. It could clearly be shown

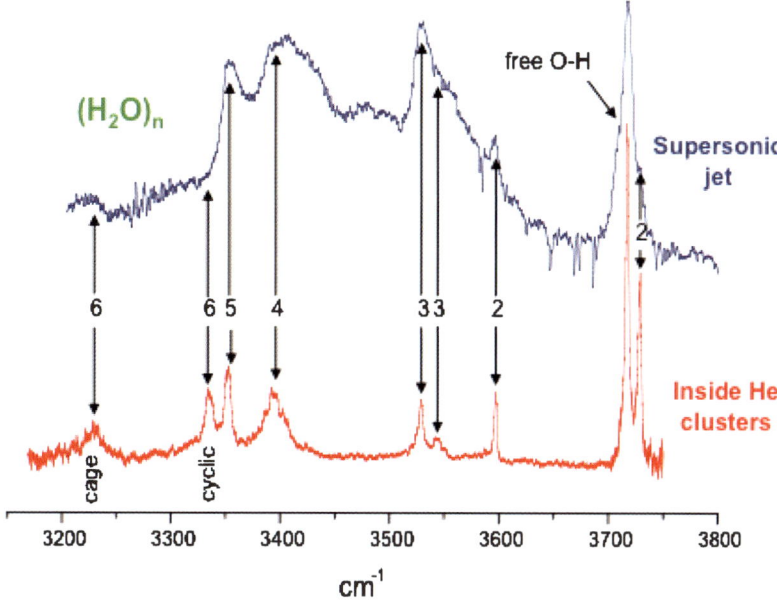

Figure 5.18 Comparison of $(H_2O)_n$ infrared absorption spectrum in a supersonic expansion and in He nanodroplets. Reproduced from Ref. 207 with permission.

that bands assigned to different isomers of the same water cluster yield the same pressure dependence, underlining the performed assignment.

If the barriers to ring insertion are large, once the cyclic trimer is formed it is expected that the addition of further water molecules would result in the formation of isomers involving water molecules bound to the trimer cyclic ring. The observation of the cyclic structures in helium implies that there are pathways between the various local minima on the PES that are low enough to be overcome under the experimental conditions, especially between structures consisting of one molecule attached to a ring of $(n–1)$ water molecules and a ring containing all n molecules.

5.12 Vibrational Energy Transfer and Predissociation

5.12.1 Aniline . . . Ar

The excitation of an intramolecular vibration (*i.e.* a vibration associated with the aniline moiety) can lead to dissociation of the aniline . . . argon complex due to the transfer of vibrational energy from the intramolecular vibration to translational motion through the bath of van der Waals vibrations into a translational dissociation motion.[208] A potential scheme to probe such S_1 dynamics is shown in Figure 5.19. The pump photon excites an intramolecular vibration in S_1 in the neutral van der Waals molecule. The ionisation rate due to

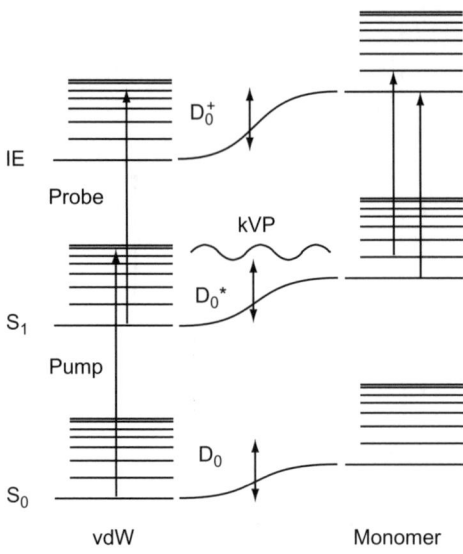

Figure 5.19   Schematic representation of an energy diagram for probing S_1 dynamics (IVR and dissociation) in van der Waals clusters. Redrawn from Ref. 79.

the probe laser (ionisation laser) then has to compete with the predissociation rate (rate constant k_{VP}). Depending on the timescale of the experiment, *i.e.* the delay between pump and probe laser, one observes either the vibrations due to the aniline argon vdW or the aniline ion monomer in the ZEKE spectrum. For small delay times (*i.e.* the time overlap of pump and probe laser) one expects mainly ionisation into the stable vdW cluster whereas vibrations of the monomer cation are observed for longer delays.

Excitation of the inversion vibration overtone in the aniline moiety (denoted I) is of particular interest.[79] The first overtone of this vibration is clearly seen in the $S_1 \leftarrow S_0$ transition at 762 cm^{-1}. Excitation of this vibration should considerably interfere with the movement of the argon sitting on the aniline aromatic moiety.[79] When pumping the I_0^2 transition, the ZEKE spectrum indeed reveals features that are attributed to both the reactant, aniline...Ar and the product, aniline monomer from Ref. 79. Figure 5.20 reproduces the ZEKE spectra for different probe laser delay times (with respect to the pump laser). An inspection of Figure 5.20 reveals that when both pump and probe laser overlap in time, the spectral features observed are mainly due to the vdW aniline... argon complex. With increasing delay time, features due to the vdW complex decrease in intensity and vibrational features due to the monomer appear (see the arrows in the middle part of Figure 5.20), becoming prominent at 4 and 12 ns delay time. Taking into account the energetics, the measured $S_1 \leftarrow S_0$ excitation spectra, and ZEKE spectra of the aniline monomer, Knee and coworkers[79] were able to assign these vibrational features in the ZEKE spectrum with 12 ns optical delay to the transition into the origin of the aniline monomer

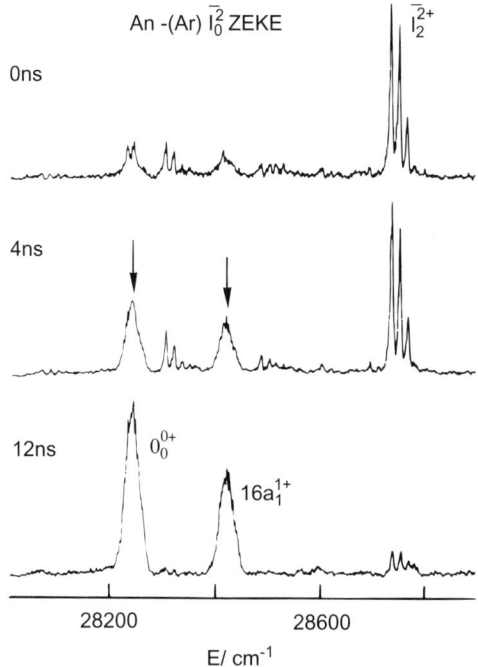

Figure 5.20 ZEKE spectra for different probe laser delay by excitation through the S_1 I_0^2 intermediate state of aniline...Ar. With increasing delay the intensity of product states (arrows) of the monomer increase, indicating dissociation and IVR in the S_1 state. Redrawn from Ref. 79.

cation (marked 0_0^{0+}) and the transition from the ν_{16a} vibration with one quantum excited into the ion also with one quantum of ν_{16a} excited (marked $16a_1^{1+}$). Therefore, these spectra lead to the conclusion that the predissociation channel can leave the monomer either vibrationless or with one quantum of the ν_{16a} excited mode. The peaks attributed to the monomer are broadened considerably compared to the ZEKE peaks not subject to predissociation. A width of approximately 28 cm^{-1} was attributed to the rotational excitation of the monomer produced in the S_1 state in the dissociation reaction.

5.12.2   Fluorobenzene ... Ar

The fluorobenzene...Ar complex was investigated by Lembach and Brutschy using MATI spectroscopy.[209] The dissociation energies in the neutral electronic ground state, the first excited state and the ionic ground state were determined to be lower than 278 cm^{-1}, 302 cm^{-1} and 502 cm^{-1}, respectively. The MATI spectrum recorded at the monomer mass via the $6b^1$ intermediate state, which is above the dissociation threshold in the first excited state, showed a very broad feature at 718 cm^{-1}. The width of this is, at 28 cm^{-1}, 5 times broader than all the other observed peaks and may therefore not represent a cationic state of the complex. Comparing the peak position with the transitions observed in the monomer spectrum this peak can be assigned as the 0^0 transition of the monomer. This observation can be interpreted as predissociation in the S_1 state, displaying that the 6b vibration couples very efficiently with the intermolecular modes. Because the delay between the two laser pulses was 4 ns, the dissociation has to occur relatively fast (< 4 ns).

Additionally, progressions of the van der Waals mode b_x, with a spacing of $13\,\text{cm}^{-1}$, could be observed in the cationic spectra of the complex and its fragment, indicating a significant structural change upon ionisation.

It is interesting that the frequency of the mode b_x decreases in the ionic ground state compared to the S_1 state even though the cationic complex is more tightly bound than the first excited state. This was explained to occur due to a flatter potential along the long axis of the aniline molecule, produced by the combination of the dispersive attraction of the ring system and the charge induced dipolar interaction with the fluorine atom, as was observed for aniline... Ar.

5.12.3   9-ethylfluorene ... Ar$_n$

Pitts and Knee have investigated the structure and dynamics of the 9-ethylfluorene... Ar$_n$ ($n = 1$–3) complex by REMPI and MATI spectroscopy.[210] The spectrum of the S_1 reveals multiple isomeric structures, see Figure 5.21, for each

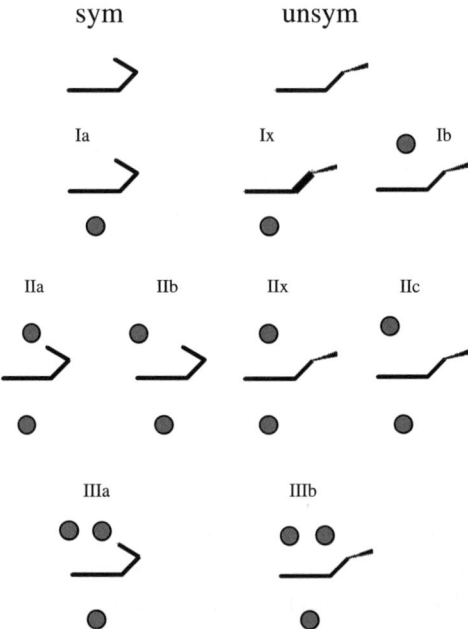

Figure 5.21   Different isomer structure of 9-ethylfluorene... Ar$_n$ ($n = 1$–3) clusters. Reproduced from Ref. 210 with permission.

of the cluster sizes associated with the two monomer conformations. The $n = 1$ cluster shows three isomers, one of the symmetric 9-ethylfluorene and two of the unsymmetric one. The $n = 2$ clusters have four possible isomers all of which are assigned to a (1/1) conformation, although each represents a unique structure with different argon-binding sites. The $n = 3$ cluster collapses down to two dominate isomers, one for each conformation of the parent. The isomer assignment in the S_1 state was assisted by the recorded MATI spectra.

MATI spectra were recorded via several intermediate states. The MATI spectra of the Ia isomer showed no evidence for redistribution (IVR) at low excess energy ($208 \, cm^{-1}$). By pumping the $398 \, cm^{-1}$ band the peak in the MATI spectrum in the $\Delta v = 0$ region is significantly broader, with an associated red-shift. The band shows significant redistribution on the nanosecond time scale.

For the IIb and IIa isomer the nanosecond MATI spectra via the $208 \, cm^{-1}$ intermediate state show a significant broadening of the peaks. While the IIa isomer behaves "normally", it shows a redistribution to the red side of the sharp reactant peak, the IIb isomer shows a blueshift. For a further investigation of this unusual phenomena picosecond MATI spectra were recorded for both isomers. A sharp peak could only be observed at very small time delays (around 0 ps). The measured temporal evolution of the reactant decay showed a rate of distribution of around 475 ps.

The anomalous blueshift for the IIb isomer was explained to occur due to the fact that the redistributed product starts to sample one or more isomeric structures. In fact, the MATI spectrum resulting from this "new" isomer is quite similar to the one from the normal red-shifted structure present in the redistribution of the IIa isomer excited to the same S_1 band. It is concluded that IIb is starting to sample the same, or similar, isomer structures as IIa. Recall that IIa and IIb are proposed to differ only in the position of an Ar on the "top", ethyl side of the ring. It is reasonable to expect that at $208 \, cm^{-1}$ the Ar is able to overcome barriers for moving on the surface of the molecule, although the surface crossing barrier is likely to be higher, as was the case in fluorene ... Ar_3.[211] Rather than considering the redistributed Ar to be a specific isomer structure it is more likely that both IIa and IIb are sampling many possible structures on the surface of the fluorene, which on average result in the broad MATI signal.

References

1. L. A. Chewter, K. Müller-Dethlefs and E. W. Schlag, *Chem. Phys. Lett.*, 1987, **135**, 219.
2. K. Kimura and M. Takahashi, *Optical Methods for Time- and State-Resolved Chemistry*, ed. C.-Y. Ng, The international Society for Optical Engineering, Washington, DC, 1992, SPIE.
3. M. Takahashi, *J. Chem. Phys.*, 1992, **96**, 2594.
4. M. Takahashi, *J. Chem. Phys.*, 1992, **96**, 6399.
5. A. M. Bush, J. M. Dyke, P. Mack, D. M. Smith and T. G. Wright, *J. Chem. Phys.*, 1998, **108**, 406.

6. J. D. Barr, J. M. Dyke, P. Mack, D. M. Smith and T. G. Wright, *J. Electron. Spectrosc. Relat. Phenom.*, 1998, **97**, 159.

7. M. C. R. J. Cockett, *Electron. Spectrosc. Relat. Phen.*, 1998, **97**, 171.

8. D. A. Beattie, M. C. R. Cockett, K. P. Lawley and R. J. Donovan, *J. Chem. Soc. Faraday Trans.*, 1997, **93**, 4245.

9. K. Müller-Dethlefs, M. Sander and E. W. Schlag, *Z. Naturforsch., A*, 1984, **39**, 1089; *Chem. Phys. Lett.*, 1984, **112**, 291.

10. K. Müller-Dethlefs and E. W. Schlag, *Angew. Chem. Int. Ed. Engl.*, 1998, **37**, 1346.

11. K. Müller-Dethlefs, O. Dopfer and T. G. Wright, *Chem. Rev.*, 1994, **94**, 1845.

12. J. M. Hutson, *Adv. Mol. Vib. Collision Dyn.*, 1991, **1**, 1.

13. H. Krause and H. J. Neusser, *J. Chem. Phys.*, 1993, **99**, 6278.

14. H. Krause and H. J. Neusser, *J. Chem. Phys.*, 1992, **97**, 5923.

15. A. Kiermeier, B. Ernstberger, H. J. Nausser and E. W. Schlag, *J. Phys. Chem.*, 1988, **92**, 3785.

16. T. Weber, A. von Bargen, E. Riedle and H. J. Neusser, *Int. J. Chem. Phys.*, 1990, **92**, 90.

17. T. Brupbacher, J. Makarewicz and A. Bauder, *J. Chem. Phys.*, 1994, **101**, 9736.

18. P. Tarakeshwar, K. S. Kim, E. Kraka and D. Cremer, *J. Chem. Phys.*, 2001, **115**, 6018.

19. R. J. Moulds, M. A. Buntine and W. D. Lawrance, *J. Chem. Phys.*, 2004, **121**, 4635.

20. P. Hobza, H. L. Selzle and E. W. Schlag, *J. Chem. Phys.*, 1991, **95**, 391.

21. E. Riedle, R. Sussmann, T. Weber and H. J. Neusser, *J. Chem. Phys.*, 1996, **104**, 865.

22. H. Krause and H. J. Neusser, *J. Chem. Phys.*, 1993, **99**, 6278.

23. P. Hobza, O. Bludský, H. L. Selzle and E. W. Schlag, *J. Chem. Phys.*, 1992, **97**, 335.

24. U. Aiger, L. Ya. Baranov, H. L. Selzle and E. W. Schlag, *J. Electron. Spectrosc. Relat. Phenom.*, 2000, **112**, 175.

25. T. L. Grebner, P. Unold and H. J. Neusser, *J. Phys. Chem. A*, 1997, **101**, 158.

26. G. Lembach and B. Brutschy, *Chem. Phys. Lett.*, 1997, **273**, 421.

27. E. D. Lipp and C. J. Seliskar, *J. Molec. Spectrosc.*, 1981, **87**, 242.

28. E. J. Bieske, M. W. Rainbird, I. M. Atkinson and A. W. Knight, *J. Chem. Phys.*, 1989, **91**, 752.

29. M. Mons, J. L. Calve, F. Piuzzi and I. Dimicoli, *J. Chem. Phys.*, 1990, **92**, 2155.

30. W. D. Sands, L. Jones and R. Moore, *J. Phys. Chem.*, 1989, **93**, 6601.

31. G. Lembach and B. Brutschy, *J. Phys. Chem.*, 1998, **102**, 6068.

32. P. Dao, S. Morgan and A. W. Castleman Jr, *Chem. Phys. Lett.*, 1985, **113**, 219.

33. M. C. Su and C. S. Parmenter, *Chem. Phys.*, 1991, **156**, 261.

34. S. Ullrich, G. Tarczay and K. Müller-Dethlefs, *J. Phys. Chem.*, 2002, **106**, 1496.

35. B. A. Jacobson, S. Humphrey and S. Rice, *J. Chem. Phys.*, 1988, **89**, 5624.

36. S. J. Baek, K. W. Choi, Y. S. Choi and S. K. Kim, *J. Phys. Chem. A*, 2003, **117**, 4826.

37. J. J. Oh, I. Park, R. J. Wilson, S. Peebles, R. L. Kuczkowski, E. Kraka and D. Cremer, *J. Chem. Phys.*, 2000, **113**, 9051.

38. C. R. Munteanu, J. L. Cacheiro, B. Fernandez and J. Makarewicz, *J. Chem. Phys.*, 2004, **121**, 1390.

39. M. Mons and J. L. Calve, *Chem. Phys.*, 1990, **146**, 195.

40. R. G. Satink, H. Piest, G. von Helden and G. Meijer, *J. Phys. Chem.*, 1999, **111**, 10750.

41. C. G. Eisenhardt and H. Baumgaertel, *Ber. Bunsen-Ges. Phys. Chem.*, 1998, **102**, 12.

42. J. Černý, X. Tong, P. Hobza and K. Müller-Dethlefs, *J. Chem. Phys.*, 2008, **128**, 114319.

43. S. Sato, T. Kojima, K. Byodo, H. Shinohara, S. Yanagihara and K. Kimura, *J. Electron. Spectrosc. Relat. Phenom.*, 2000, **112**, 247.

44. T. Wright, S. Panov and T. Miller, *J. Chem. Phys.*, 1995, **102**, 4793; G. Lembach and B. Brutschy, *Chem. Phys. Lett.*, 1997, **273**, 421.

45. S. M. Bellm, J. R. Gascooke and W. D. Lawerence, *Chem. Phys. Lett.*, 2000, **330**, 103.

46. S. Georgiev, T. Chakraborty and H. J. Neusser, *J. Phys. Chem. A*, 2004, **108**, 3304.

47. T. L. Grebner and H. J. Neusser, *Chem. Phys. Lett.*, 1995, **245**, 578.

48. A. Gaber, M. Riese, F. Witte and J. Grotemeyer, *Phys. Chem. Chem. Phys.*, 2009, **11**, 1628.

49. A. Gaber, M. Riese and J. Grotemeyer, *J. Phys. Chem. A*, 2008, **112**, 425.

50. X. Tong, J. Černý and K. Müller-Dethlefs, *J. Phys. Chem. A*, 2008, **112**, 5872.

51. X. Tong, M. S. Ford, C. E. H. Dessent and K. Müller-Dethlefs, *J. Chem. Phys.*, 2003, **119**, 12908.

52. L. A. Chewter, M. Sander, K. Müller-Dethlefs and E. W. Schlag, *J. Chem. Phys.*, 1987, **86**, 4737.

53. K. T. Lu, G. C. Eiden and J. C. Weisshaar, *J. Phys. Chem.*, 1992, **96**, 9742.

54. M. Takahashi and K. Kimura, *J. Chem. Phys.*, 1992, **97**, 2920.

55. S. Leutwyler and J. Bosiger, *Chem. Rev.*, 1990, **90**, 489.

56. R. Leist, J. A. Frey, P. Ottiger, H. M. Frey, S. Leutwyler, R. A. Bachorz and W. Klopper, *Angew. Chem. Int. Ed.*, 2007, **46**, 7449.

57. T. M. Trnka and R. H. Grubbs, *Acc. Chem. Res.*, 2001, **34**, 18.

58. S. S. Kaye and J. R. Long, *J. Am. Chem. Soc.*, 2008, **130**, 806.

59. X. Hu, M. Trudeau and D. M. Antonelli, *Chem. Mater.*, 2007, **19**, 1388.

60. K.-W. Choi, S. Choi, D.-S. Ahn, S. Han, T. Y. Kang, S. J. Baek and S. K. Kim, *J. Phys. Chem. A*, 2008, **112**, 7125.

61. D. W. Pratt, *Annu. Rev. Phys. Chem.*, 1999, **49**, 481.
62. M. R. Hockbridge, S. M. Knight, E. G. Robertson, J. P. Simons, J. McComie and M. Walker, *Phys. Chem. Chem. Phys.*, 1999, **1**, 407.
63. S. J. Martinez, J. C. Alafano and D. H. Levy, *J. Mol. Spectrosc.*, 1989, **137**, 420.
64. B. D. Howells, J. McCombie, T. F. Palmer, J. P. Simons and A. Walters, *J. Chem. Soc. Faraday Trans.*, 1992, **88**, 2603.
65. A. Bacon and J. M. Hollas, *Faraday Discuss. Chem. Soc.*, 1988, **86**, 129.
66. K. Wang and V. McKoy, *Annu. Rev. Phys. Chem.*, 1995, **46**, 275.
67. K. Müller-Dethlefs, *J. Chem. Phys.*, 1991, **95**, 4821.
68. I. Fischer, R. Linder and K. Müller-Dethlefs, *J. Chem. Soc. Faraday Trans.*, 1994, **90**, 2425.
69. M. S. Ford and K. Müller-Dethlefs, *Phys. Chem. Chem. Phys.*, 2004, **6**, 23.
70. J. L. Lyman, G. Müller, P. L. Houston, M. Piltch, W. E. Schmid and K. L. Kompa, *J. Chem. Phys.*, 1985, **82**, 810.
71. T. Brupbacher and A. Bauder, *Chem. Phys. Lett.*, 1990, **173**, 435.
72. R. K. Sampson and W. D. Lawrance, *Aust. J. Chem.*, 2003, **56**, 275.
73. U. Boesl, *J. Phys. Chem.*, 1991, **95**, 2949.
74. M. Schmidt, F. Piuzzi, M. Mons, J. Lecalve and I. Dimicoli, Resonance Ionization Spectroscopy 1990, *Inst. Phys. Conf. Ser.*, 1991, **199**, 114.
75. R. Knochenmuss, D. Ray and W. P. Hess, *J. Chem. Phys.*, 1994, **100**, 44.
76. N. Gonohe, H. Abe and N. Mikami, *J. Phys. Chem.*, 1985, **89**, 3642.
77. N. Gonohe, N. Suzuki, H. Abe, N. Mikami and M. Ito, *Chem. Phys. Lett.*, 1983, **94**, 549.
78. E. J. Bieske, M. W. Rainbird, I. M. Atkinson and A. E. W. Knight, *J. Chem. Phys.*, 1989, **91**, 752.
79. X. Zhang, J. M. Smith and J. L. Knee, *J. Chem. Phys.*, 1992, **97**, 2843.
80. H. Piest, G. von Helden and G. Meijer, *J. Chem. Phys.*, 1999, **110**, 2010.
81. Q. Gu and J. L. Knee, *J. Chem. Phys.*, 2008, **128**, 064311.
82. K. Rademann, B. Brutschy and H. Baumgärtel, *Chem. Phys.*, 1983, **80**, 129.
83. T. Ebata, A. Iwasaki and N. Mikami, *J. Phys. Chem. A.*, 2000, **104**, 7974.
84. G. V. Hartland, B. F. Henson, V. A. Venturo and P. M. Felker, *J. Phys. Chem.*, 1992, **96**, 1164.
85. S. R. Haines, C. E. H. Dessent and K. Müller-Dethlefs, *J. Electron Spectrosc. Relat. Phenom.*, 2000, **108**, 1.
86. J. Černý, X. Tong, P. Hobza and K. Müller-Dethlefs, *J. Chem. Phys.*, 2008, **128**, 114319.
87. M. S. Ford, S. R. Haines, I. Pugliesi, C. E. H. Dessent and K. Müller-Dethlefs, *J. Electron. Spectrosc. Relat. Phenom.*, 2000, **112**, 231.
88. Y. H. Lee, J. W. Jung, B. Kim, P. Butz, L. C. Snoek, R. T. Kroemer and J. P. Simons, *J. Phys. Chem. A.*, 2004, **108**, 69.
89. D. R. Borst, P. W. Joireman, D. W. Pratt, E. G. Robertson and J. P. Simons, *J. Chem. Phys.*, 2002, **116**, 7057.

90. M. S. Ford, S. R. Haines, I. Pugliesi, C. E. H. Dessent and K. Müller-Dethlefs, *J. Electron. Spectrosc. Relat. Phenom.*, 2000, **112**, 231.
91. C. E. H. Dessent, S. R. Haines and K. Müller-Dethlefs, *Chem. Phys. Lett.*, 1999, **315**, 103.
92. N. Solcá and O. Dopfer, *J. Phys. Chem. A.*, 2001, **105**, 5637.
93. N. Solcá and O. Dopfer, *Chem. Phys. Lett.*, 2003, **369**, 68.
94. O. Dopfer, *Z. Phys. Chem.*, 2005, **219**, 125.
95. N. Solcá and O. Dopfer, *Eur. Phys. J. D*, 2002, **20**, 469.
96. J. Černý, X. Tong, P. Hobza and K. Müller-Dethlefs, *Phys. Chem. Chem. Phys.*, 2008, **10**, 2780.
97. E. J. Bieske, A. Z. Vichanco, M. W. Rainbird and A. E. W. Knight, *J. Chem. Phys.*, 1991, **94**, 7029.
98. P. Hermine, P. Parneix, B. Coutant, F. G. Amar and P. Bréchignac, *Z. Phys. D*, 1992, **22**, 529.
99. P. Parneix, P. Bréchignac and F. G. Amar, *J. Chem. Phys.*, 1996, **104**, 983.
100. S. Douin, S. Piccirillo and P. Bréchignac, *Chem. Phys. Lett*, 1997, **273**, 389.
101. S. Douin, P. Parneix, F. G. Amar and P. Bréchignac, *J. Phys. Chem. A*, 1997, **101**, 122.
102. S. Ishiuchi, Y. Tsuchida, O. Dopfer, K. Müller-Dethlefs and M. Fujii, *J. Phys. Chem. A*, 2007, **111**, 7569.
103. M. Schmitt, M. Mons and J. Le Calvé, *Z. Phys. D*, 1990, **17**, 153.
104. X. Tong, PhD Thesis, *Non-Covalent Interactions in Aromatic Molecules and Clusters Studies by Laser Spectroscopy*, University of York, York, 2005.
105. S. Ishiuchi, M. Sakai, Y. Tsuchida, A. Takeda, Y. Kawaschima, O. Dopfer, K. Müller-Dethlefs and M. Fujii, *J. Chem. Phys.*, 2007, **127**, 114307.
106. S. Ishiuchi, M. Sakai, Y. Tsuchida, A. Takeda, Y. Kawaschima, O. Dopfer, K. Müller-Dethlefs and M. Fujii, *Angew. Chem. Int. Ed*, 2005, **44**, 6149.
107. I. Kalkman, C. Brand, T. Chau Vu, W. Meerts, Y. Svartsov, O. Dopfer, X. Tong, K. Müller-Dethlefs, S. Grimme and M. Schmitt, *J. Chem. Phys.* 2009, **130**, 224303.
108. T. Nakanaga, F. Ito, J. Miyawaki, K. Sugawara and H. Takeo, *Chem. Phys. Lett.*, 1996, **261**, 414.
109. T. Nakanaga and F. Ito, *Chem. Phys. Lett.*, 2002, **355**, 109.
110. M. S. Ford, S. R. Haines, I. Pugliesi, C. E. H. Dessent and K. Müller-Dethlefs, *J. Electron Spectrosc. Relat. Phenom.*, 2000, **112**, 231.
111. X. Tong, A. Armentano, M. Riese, M. Ben Yezzar, S. Pimblott, K. Müller-Dethlefs, O. Dopfer, S. Ishiuchi, M. Sakai and M. Fujii, in preparation 2009.
112. P. Parneix, F. G. Amar and P. Bréchingac, *Z. Phys. D.*, 1993, **26**, 217.
113. S. Douin, P. Parneix, F. G. Amar and P. Bréchingac, *J. Phys. Chem. A*, 1997, **101**, 122.

114. S. Douin, P. Hermine, P. Parneix, F. G. Amar and P. Brechingac, *J. Chem. Phys.*, 1992, **97**, 2160.

115. A. Armentano, M. Riese and K. Müller-Dethlefs, in prep. 2009.

116. O. Dopfer, M. Melf and K. Müller-Dethlefs, *Chem. Phys.*, 1996, **207**, 437.

117. M. Miyazaki, A. Fujii, T. Ebata and N. Mikami, *Chem. Phys. Lett.*, 2001, **349**, 431.

118. M. Miyazaki, A. Fujii, T. Ebata and N. Mikami, *Phys. Chem. Chem. Phys.*, 2003, **5**, 1137.

119. N. Solcá and O. Dopfer, *Chem. Phys. Lett.*, 2001, **347**, 59.

120. N. Solcá and O. Dopfer, *J. Phys. Chem. A*, 2003, **107**, 4046.

121. A. Courty, M. Mons, I. Dimicoli, F. Piuzzi, M.-P. Gaigeot, V. Brenner, P. de Pujo and P. Millie, *J. Phys. Chem. A*, 1998, **102**, 6590.

122. H. Tachikawa and M. Igarashi, *J. Phys. Chem. A*, 1998, **102**, 8648.

123. H. Tachikawa, M. Igarashi and T. Ishibashi, *Phys. Chem. Chem. Phys.*, 2001, **3**, 3052.

124. R. N. Pribble, A. W. Garrett, K. Haber and T. S. Zwier, *J. Chem. Phys.*, 1995, **103**, 531.

125. J. R. Gord, A. W. Garrett, R. E. Bandy and T. S. Zwier, *Chem. Phys. Lett.*, 1990, **171**, 443.

126. A. J. Gotch and T. S. Zwier, *J. Chem. Phys.*, 1990, **93**, 6977.

127. A. J. Gotch and T. S. Zwier, *J. Chem. Phys.*, 1992, **96**, 3388.

128. M. Miyazaki, A. Fujii and N. Mikami, *J. Phys. Chem. A*, 2004, **108**, 8269.

129. W. Scherzer, O. Krätzschmar, H. L. Selzle and E. W. Schlag, *Z. Naturforsch. Phys. Sci.*, 1992, **47**, 1248.

130. E. Arunan and H. S. Gutowsky, *J. Chem. Phys.*, 1993, **98**, 4294.

131. U. Erlekam, M. Frankowski, G. Meijer and G. von Helden, *J. Chem. Phys.*, 2006, **124**, 171101.

132. M. Pavone, N. Rega and V. Barone, *Chem. Phys. Lett.*, 2008, **452**, 333.

133. J. Řezáč and P. Hobza, *J. Chem. Theory Comput.*, 2008, **4**, 1835.

134. P. Hobza, V. Špirko, H. L. Selzle and E. W. Schlag, *J. Phys. Chem. A.*, 1998, **102**, 2501.

135. P. Hobza and Z. Havlas, *Chem. Rev.*, 2000, **100**, 4253.

136. W. Z. Wang, M. Pitoňák and P. Hobza, *ChemPhysChem*, 2007, **8**, 2107.

137. J. Braun, H. J. Neusser and P. Hobza, *J. Phys. Chem. A*, 2003, **107**, 3918.

138. I. K. Yanson, A. B. Teplitsky and L. F. Sukhodub, *Biopolymers*, 1979, **18**, 1149.

139. E. Nir, C. Plutzer, K. Kleinermanns and M. de Vries, *Eur. Phys. J. D*, 2002, **20**, 317.

140. E. Nir, Ch. Janzen, P. Imhof, K. Kleinermanns and M. S. de Vries, *Phys. Chem. Chem. Phys.*, 2002, **4**, 740.

141. E. Nir, K. Kleinermanns and M. S. de Vries, *Nature*, 2000, **408**, 949.

142. E. Nir, C. Janzen, P. Imhof, K. Kleinermanns and M. S. de Vries, *J. Chem. Phys.*, 2001, **115**, 4604.

143. E. Nir, C. Janzen, P. Imhof, K. Kleinermanns and M. S. de Vries, *Phys. Chem. Chem. Phys.*, 2002, **4**, 732.

144. Ch. Plützer, P. Hünig and K. Kleinermanns, *Phys. Chem. Chem. Phys.*, 2003, **5**, 1158.

145. I. R. Gould and P. A. Kollman, *J. Am. Chem. Soc.*, 1994, **116**, 2493.

146. J. Šponer, J. Leszczynski and P. Hobza, *J. Phys. Chem.*, 1996, **100**, 5590.

147. P. Jurečka and P. Hobza, *J. Am. Chem. Soc.*, 2003, **125**, 15608.

148. M. Kabeláč and P. Hobza, *J. Phys. Chem. B*, 2001, **105**, 5804.

149. P. Cieplak and P. Kollman, *J. Am. Chem. Soc.*, 1988, **110**, 3734.

150. R. A. Friedman and B. Honig, *Biophys. J.*, 1995, **69**, 1528.

151. J. Norberg and L. Nilsson, *Biophys. J.*, 1995, **69**, 2277.

152. K. Murata, Y. Sugita and Y. Okamoto, *Chem. Phys. Lett.*, 2004, **385**, 1.

153. H. Kang, B. Jung and S. K. Kim, *J. Chem. Phys.*, 2003, **118**, 6717.

154. G. A. Pino, C. Dedonder-Lardeux, G. Gregoire, C. Jouvet, S. Martrenchard and D. Solgadi, *J. Chem. Phys.*, 1999, **111**, 10747.

155. G. Pino, G. Gregoire, C. Dedonder-Lardeux, C. Jouvet, S. Martrenchard and D. Solgadi, *Phys. Chem. Chem. Phys.*, 2000, **2**, 893.

156. C. Dedonder-Lardeux, G. Gregoire, C. Jouvet, S. Martrenchard and D. Solgadi, *Chem. Rev.*, 2000, **100**, 4023.

157. A. L. Sobolewski, W. Domcke, C. Dedonder-Lardeux and C. Jouvet, *Phys. Chem. Chem. Phys.*, 2002, **4**, 1093.

158. S. Perun, A. L. Sobolewski and W. Domcke, *J. Am. Chem. Soc.*, 2005, **127**, 6257.

159. T. Schultz, E. Samoylova, W. Radloff, I. V. Hertel, A. L. Sobolewski and W. Domcke, *Science*, 2004, **306**, 1765.

160. A. L. Sobolewski and W. Domcke, *Chem. Phys.*, 2003, **294**, 73.

161. A. L. Sobolewski and W. Domcke, *J. Chem. Phys.*, 2005, **122**.

162. E. Nir, T. Hünig, K. Kleinermanns and M. S. de Vries, *ChemPhysChem.*, 2004, **5**, 131.

163. H. Kang, K. T. Lee, B. Jung, Y. J. Ko and S. K. Kim, *J. Am. Chem. Soc.*, 2002, **124**, 12958.

164. F. Piuzzi, M. Mons, I. Dimicoli, B. Tardivel and Q. Zhao, *Chem. Phys.*, 2001, **270**, 2005.

165. G. A. Pino, C. Dedonder-Lardeux, G. Gregoire, C. Jouvet, S. Martrenchard and D. Solgadi, *J. Chem. Phys.*, 1999, **111**, 10747.

166. G. Pino, G. Gregoire, C. Dedonder-Lardeux, C. Jouvet, S. Martrenchard and D. Solgadi, *Phys. Chem. Chem. Phys.*, 2000, **2**, 893.

167. C. Dedonder-Lardeux, G. Gregoire, C. Jouvet, S. Martrenchard and D. Solgadi, *Chem. Rev.*, 2000, **100**, 4023.

168. A. L. Sobolewski, W. Domcke, C. Dedonder-Lardeux and C. Jouvet, *Phys. Chem. Chem. Phys.*, 2002, **4**, 1093.

169. S. Perun, A. L. Sobolewski and W. Domcke, *J. Am. Chem. Soc.*, 2005, **127**, 6257.

170. T. Schultz, E. Samoylova, W. Radloff, I. V. Hertel, A. L. Sobolewski and W. Domcke, *Science*, 2004, **306**, 1765.

171. A. L. Sobolewski and W. Domcke, *Chem. Phys.*, 2003, **294**, 73.

172. K. Daigoku, S. Ishiuchi, M. Sakai, M. Fujii and K. Hashimoto, *J. Chem. Phys.*, 2003, **119**, 5149.

173. S. Ishiuchi, K. Daigoku, K. Hashimoto and M. Fujii, *J. Chem. Phys.*, 2004, **120**, 3215.
174. S. Ishiuchi, K. Daigoku, M. Saeki, M. Sakai, K. Hashimoto and M. Fujii, *J. Chem. Phys.*, 2002, **117**, 7077.
175. S. Ishiuchi, K. Daigoku, M. Saeki, M. Sakai, K. Hashimoto and M. Fujii, *J. Chem. Phys.*, 2002, **117**, 7083.
176. A. Stolow, *Annu. Rev. Phys. Chem.*, 2003, **54**, 89.
177. A. Abo-Riziq, T. Grace, E. Nir, M. Kabeláč, P. Hobza and M. S. de Vries, *Proc. Natl. Acad. Sci. USA*, 2005, **102**, 20.
178. M. Meuwly, A. Bach and S. Leutwyler, *J. Am. Chem. Soc.*, 2001, **123**, 11446.
179. C. Tanner, C. Manca and S. Leutwyler, *Science*, 2003, **302**, 1736.
180. O. Cheshnovsky and S. Leutwyler, *J. Chem. Phys.*, 1988, **88**, 4127.
181. S. K. Kim, J. J. Breen, D. M. Willberg, L. W. Peng, A. Heikal, J. A. Syage and A. H. Zewail, *J. Phys. Chem.*, 1995, **99**, 7421.
182. O. David, C. Dedonder-Lardeux, C. Jouvet and A. L. Sobolewski, *J. Phys. Chem. A*, 2006, **110**, 9383.
183. C. Dedonder-Lardeux, D. Grosswasser, C. Jouvet and S. Martrenchard, *Phys. Chem. Comm.*, 2001, **4**, 21.
184. G. Pino, G. Gregoire, C. Dedonder-Lardeux, C. Jouvet, S. Martrenchard and D. Solgadi, *Phys. Chem. Chem. Phys.*, 2000, **2**, 893.
185. G. A. Pino, C. Dedonder-Lardeux, G. Gregoire, C. Jouvet, S. Martrenchard and D. Solgadi, *J. Chem. Phys.*, 1999, **111**, 10747.
186. M. N. R. Ashfold, B. Cronin, A. L. Devine, R. N. Dixon and M. G. D. Nix, *Science*, 2006, **312**, 1637.
187. K. Daigoku, S. Ishiuchi, M. Sakai, M. Fujii and K. Hashimoto, *J. Chem. Phys.*, 2003, **119**, 5149.
188. S. Ishiuchi, M. Saeki, M. Sakai and M. Fujii, *Chem. Phys. Lett.*, 2000, **322**, 27.
189. S. Ishiuchi, M. Sakai, K. Daigoku, K. Hashimoto and M. Fujii, *J. Chem. Phys.*, 2007, **127**, 234304.
190. S. Ishiuchi, M. Sakai, K. Daigoku, T. Ueda, T. Yamanaka, K. Hashimoto and M. Fujii, *Chem. Phys. Lett.*, 2001, **347**, 87.
191. N. Tsuji, S. Ishiuchi, M. Sakai, M. Fujii, T. Ebata, C. Jouvet and C. Dedonder-Lardeux, *Phys. Chem. Chem. Phys.*, 2006, **8**, 114.
192. W. Domcke and A. L. Sobolewski, *Science*, 2003, **302**, 1693.
193. A. L. Sobolewski, W. Domcke, C. Dedonder-Lardeux and C. Jouvet, *Phys. Chem. Chem. Phys.*, 2003, **4**, 1093.
194. S. Ishiuchi, K. Daigoku, M. Saeki, M. Sakai, K. Hashimoto and M. Fujii, *J. Chem. Phys.*, 2002, **117**, 7083.
195. S. Ishiuchi, K. Daigoku, K. Hashimoto and M. Fujii, *J. Chem. Phys.*, 2004, **120**, 3215.
196. S. Ishiuchi, K. Daigoku, M. Saeki, M. Sakai, K. Hashimoto and M. Fujii, *J. Chem. Phys.*, 2002, **117**, 7077.
197. H. Yokoyama, H. Watanabe, T. Omi, S. Ishiuchi and M. Fujii, *J. Phys. Chem. A*, 2001, **105**, 9366.

198. F. Yokoyama, H. Watanabe, T. Omi, S. Ishiuchi and M. Fujii, *J. Phys. Chem. A*, 2002, **106**, 854.
199. A. Hara, K. Sakota, M. Nakagaki and H. Sekiya, *Chem. Phys. Lett.*, 2005, **407**, 30.
200. K. Sakota, N. Inoue, Y. Komoto and H. Sekiya, *J. Phys. Chem. A*, 2007, **111**, 4596.
201. S. K. Kim and E. R. Bernstein, *J. Phys. Chem.*, 1990, **94**, 3531.
202. Y. Koizumi, C. Jouvet, T. Norihiro, S. Ishiuchi, C. Dedonder-Lardeux and M. Fujii, *J. Chem. Phys.*, 2008, **129**, 104311.
203. J. Chocholoušová, J. Vacek and P. Hobza, *Phys. Chem. Chem. Phys.*, 2002, **4**, 2119.
204. F. Madeja, M. Havenith, K. Nauta, R. E. Miller, J. Chocholoušová and P. Hobza, *J. Chem. Phys.*, 2004, **120**, 10554.
205. J. Chocholoušová, J. Vacek, F. Huisken, O. Werhahn and P. Hobza, *J. Phys. Chem. A*, 2002, **106**, 11540.
206. K. Nauta and R. E. Miller, *Science*, 2000, **287**, 293.
207. C. J. Burnham, S. S. Xantheas, M. A. Miller, B. E. Applegate and R. E. Miller, *J. Chem. Phys.*, 2002, **117**, 1109.
208. X. Zhang and J. L. Knee, *Faraday Discuss.*, 1994, **97**, 299.
209. G. Lembach and B. Brutschy, *J. Phys. Chem.*, 1996, **100**, 19758.
210. J. D. Pitts and J. L. Knee, *J. Chem. Phys.*, 1999, **110**, 3389.
211. X. Zhang, J. D. Pitts, R. Nadarajah and J. L. Knee, *J. Chem. Phys.*, 1997, **107**, 8239.

CHAPTER 6

Extended Molecular Clusters in Chemistry, the Atmosphere and Stereospecific Molecular Recognition

In the preceding chapter we presented experimental and theoretical results of relatively small clusters in order to gain an understanding of the basic principles of non-covalent interactions. In the present chapter we will present some extended examples.

Solvent-cluster research is a link between the gas phase and the solution phase. Due to the unique hydrogen-bonding capabilities of water and its importance in chemistry and biology[1,2] the investigation of water clusters is of special interest. Examples of water clusters were already given in Section 5.11.1, focusing on the neutral clusters. Here, we will present some results obtained for protonated water clusters. Protonated water clusters do not only play fundamental role in Earth chemistry and biology but have also be found to be an influential element in the upper atmosphere.[3] Directly connected to atmospheric chemistry and physics are aerosols, which will be discussed as a second example.

As a third example, which is only indirectly related to the former ones but of not minor importance, we will discuss chirality. Chirality is most important for the building blocks in nature and its unique significance for life.

6.1 Protonated Water Clusters $[H(H_2O)_n]^+$

6.1.1 Magic Numbers

6.1.1.1 Observation of Magic Numbers $n = 21$ and 28

Lin[4] observed in 1973 that the intensity of water clusters $[H(H_2O)_n]^+$ decreases with increasing n, except for $n = 21$, 22. The intensity of the $n = 21$ cluster was

RSC Theoretical and Computational Chemistry Series No. 2
Non-covalent Interactions: Theory and Experiment
By Pavel Hobza and Klaus Müller-Dethlefs
© Pavel Hobza and Klaus Müller-Dethlefs 2010
Published by the Royal Society of Chemistry, www.rsc.org

higher and the intensity of the $n = 22$ cluster was lower than expected. In a variety of different experiments[5–9] an anomalous intensity was observed for the $[H(H_2O)_{21}]^+$ cluster, indicating that it has a particular stable structure. Therefore, it has been described as "magic". Another cluster that seems to have a special stability is the $[H(H_2O)_{28}]^+$ cluster.[5–7,10] The behaviour and existence of these magic number clusters are independent of the source of ionisation and many different techniques have been successful in generating these ensembles of water clusters.[2,10–14]

6.1.1.2 Structure of the Magic Number Clusters

To account for the high stability of the $[H(H_2O)_{21}]^+$ cluster a pentagonal dodecahedral cage structure has been suggested by Searcy and Fenn.[15]

In vibrational predissociation experiments using a tandem mass spectrometer the structure of a range of protonated water clusters $[H(H_2O)n]^+$ has been probed.[12,16–18] It was shown that the spectra depended on the size and structure of the water cluster being investigated.[16,18] The O–H stretching vibrational spectra of the $[H(H_2O)_{21}]^+$ and $[H(H_2O)_{22}]^+$ differed from the other $[H(H_2O)_n]^+$ clusters ($6 \leq n \leq 27$) as recorded by Park *et al.*[16] While the spectra of the neighbouring cluster displayed a doublet peak, the spectra of the $n = 21$ and 22 cluster yield only a single peak, implying that *all* of the dangling OH groups on the $[H(H_2O)_{21}]^+$ cluster are due to water molecules that are bound in a similar way, providing support for the pentagonal dodecahedral cage structure.

That the stability of the $[H(H_2O)_{21}]^+$ cluster is due to a distorted pentagonal dodecahedral cage structure (Figure 6.1) is supported by different computational methods such as *ab initio* calculations,[19–22] Monte Carlo simulations,[23–25] and molecular dynamics simulations[26] that have been used to calculate the structure of the $[H(H_2O)_{21}]^+$ cluster. The stability of this structure arises from the fact that each water molecule is hydrogen bonded to three other water molecules within the cage. Whether there is a single water molecule or a hydronium ion in the centre has been widely discussed.[27] The location of a hydronium in the centre of the cage was suggested by different calculations,[21,24,28,29] while others suggest it is found on the surface of the cage.[19,30,31] The latter result is supported by recent computational studies, which suggest that there is a water molecule within the cage and the hydronium ion is found on the surface of the cage.[6,20] Also for the $[H(H_2O)_{28}]^+$ cluster a similar clathrate-like structure is proposed.[6]

6.1.1.3 Investigation of the Stability of the Magic Clusters

Wu *et al.*[6] applied vibrational predissociation spectroscopy in an ion-trap tandem mass spectrometer to investigate the stability of $[H(H_2O)_{21}]^+$ (generated in a continuous corona-discharge). Anomalous intensities between the $n = 21$ and 22 clusters were observed at different backing pressures. At higher backing pressures (340 and 200 Torr) it was observed that $[H(H_2O)_{21}]^+$, which

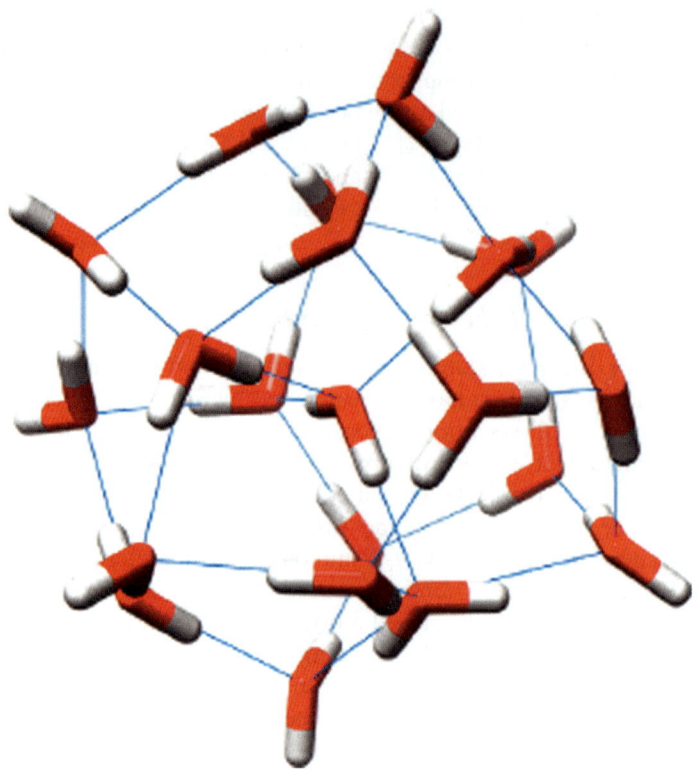

Figure 6.1 An energy-minimised model of the $[H(H_2O)_{21}]^+$ ion, a distorted dodecahedral cage in which each edge is a hydrogen bond. Redrawn from Ref. 15.

is the most intense peak, has significantly smaller dissociation rates than the
$n = 20$ and 22 clusters.

Echt et al.[10] generated and studied $[H(H_2O)_{21}]^+$ clusters using electron
impact-ionisation time-of-flight (TOF) mass spectrometry. The TOF spectra
shows that after a time delay, anomalous intensities for the $[H(H_2O)_{21}]^+$ and
$[H(H_2O)_{28}]^+$ clusters are evident. They investigated the amount of decomposition occurring in the drift tube of the TOF mass analyser. While only 27%
of the $[H(H_2O)_{21}]^+$ cluster decomposed, almost twice as much, about 50%, of
the $[H(H_2O)_{22}]^+$ ion decomposed under the same conditions.

Magnera et al.[32] calculated the proton hydration energies for clusters with
1–28 water molecules. They investigated the binding energies of clusters generated by fast-atom bombardment using collision-induced dissociation (CID)
mass spectrometry. A significant decrease of the binding energy was observed
up to a cluster sizes $n = 9$, while it starts to increases slowly for $n > 9$. For the
$[H(H_2O)_{21}]^+$ cluster it was found that its binding energy is about 2 kcal/mol
higher than its neighbouring clusters due to its high stability.

Schindler et al.[33] investigated the fragmentation of the $[H(H_2O)_{58}]^+$ water
cluster using FTICR-MS. The fragmentation of this cluster led to the

formation of clusters of $n = 57–51$, where the clusters $n = 55$ and 53 had particularly long lifetimes. A further fragmentation of these clusters to smaller clusters was not performed but their investigation showed that the $n = 21$ cluster had a longer lifetime than other clusters.

McQuinn et al.[34] used an energy-dependent electrospray ionisation tandem mass spectrometer (EDESI–MS/MS) for the investigation of $[H(H_2O)n]^+$ where $n = 26–76$. Collision-induced dissociation was performed in an argon-filled collision cell. Increasing the collision voltage leads to the sequential loss of water molecules:

$$[H(H_2O)_n]^+ + Ar \ldots [H(H_2O)_{n-1}]^+ + H_2O + Ar$$

$$[H(H_2O)_{n-1}]^+ + Ar \ldots [H(H_2O)_{n-2}]^+ + H_2O + Ar, \text{ etc.}$$

For all the EDESI-MS experiments it could be shown, that the selected cluster $[H(H_2O)_n]^+$ $n = 26–76$, preferentially fragments to $[H(H_2O)_{21}]^+$ clusters.

In a 3D plot, a fairly steady intensity of water clusters is observed with the exception of the anomalously strong peak at $m/z = 379.24$ corresponding to the $[H(H_2O)_{21}]^+$ cluster. What appears as a fairly uniform "wave" of ions is interrupted by a distinct spike in intensity. This peak is also broader than all other peaks in the plot, indicating that this cluster is a predominant feature at more than one collision voltages. In fact, the $[H(H_2O)_{21}]^+$ cluster is observed to dominate the MS/MS spectrum at lower collision voltages than expected and maintains its strong intensity for a larger range of collision voltages than other clusters. The magic $[H(H_2O)_{21}]^+$ cluster dominates the MS/MS spectra for more than 15 consecutive collision voltage settings, while normally a single cluster is the base peak in the MS/MS spectra only for one or two different values of the collision voltage. At higher collision voltages the intensities of the cluster decrease steadily as the collision voltage increases, with the exception of $[H(H_2O)_{21}]^+$.

Also for the $[H(H_2O)_{28}]^+$ cluster a slightly higher intensity than its neighbouring clusters could be observed, while the intensity of the $n = 29$ cluster is anomalous low, indicating the preferential formation of the magic $[H(H_2O)_{28}]^+$ cluster. But compared to the $[H(H_2O)_{21}]^+$ cluster a prolonged existence over a large range of collision voltages could not be observed.

6.1.1.4 Clusters of $n = 12$, 13

One unexpected feature of all the EDESI–MS/MS spectra as recorded by McQuinn et al.[34] is the consistent appearance of broad intensity maxima centred at about the 13-mer, 25-mer and 36-mer clusters, and a distinct minimum at the 32-mer. These features are quite different from the "spike" apparent for the 21-mer, or the more subtle increased intensity of the 28-mer; the increased intensity is not at the expense of the neighbouring peaks, instead it manifests itself as the superposition of an approximately sinusoidal wave upon the background intensity. Interestingly, the "wave" has maxima every

12–13 water molecules, and the undulation can be seen to continue out further for the larger clusters. Theoretical studies of the energies of global minima of protonated water clusters (to the 20-mer) reveal a very consistent energy change as each water molecule is added, with no notable discontinuities.[23] Interestingly, experimental and theoretical studies of some rare-gas clusters indicate particularly stable icosahedral structures for the 13-mer among others.[35–37] The mass spectra of these species display sharply higher intensities for the magic numbers.

6.1.2   Formation of Nanoscale Cages

Miyazaki *et al.*[18] recorded IR spectra of protonated water clusters, $H^+(H_2O)_n$, from $n = 4$ to 27, to investigate the development of the hydrogen-bond network structure. The spectra for $n = 4–8$ show gradual changes in the free as well as the hydrogen-bond OH-stretch region, reflecting a general trend towards a hydrogen-network structure that strongly depends on the cluster size. For clusters up to $n = 6$ it could be shown by Jiang *et al.*[38] that the H_3O^+ or $H_5O_2^+$ ion core is located at the centre of the cluster and radial hydrogen-bond chains originate from this 3- or 4-coordinated ion core (Figure 6.2(A)). Miyazaki *et al.* observed that an obvious spectral change occurs in the free OH-stretch region. The v_1 and v_3 bands show a gradual high-frequency shift with increasing cluster size, reflecting the increase in distance from the charge centre. Moreover, the intensity of these bands decreases with cluster size, and the bands finally disappear at $n \sim 10$. At the same time as this decrease, a new band starts to appear at $\sim 3690 \, cm^{-1}$ that has been assigned to the dangling OH-stretching vibration in a 3-coordinated water molecule.[38–41] Such a 3-coordinated water molecule occurs at a bridging site of hydrogen-bond chains. The disappearance of the v_1 and v_3 bands and simultaneous appearance of the dangling OH stretching bands for clusters $n = 7$ to 10 indicates the development of the hydrogen network from the *chain* type to a 2D *net* structure (Figures 6.2(A) and (B)).

From the recorded spectra it was shown that the hydrogen-bond network preserves a 2D net structure for $n = 10–19$. From $n > \sim 19$, new spectral features representing the next stage of network growth were observed. The spectra for $n \geq 21$ clearly indicate that the 2D net types are converted to 3D cage structures in this size region (Figures 6.2(B) and (C)). Such nanometer-sized cage structures have not previously been experimentally confirmed for hydrogen-bonded cluster systems. The central H_3O^+ or $H_2O_5^+$ ion core prefers the planar coordination by nature, and it allows 3D cage formation only in such large-sized clusters.

6.2   Aerosols

Generally, an aerosol is a multiphase system composed of solid or liquid particles suspended within a gas phase. The two phases are coupled, with changes in the composition or temperature of the gas phase affecting the behaviour of

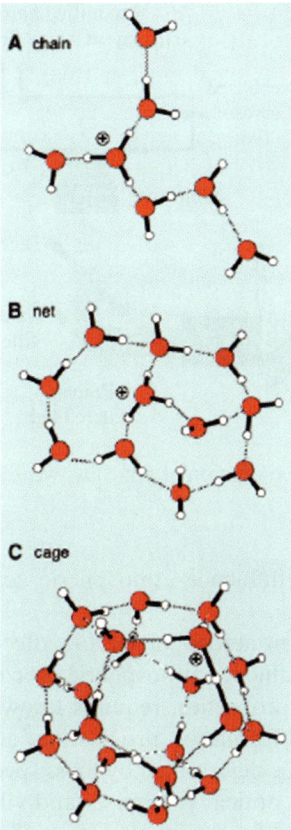

Figure 6.2   IR spectra of protonated water clusters, $H^+(H_2O)_n$, from $n=4$ to $n=27$. Redrawn from Ref. 18.

the condensed phase particles. Aerosols play an important role for planetary atmospheres, including the Earth's, nanoparticle production and health (*e.g.* drug delivery).

A significant fraction of the total number of particles present in the atmosphere is formed originally by nucleation from the gas phase, creating secondary aerosols. The nucleation involves multiple compounds. Binary nucleation of sulfuric acid and water, ternary nucleation of sulfuric acid, water and ammonia and ion-induced nucleation are thought to be the most important aerosol-nucleation processes in the atmosphere. The nucleation competes with growth and coagulation of existing particles and couples to oxidation reactions that convert volatile gas-phase molecules to low-volatility molecules.

A detailed understanding of atmospheric aerosol-nucleation processes is needed as the freshly formed particles directly influence the number concentration and size distribution of the atmospheric aerosol. The formation of clouds and precipitation is affected and influences on climate are anticipated.

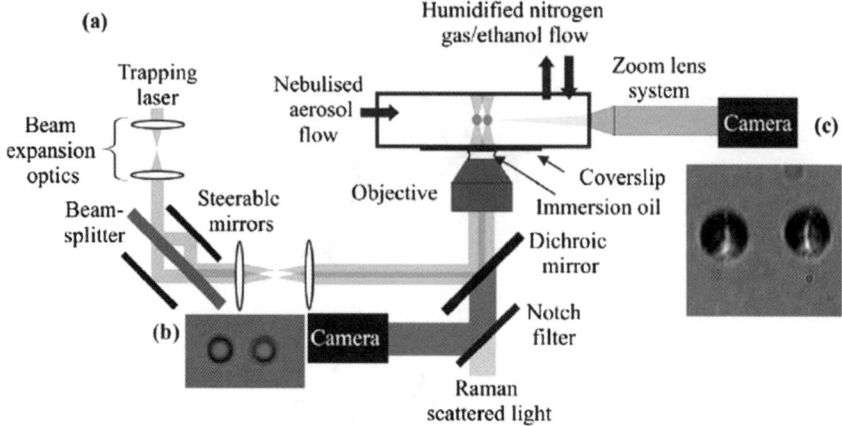

Figure 6.3 Setup for the investigation of optically tweezered aerosol droplets. Redrawn from Ref. 47.

Anthropogenic emissions influence atmospheric aerosol-nucleation processes considerably.

The understanding of how aerosol properties affect scattering of sunlight and activation, how aerosols influence atmospheric processes, and how atmospheric processes change aerosol properties, requires knowledge of chemical composition, size, reactivity, and formation processes of aerosols.

Aerosol properties can be determined by mass spectrometric methods[42–46] or optical techniques such as optical tweezers[47] and vibrational spectroscopy.[48]

Figure 6.3 shows a setup for the investigation of optically tweezered aerosol droplets. A focused laser beam is used to trap the droplets. An in-plane image of the droplet is acquired using conventional bright-field microscopy and a side image using a zoom-lens system. Additionally, the light of the laser is scattered by the droplets and this scattered light can by analysed.

Adjusting the flow rate of humidified nitrogen gas into the cell controls the relative humidity. The trapped droplets can be held for indefinitely long periods in the cell and investigated with respect to the conditions of the environment.

A typical Raman spectrum for optical tweezered aqueous/NaCl droplets of different size is displayed in Figure 6.4.

6.2.1   Spontaneous Raman Scattering

The intramolecular OH-stretching vibration is the origin of the spontaneous Raman scattering (broad background) observed from $2900\,cm^{-1}$ to $3700\,cm^{-1}$. It reflects the distribution of intermolecular hydrogen environments of water molecules.[50] Thus, the band shape shows a systematic dependence on ionic strength due to changes in the hydrogen-bonding network with ion concentration.[50,51] The band in the spectrum presented here has a maximum at $3480\,cm^{-1}$

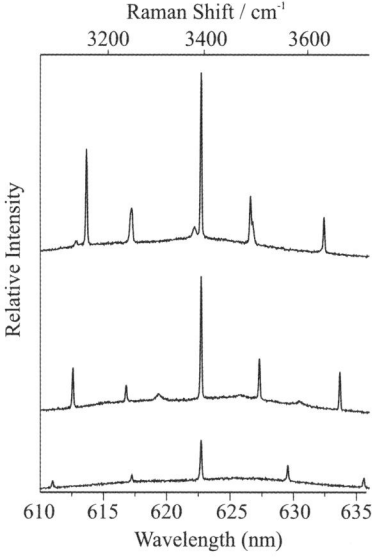

Figure 6.4 A typical Raman spectrum for optical tweezered aqueous/NaCl droplets of different size. Redrawn from Ref. 49.

and a shoulder at $3290\,\mathrm{cm}^{-1}$. An increase in the NaCl concentration of the droplet results in a greater disruption of the hydrogen-bonding environment within the water phase, and as a consequence the OH-stretching band peak maximum is observed to shift to higher frequencies, whilst the low-frequency shoulder at $3290\,\mathrm{cm}^{-1}$ diminishes in intensity.[50,51] The spontaneous Raman scattering therefore provides a fingerprint of the chemical composition.

6.2.3 Stimulated Raman Scattering

At discrete wavelengths, the inelastically scattered Raman light (sharp peaks) can become trapped by total internal reflection, forming a standing wave, which circulates the droplet circumference.[52,53] At these wavelengths, which are commensurate with whispering-gallery modes (WGMs), the droplet behaves as a low-loss optical cavity providing a mechanism for optical feedback leading to the amplification of the Raman signal. The threshold intensity for the occurrence of SRS is observed at much lower incident laser intensities for droplets than bulk samples, as a result of the ability of the droplet to act as a low-loss cavity.[52,53] The WGM wavelengths can be expressed in terms of the size parameter, x, which is determined from the radius of the particle and the wavelength of the WGM λ.[54,55]

By comparison of the WGM wavelengths with Mie scattering calculations the size of the optically trapped aerosol droplet can be calculated with nanometre accuracy, dependent upon the accuracy with which the WGM wavelengths and the refractive index of the droplet are known.

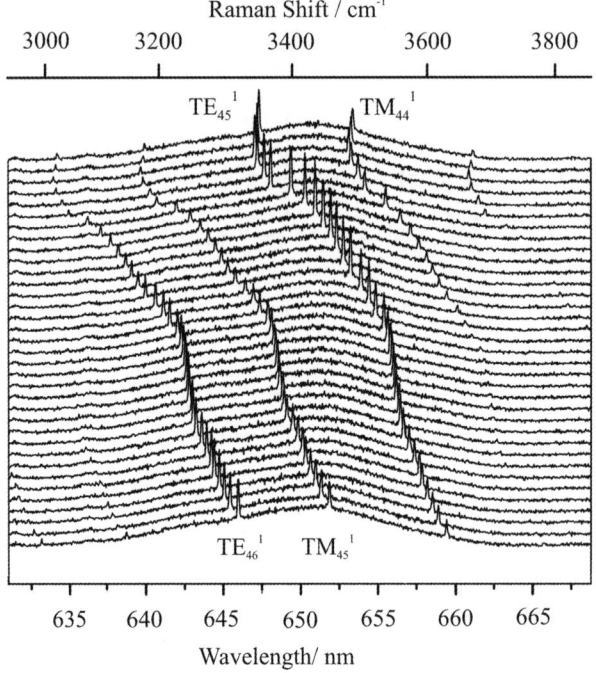

Figure 6.5 Variation in WGM wavelengths observed during equilibration of a NaCl/ aqueous aerosol droplet with the RH of the aerosol cell. Redrawn from Ref. 50.

Figure 6.5 shows the variation in WGM wavelengths observed during equilibration of a NaCl/aqueous aerosol droplet with the RH of the aerosol cell, illustrating the achievable time resolution and size precision.[52] Initially, the droplet radius equals 3.986 mm and over a period of 30 min the droplet evaporates by 73 nm to a size of 3.913 mm. The shift in the resonance wavelengths to the blue is indicative of a decrease in droplet size. The change in droplet refractive index arising from the evolving composition during the evaporation is accounted for when calculating the variation in droplet size.

6.3 Chirality and Molecular Complexes

Chirality plays an important role in a variety of fields such as biology, chemistry and pharmacy. Biological chiral discrimination is an important phenomenon in amino acids or sugars, where one enantiomer is exclusively selected by nature over the other form. All amino acids in proteins are of the L-form, while all sugars in DNA and RNA, and in the metabolic pathways, are of the D-form. Contrarily, amino acids produced in the laboratory are a 1:1 mixture of both forms. The understanding of homochirality in Nature is therefore important for the understanding of life. Later, chiral discrimination is very

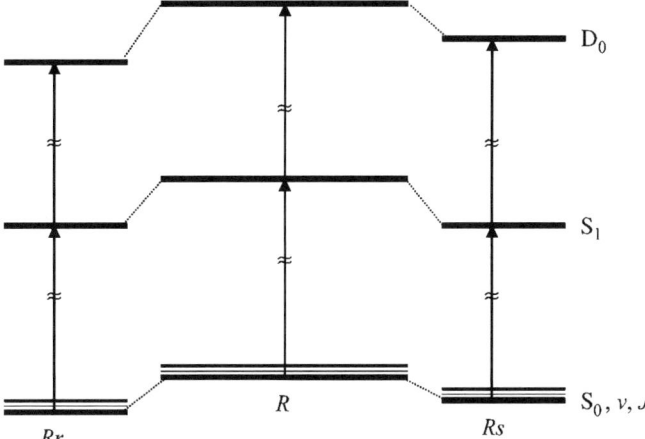

Figure 6.6 Energy-level scheme for chiral recognition. Redrawn from Ref. 59.

important in the pharmaceutical industry, where one enantiomer may have a different pathway in the body, or it may be toxic. For chiral discrimination molecular recognition plays a major role. Chiral recognition is the ability of a chiral molecule to distinguish between the two enantiomeric forms of another chiral molecule and is thought to happen through formation of non-covalently bound contact pairs. The knowledge of the structure, the stability and the dynamics of molecule/receptor pairs is needed to understand the phenomena of molecular recognition. In the last few years, the development of new gas-phase techniques allowed the study of intrinsic interactions in tailor-made molecular clusters mimicking receptor/molecule systems in the isolated state without any interference from the environment.[56–58]

The principle of chiral recognition is shown in Figure 6.6. The complexation of a chiral selection R by enantiomorphs r or s leads to enantiomer selective shifts of the neutral ground state, electronic excited states and the ionic ground state.

6.3.1 Theoretical Approaches to Chiral Recognition

Theoretical approaches to chirality recognition are also subject to severe limitations, owing to the size and the complexity of the systems at play. Numerous attempts have been made for modelling the selectivity of interactions between chiral molecules to mimic similar selectivity in enzyme-substrate systems. Empirical approaches often derive from the famous "lock-and-key" principle proposed by Fischer at the end of the 19th century to account for the specificity of enzyme reactions.[60] The "three-point rule"[61] has been applied to explain the enantiospecificity of chiral-phase chromatography.[62] Similar approaches discuss the mechanism of chirality recognition in terms of the number of interaction points or stereochemical factors.[63-66] Molecular-dynamics calculations have been used to assess the role of hydrogen bonding in chirality recognition between cyclodextrins and amino acids.[67] Molecular modelling describes molecular interactions in macromolecular systems, making rational drug design possible.[68]

6.3.2 Experiments in the Gas Phase and Supersonic Jets

A large body of literature has been published concerning the thermodynamics of ionic clusters of chiral molecules using, in particular, the kinetic method of Cooks *et al.*[69,70] Special attention has been paid to interactions involving amino acids.[71,72] Ionic clusters have been the subject of several review articles.[73-75] Because of the large binding energy of ionic complexes, the mass spectrometry experiments are usually done at room temperature, which makes spectroscopic measurements difficult. This is why direct structural information is rarely available. Vibrational spectra of protonated homochiral serine clusters have been reported,[76,77] in particular that of the octamer, the outstanding stability of which is proposed to play a role in the homochirality of life.[78] Several studies have described the reactivity of simple aromatic ions produced by photoionisation of jet-cooled neutral complexes.[79] In contrast to ions and zwitterions, neutral molecular complexes usually involve low binding energies. That is why their study mostly resorts to supersonic expansions, taking advantage of their capability to produce cold, weakly bound complexes in isolated conditions. A very interesting study comes from the group of Zehnacker on chiral recognition in jet-cooled complexes of (1R,2S)-(+)-cis-1-amino-2-indanol with methyl-lactate confirming the importance of the CH ... π interaction.[80] This work demonstrates chiral-sensitive and -non-sensitive insertion for two different cluster distereomers (Figures 6.7 and 6.8).

6.3.3 Neurotransmitters: (1S, 2S)-N-Methylpseudoephedrine

Neurotransmitters are molecules that interpose between neurons (nerve cells), or between neurons and muscle cells. The simplest neurotransmitters such as amphetamine, dopamine, noradrenaline, *etc.* have been studied regarding their different conformation and conformational changes due to monosolvation.

Only a few investigations deal with non-covalent interactions between a chiral solvent and neurotransmitters, which represents the basis of the receptor recognition of a chiral neurotransmitter molecule.

R2PI spectroscopic investigations of the neurotransmitter (1S,2S)-N-methylpseudoephedrine (MPE) and its complexes with chiral solvents methyl lactate (L) and 2-butanol (B) have been performed by Guidoni *et al.*[81] The results have been compared with those of the complex with water (W).

Insertion <u>not sensitive</u> to chirality

Two H bonds (dominant interactions) ➡ rigid system with
poor chiral discrimination

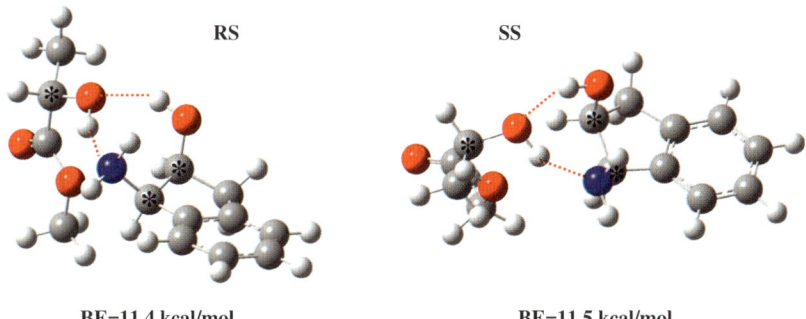

RS   SS

BE=11.4 kcal/mol   **BE=11.5 kcal/mol**

Figure 6.7   1-amino-2-indanol/methyl lactate, nonchiral-sensitive insertion. Redrawn from Ref. 80.

Addition or cycle <u>sensitive</u> to chirality

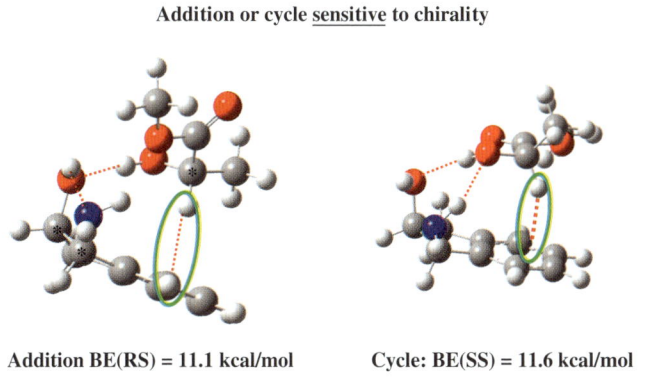

Addition BE(RS) = 11.1 kcal/mol   **Cycle: BE(SS) = 11.6 kcal/mol**

Figure 6.8   Chiral-sensitive insertion. Redrawn from Ref. 80.

Like ephedrine, pseudoephedrine and their complexes MPE ($m/z = 179$) undergo efficient dissociation ($> 99\%$), therefore the R2PI spectrum of the MPE monomer was recorded at $m/z = 72$ ($CH_3CHNHCH_3$). Two conformers could be observed in the spectrum.

The spectral patterns of the selected complexes have been interpreted in terms of the specific hydrogen-bond interactions operating in the diastereomeric complexes, whose nature in turn depends on the structure and the configuration of the solvent molecule. The obtained results confirm the view that a representative neurotransmitter molecule, such as MPE, communicates with the enantiomers of a chiral substrate through different, specific interactions. These findings can be regarded as a further contribution to modelling neurotransmitter functions in biological systems.

References

1. P. Ball, *Chem. Rev.*, 2008, **108**, 74.
2. F. Dong, S. Heinbuch, J. J. Rocca and E. R. Bernstein, *J. Chem. Phys.*, 2006, **124**, 224319.
3. R. P. Wayne, *Chemistry of Atmospheres*, Oxford University Press, Oxford, UK, 1991.
4. S.-S. Lin, *Rev. Sci. Instrum.*, 1973, **44**, 516.
5. X. Yang and A. W. Castleman Jr, *J. Am. Chem. Soc.*, 1989, **111**, 6845.
6. C. C. Wu, C.-K. Lin, H.-C. Chang, J.-C. Jiang, J.-L. Kuo and M. L. Klein, *J. Chem. Phys.*, 2005, **122**, 074315.
7. M. Tsuchiya, T. Tashiro and A. Shigihara, *J. Mass Spectrom. Soc. Jpn.*, 2004, **52**, 1.
8. G. Hulthe, G. Stenhagen, O. Wennerstroem and C.-H. Ottosson, *J. Chromatogr. A*, 1997, **777**, 155.
9. P. P. Radi, P. Beaud, D. Franzke, H. M. Frey, T. Gerber, B. Mischler and A. P. Tzannis, *J. Chem. Phys.*, 1999, **111**, 512.
10. O. Echt, D. Kreisle, M. Knapp and E. Recknagel, *Chem. Phys. Lett.*, 1984, **108**, 401.
11. D. W. Ledman and R. O. Fox, *J. Am. Soc. Mass Spectrom.*, 1997, **8**, 1158.
12. S. Y. Huang, C. D. Huang, B. T. Chang and C. T. Yeh, *J. Phys. Chem. B*, 2006, **110**, 21783.
13. Z. Shi, J. V. Ford, S. Wei and A. W. Castleman Jr, *J. Chem. Phys.*, 1993, **99**, 8009.
14. K. Mori, D. Asakawa, J. Sunner and K. Hiraoka, *Rapid Commun. Mass Spectrom.*, 2006, **20**, 2596.
15. J. C. Searcy and J. B. Fenn, *J. Chem. Phys.*, 1974, **61**, 5282.
16. S. J. Park, D. M. Shin, S. Sakamoto, K. Yamaguchi, Y. K. Chung, M. S. Lah and J. I. Hong, *Chem. Eur. J.*, 2004, **11**, 235.
17. J. M. Headrick, E. G. Diken, R. S. Walters, N. I. Hammer, R. A. Christie, J. Cui, E. M. Myshakin, M. A. Duncan, M. A. Johnson and K. D. Jordan, *Science*, 2005, **308**, 1765.
18. M. Miyazaki, A. Fujii, T. Ebata and N. Mikami, *Science*, 2004, **304**, 1134.
19. A. Khan, *Chem. Phys. Lett.*, 2000, **319**, 440.
20. S. S. Iyengar, M. K. Petersen, T. J. F. Day, C. J. Burnham, V. E. Teige and G. A. Voth, *J. Chem. Phys.*, 2005, **123**, 084309.
21. H. Shinohara, U. Nagashima, H. Tanaka and N. Nishi, *J. Chem. Phys.*, 1985, **83**, 4183.
22. D. J. Wales and M. P. Hodges, *Chem. Phys. Lett.*, 1998, **286**, 65.
23. M. P. Hodges and D. J. Wales, *Chem. Phys. Lett.*, 2000, **324**, 279.
24. M. Svanberg and J. B. C. Pettersson, *J. Phys. Chem. A*, 1998, **102**, 1865.
25. C. C. Wu, C. K. Lin, H. C. Chang, J. C. Jiang, J. L. Kuo and M. L. Klein, *J. Chem. Phys.*, 2005, **122**, 074315.
26. E. Brodskaya, A. P. Lyubartsev and A. Laaksonen, *J. Phys. Chem. B*, 2002, **106**, 6479.
27. T. S. Zwier, *Science*, 2004, **304**, 1119.

28. R. E. Kozack and P. C. Jordan, *J. Chem. Phys.*, 1993, **99**, 2978.
29. K. Laasonen and M. L. Klein, *J. Phys. Chem.*, 1994, **98**, 10079.
30. P. M. Holland and A. W. Castleman Jr, *J. Chem. Phys.*, 1980, **72**, 5984.
31. R. Kelterbaum and E. Kochanski, *J. Phys. Chem.*, 1995, **99**, 12493.
32. T. F. Magnera, D. E. David and J. Michl, *Chem. Phys. Lett.*, 1991, **182**, 363.
33. T. Schindler, C. Berg, G. Niedner-Schatteburg and V. E. Bondybey, *Chem. Phys. Lett.*, 1996, **250**, 301.
34. K. McQuinn, F. Hof and J. S. McIndoe, *Int. J. Mass Spect.*, 2009, **279**, 32.
35. K. McQuinn, J. S. McIndoe and F. Hof, *Chem. Eur. J.*, 2008, **14**, 6483.
36. S. A. McLuckey, *J. Am. Soc. Mass Spectrom.*, 1992, **3**, 599.
37. C. P. G. Butcher, B. F. G. Johnson, J. S. McIndoe, X. Yang, X.-B. Wang and L.-S. Wang, *J. Chem. Phys.*, 2002, **116**, 6560.
38. J.-C. Jiang, *et al. J. Am. Chem. Soc.*, 2000, **122**, 1398.
39. B. Rowland, M. Fisher and J. P. Devlin, *J. Chem. Phys.*, 1991, **95**, 1378.
40. C. J. Gruenloh, *et al. Science*, 1997, **276**, 1678.
41. U. Buck, I. Ettischer, M. Melzer, V. Buch and J. Sadlej, *Phys. Rev. Lett.*, 1998, **80**, 2578.
42. M. R. Canagaratna, *et al. Mass Spectrom. Rev.*, 2007, **26**, 185.
43. D. M. Murphy, *Mass Spectrom. Rev.*, 2007, **26**, 150.
44. E. S. Cross, *et al. Aerosol Sci. Technol.*, 2007, **41**, 343.
45. R. C. Moffet and K. A. Prather, *Analytical Chem.*, 2005, **77**, 6535.
46. A. Zelenyuk, Y. Cai and D. Imre, *Aerosol Sci. Technol.*, 2006, **40**, 197.
47. L. Mitchem and J. P. Reid, *Chem. Soc. Rev.*, 2008, **37**, 756.
48. R. Signorell and M. Jetzki, *Faraday Discuss.*, 2008, **137**, 51.
49. R. J. Hopkins, L. Mitchem, A. D. Ward and J. P. Reid, *Phys. Chem. Chem. Phys.*, 2004, **6**, 4924.
50. L. Mitchem, J. Buajarern, R. J. Hopkins, A. D. Ward, R. J. J. Gilham, R. L. Johnston and J. P. Reid, *J. Phys. Chem. A*, 2006, **110**, 8116.
51. R. Symes, R. M. Sayer and J. P. Reid, *Phys. Chem. Chem. Phys.*, 2004, **6**, 474.
52. S. C. Hill and R. E. Benner, *Morphology-Dependent Resonances in Optical Effects Associated with Small Particles*, ed. P. W. Barber, R. K. Chang, World Scientific, Singapore, 1988.
53. A. Campillo, J. D. Eversole and H. B. Lin, *Optical Processes in Micro-cavities*, World Scientific, Singapore, 1996.
54. J. P. Reid and L. Mitchem, *Annu. Rev. Phys. Chem.*, 2006, **57**, 245.
55. R. Symes, R. M. Sayer and J. P. Reid, *Phys. Chem. Chem. Phys.*, 2004, **6**, 474.
56. E. Nir, K. Kleinermanns and M. S. De Vries, *Nature*, 2000, **408**, 949.
57. M. S. De Vries and P. Hobza, *Annu. Rev. Phys. Chem.*, 2007, **58**, 585.
58. J. P. Simons, P. Carcabal, B. G. Davis, D. P. Gamblin, I. Hunig, R. A. Jockusch, R. T. Kroemer, E. M. Marzluff and L. C. Snoek, *Int. Rev. Phys. Chem.*, 2005, **24**, 489.
59. A. Zehnacker-Retien and M. A. Suhm, *Angew. Chem. Int. Ed.*, 2008, **47**, 6970.

60. E. Fischer, *Ber. Dtsch. Chem. Ges.*, 1894, **27**, 2985.
61. A. G. Ogstron, *Nature*, 1948, **162**, 963.
62. W. H. Pirkle and T. C. Pochapsky, *Chem. Rev.*, 1989, **89**, 347.
63. S. Garten, P. U. Biedermann, I. Agranat and S. Topiol, *Chirality*, 2005, **17**, S159.
64. T. P. Radhakrishnan, S. Topiol, P. U. Biedermann, S. Garten and I. Agranat, *Chem. Commun.*, 2002, 2664.
65. S. Topiol, *Chirality*, 1989, **1**, 69.
66. S. Topiol and M. Sabio, *J. Am. Chem. Soc.*, 1989, **111**, 4109.
67. K. B. Lipkowitz, S. Raghothama and J. Yang, *J. Am. Chem. Soc.*, 1992, **114**, 1554.
68. G. Subramanian, *Chiral Separation Techniques*, 3rd edn, Wiley-VCH, Weinheim, 2001.
69. R. G. Cooks and T. L. Kruger, *J. Am. Chem. Soc.*, 1977, **99**, 1279.
70. W. A. Tao and R. G. Cooks, *Anal. Chem.*, 2003, **75**, 25A.
71. J. Ramirez, F. He and C. B. Lebrilla, *J. Am. Chem. Soc.*, 1998, **120**, 7387.
72. Y. J. Liang, J. S. Bradshaw and D. V. Dearden, *J. Phys. Chem. A*, 2002, **106**, 9665.
73. C. A. Schalley, *Mass Spectrom. Rev.*, 2001, **20**, 253.
74. M. Sawada, *Mass Spectrom. Rev.*, 1997, **16**, 73.
75. M. Speranza, *Int. J. Mass Spectrom.*, 2004, **232**, 277.
76. H. A. Cox, R. Hodyss and J. L. Beauchamp, *J. Am. Chem. Soc.*, 2005, **127**, 4078.
77. X. L. Kong, I. A. Tsai, S. Sabu, C. C. Han, Y. T. Lee, H. C. Chang, S. Y. Tu, A. H. Kung and C. C. Wu, *Angew. Chem. Int. Ed.*, 2006, **45**, 4130.
78. P. Yang, R. Xu, S. C. Nanita and R. G. Cooks, *J. Am. Chem. Soc.*, 2006, **128**, 17074.
79. M. Speranza, M. Satta, S. Piccirillo, F. Rondino, A. Paladini, A. Giardini, A. Filippi and D. Catone, *Mass Spectrom. Rev.*, 2005, **24**, 588.
80. K. Le Barbu-Debus, M. Broquier, A. Mahjoub and A. Zehnacker-Rentien, *J Chem. Phys.*, 2009, DOI: 10.1039/b906834a.
81. A. G. Guidoni, A. Paladini, S. Piccirillo, F. Rondino, M. Satta and M. Speranza, *Org. Biomol. Chem.*, 2006, **4**, 2012.

Subject Index

Page references to *figures*, *tables* and *text boxes* are shown in *italics*.